Jie Lu, Lakhmi C. Jain, and Guangquan Zhang (Eds.)

Handbook on Decision Making

Intelligent Systems Reference Library, Volume 33

Editors-in-Chief

Prof. Janusz Kacprzyk
Systems Research Institute
Polish Academy of Sciences
ul. Newelska 6
01-447 Warsaw
Poland
E-mail: kacprzyk@ibspan.waw.pl

Prof. Lakhmi C. Jain
University of South Australia
Adelaide
Mawson Lakes Campus
South Australia 5095
Australia
E-mail: Lakhmi.jain@unisa.edu.au

Further volumes of this series can be found on our homepage:
springer.com

Jie Lu, Lakhmi C. Jain, and Guangquan Zhang (Eds.)

Handbook on Decision Making

Vol 2: Risk Management in Decision Making

 Springer

Editors

Prof. Jie Lu
University of Technology, Sydney (UTS)
Australia

Prof. Guangquan Zhang
University of Technology, Sydney (UTS)
Australia

Prof. Lakhmi C. Jain
University of South Australia
Australia

ISSN 1868-4394
ISBN 978-3-662-50666-0
DOI 10.1007/978-3-642-25755-1
Springer Heidelberg New York Dordrecht London

e-ISSN 1868-4408
e-ISBN 978-3-642-25755-1

Preface

This book provides a systematic overview of developments in the field of risk management in decision making and outlines state-of-the-art research in fundamental approaches, methodologies, software systems, and applications in this area. It demonstrates how adaptations of intelligent methodologies and technologies benefit the study of risk management in decision making. The book promotes new research development through collaboration with research groups and researchers throughout eleven countries/regions in the world.

Risk management in decision making is more difficult in today's complex and rapidly changing decision environment than ever before. In recent years, both decision optimization under risk, and risk management in decision making have had unimaginable improvements. Decision support systems, risk analysis systems and emergency response systems are playing significantly more important roles in organizations in every discipline, including health, business, engineering, education and finance. At the same time, organizational decision makers and risk response officers are experiencing increasing requirements for advanced knowledge, previously successful experiences, and intelligent technical conditions to enable and support better risk analysis and decision making. As a result, the applications of intelligent methodologies and technologies are improving the functions and performance of these systems. This book, as the title suggests, aims to offer a thorough introduction and systematic overview of various aspects of the field, including both theorems and applications.

The book is organized in three parts, with 21 chapters: (1) Decision making under risk; (2) Risk management in business decision making; (3) Risk assessment and response systems. It provides a comprehensive research record of the theories, methodologies and technologies of risk management in decision making by outlining the application of various intelligence technologies such as fuzzy logic and similarities, agents, and bi-level optimization. It also includes various application-oriented chapters from water-related risk management, real estate investment prediction, road safety management, and supply chain risk management from the practical point of view.

Academic and applied researchers working on risk management, decision making, and management information systems areas will find the book to be a valuable reference resource. The methods, models and systems proposed in this book can be used by a large number of organizations in related applications. Business managers will also directly benefit from the information outlined in this book. Also, final year undergraduate, Masters and PhD students in computer science, information systems, industry engineering, business management, and many other related areas will find that the book is an excellent reference text for their studies.

We wish to thank all the contributors and referees for their excellent work and assistance in producing this publication.

Jie Lu

Decision Systems & e-Service Intelligence lab
Centre for Quantum Computation & Intelligent Systems
School of Software, Faculty of Engineering and Information Technology
University of Technology, Sydney
PO Box 123, Broadway, NSW 2007 Australia

Lakhmi Jain

School of Electrical and Information Engineering
Division of Information Technology, Engineering and the Environment
University of South Australia
Adelaide, Mawson Lakes Campus, SA 5095,
Australia

Guangquan Zhang

Decision Systems & e-Service Intelligence lab
Centre for Quantum Computation & Intelligent Systems
Faculty of Engineering and Information Technology
University of Technology, Sydney
PO Box 123, Broadway, NSW 2007 Australia

Contents

Part I: Decision Making under Risk

Chapter 1: Risk Management in Decision Making
Jie Lu, Lakhmi C. Jain, Guangquan Zhang

Chapter 2: Computational Intelligence Techniques for Risk Management
in Decision Making
İhsan Kaya, Cengiz Kahraman, Selçuk Çebi

Chapter 3: Using Belief Degree Distributed Fuzzy Cognitive Maps for Energy
Policy Evaluation
Lusine Mkrtchyan, Da Ruan

Chapter 4: The Risk of Comparative Effectiveness Analysis for Decision Making Purposes
Patricia Cerrito

Chapter 5: Portfolio Risk Management Modelling by Bi-level Optimization
Todor Stoilov, Krasimira Stoilova

Chapter 6: Possibilistic Decision-Making Models for Portfolio Section Problems
Peijun Guo

Chapter 7: Searching Musical Representative Phrases Using Decision Making Based on Fuzzy Similarities

Emerson Castañeda, Luis Garmendia, Matilde Santos

Chapter 8: A Risk-Based Multi-criteria Decision Support System for Sustainable Development in the Textile Supply Chain

Besoa Rabenasolo, Xianyi Zeng

Part II: Risk Management in Business Decision Making

Chapter 12: Supply Chain Risk Management: Resilience and Business Continuity
Mauricio F. Blos, Hui Ming Wee, Wen-Hsiung Yang

Chapter 13: A Fuzzy Decision System for an Autonomous Car Parking
Carlos Martín Sánchez, Matilde Santos Peñas, Luis Garmendia Salvador

Chapter 14: Risk-Based Decision Making Framework for Investment in the Real Estate Industry
Nur Atiqah Rochin Demong, Jie Lu

Chapter 15: Risk Management in Logistics
Hui Ming Wee, Mauricio F. Blos, Wen-Hsiung Yang

Part III: Risk Assessment and Response Systems

Chapter 16: Natural Disaster Risk Assessment Using Information Diffusion and Geographical Information System
Zhang Jiquan, Liu Xingpeng, Tong Zhijun

Chapter 17: Applications of Social Systems Modeling to Political Risk Management

Gnana K. Bharathy, Barry Silverman

Chapter 18: An Integrated Intelligent Cooperative Model for Water-Related Risk Management and Resource Scheduling

Yong-Sheng Ding, Xiao Liang, Li-Jun Cheng, Wei Wang, Rong-Fang Li

Chapter 19: Determining the Significance of Assessment Criteria for Risk Analysis in Business Associations

Omar Hussain, Khresna Bayu Sangka, Farookh Khadeer Hussain

Chapter 20: Artificial Immune Systems Metaphor for Agent Based Modeling of Crisis Response Operations

Khaled M. Khalil, M. Abdel-Aziz, Taymour T. Nazmy, Abdel-Badeeh M. Salem

Chapter 21: Mobile-Based Emergency Response System Using Ontology-Supported Information Extraction

Khaled Amailef, Jie Lu

Part I

Decision Making under Risk

Chapter 1
Risk Management in Decision Making

Jie Lu[1], Lakhmi C. Jain[2], and Guangquan Zhang[3]

[1] Decision Systems & e-Service Intelligence Laboratory
Centre for Quantum Computation & Intelligent Systems
School of Software, Faculty of Engineering and Information Technology
University of Technology, Sydney (UTS)
P.O. Box 123, Broadway, NSW 2007
Australia
[2] School of Electrical and Information Engineering
University of South Australia
Adelaide, Mawson Lakes Campus, SA 5095
Australia
[3] Decision Systems & e-Service Intelligence Laboratory
Centre for Quantum Computation & Intelligent Systems
School of Software, Faculty of Engineering and Information Technology
University of Technology, Sydney (UTS)
P.O. Box 123, Broadway, NSW 2007, Australia

1 Risk Management and Decision Making

Organizational decision making often occurs in the face of uncertainty about whether a decision maker's choices will lead to benefit or disaster. Risk is the potential that a decision will lead to a loss or an undesirable outcome. In fact, almost any human decision carries some risk, but some decisions are much more risky than others. Risk and decision making are two inter-related factors in organizational management, and they are both related to various uncertainties.

There are several methods used to assess, evaluate, or measure risk in order to support better decision making. Some of them are quantitative and some are more subjective. They are all used in support human decision making.

Risk management is defined as "coordinated activities to direct and control an organization with regard to risk [1]" by the International Organization for Standardization (ISO), where a risk refers to "an uncertain event or set of events which, should it occur, will have an effect on the achievement of objectives [4]". Risk management is an important organisational activity used to identify and assess potential risks, and make appropriate decisions in response to, and to control, those risks.

A practical implementation of the risk management process follows a series of principles and guidelines. A well-established relationship diagram between the risk management process, the implementation principle, and the process framework is defined in [1] (see Figure 1).

Typical application fields of risk management include engineering, finance and banking. In recent years, risk management techniques and methodologies have been

J. Lu, L.C. Jain, and G. Zhang: Handbook on Decision Making, ISRL 33, pp. 3–7.
springerlink.com   © Springer-Verlag Berlin Heidelberg 2012

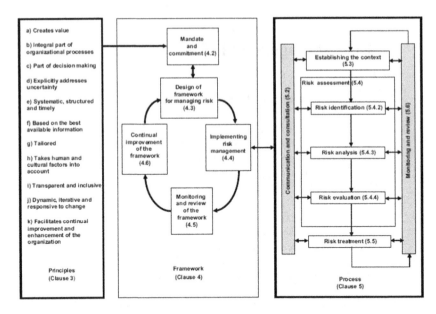

Fig. 1. Relationships between the risk management principles, framework and process [1]

extensively and successfully used in areas such as social management [5], energy management [3] and sustainable development [6], road and food safety [2], disaster forecasting and prediction, industrial and business competitions, and politics. Meanwhile, new challenges to the theories, the methodologies, and the techniques for risk management have emerged from the developments and deployments of risk management implementations.

Decision making is ubiquitous, and is closely related to risk management. On one hand, appropriate decision making is an important task in risk management implementation; on the other hand, a risk management process in decision making is an important step for better decision making. In practice, a decision environment for a real application becomes more and more complex and uncontrollable. Making an appropriate decision is not easy. Various uncertainties, which occur in natural environments and human societies increases the possibility of making inappropriate decisions. How to effectively make a decision, therefore, is a challenging issue for organisations needing to reach their targeted achievements.

Research into combining risk management with decision making has drawn considerable attention from researchers in a variety of disciplines. It includes decision making under risk, as well as risk decision making. In particular, research about how to respond to an emergency and reduce risk in a disaster has recently developed.

This book aims to present innovative theories, methodologies, and techniques in the risk management and decision making field. It introduces new research and technology development to readers interested in the area and provides a comprehensive image of their potential applications. The collected works cover: computational intelligence for risk management, multi-criteria decision making, healthcare modelling, risk forecasting and evaluation, public security and community safety, supply chain optimisation and resource allocation, and business and political risk management.

2   Chapter Outlines

Following Chapter 1, the book has 20 chapters in three parts that cover state-of-the art of research and development of various aspects of risk management in decision making, including both theories and applications.

Part one contains Chapters 2-9, which focus on decision making under risk.

Chapter 2 discusses Computational Intelligence (CI) techniques for risk management in decision making. A detailed classification of existing techniques is presented. Future directions of CI for risk management are presented.

Chapter 3 outlines the use of belief degree in distributed fuzzy cognitive maps for energy policy evaluation. The authors present a tool for dealing with casual reasoning.

Chapter 4 presents the comparative effectiveness analysis for decision making purposes. With reference to cancer drugs, comparative effectiveness analysis tends to compare quality threshold values. The author has used data mining to compare different aspects of comparative analysis for the treatment of osteoporosis with the overall cost of healthcare.

Chapter 5 outlines portfolio risk management modelling by bi-level optimization for investment. The authors present a formal model of optimization for the portfolio problem. They state that the risk of investment can be minimized twice by optimal content of portfolio securities and optimal assessment of the parameter of risk preference.

Chapter 6 presents a set of possibilitic decision making models for portfolio problems. The authors state that, since portfolio experts' knowledge is characterized by the upper and lower possibility distributions, the obtained portfolio will reflect portfolio experts' judgment.

Chapter 7 considers searching musical representation phrases using decision making based on fuzzy similarities. The authors present a method of locating representative phrases from a musical score and validating its superiority by use of examples.

Chapter 8 focuses on a risk-based multi-criteria decision support system for sustainable development in the textile supply chain. The authors have used a method of data aggregation with multiple fuzzy criteria for selecting the most appropriate textile material and the most suitable supplier.

Chapter 9 reports a fuzzy decision system for road safety. This is a significant contribution towards reducing road accidents by proposing corrective actions that should be taken by drivers.

Part two focuses on risk management in business decision making and includes Chapters 10 to15.

Chapter 10 describes a latex price forecasting model to reduce the risk of rubber over-production in Thailand. The model is validated using real rubber latex prices trend data, which in turn is compared with experimental forecasting results to determine forecasting accuracy.

Chapter 11 develops an agent-based model for pandemic influenza in Egypt. The results help us to understand the characteristics of a pandemic and the conditions under which an outbreak occurs.

Chapter 12 focuses on supply chain risk management. The authors present an overview of how better supply chain decision making with risk can be made so that it achieves supply chain resilience and business continuity.

Chapter 13 outlines the development of a fuzzy decision system for autonomous car parking. The authors validated this system using a simulation applet to park successfully in most of the initial conditions.

Chapter 14 presents a risk-based decision making framework for investment in the real estate industry. The proposed framework provides a comprehensive analysis of risk-based decision making for optimal decisions. The framework can be applied to problem solving involving different issues in the decision making process where risk is a factor.

Chapter 15 provides a compressive review of literature on risk management in logistics in order to provide valuable insights for enterprise by understanding the essential logistical risk management concepts.

Part 3 focuses on risk assessment and response systems in Chapters 16 to 21.

Chapter 16 studies natural disaster risk assessment using the information diffusion technique and a geographical information system. It takes grassland fire disasters in Northern China as the case study and tests the reliability of the proposed approach.

Chapter 17 applies a social systems modelling technique to political risk management. The authors present a social system modelling technique and its associated intelligent system tools; these are applied to assess and manage political risk. The agent models are subjected to real-world examples to establish the validity.

Chapter 18 outlines the development of an integrated intelligent cooperative model for water-related risk management and resource scheduling. Two of models are proposed by the authors. Simulation results demonstrate that the first model makes full use of the spatial and time data of a drought so that high accuracy of evaluation and classification of the drought severity can therefore be acquired. The second model distributes water storage between reservoirs timely and efficiently.

Chapter 19 discusses the significance of assessment criteria for risk analysis in business associations. The authors state that the proposed approach is applicable to any activity in any domain to determine the significance of the assessment criteria, while performing risk assessment and management.

Chapter 20 focuses on artificial immune systems for agent based modelling of crisis response operations. The proposed model is applied to the spread of pandemic influenza in Egypt.

Chapter 21 discusses a mobile-based emergency response system using an ontology supported information extraction technique. The proposed scheme can extract many kinds of semantic elements of emergency situation information such as disaster location, disaster event, and status of a disaster.

3   Summary

Organizational decision is usually made under a certain degree of risk. Risk management has been widely and successfully used in many decision problems. This book presents an overview of new developments of risk management and decision making theories and techniques and their applications. Research has shown that many newly emerging problems are still facing managers, users, and organisations in the areas of risk management and decision making. This book aims to address some of those problems and provide innovative, yet practical solutions for organisational decision making.

References and Further Readings

1. AS/NZA ISO 31000: Risk management – Principles and guidelines, Standards Australia (2009)
2. Chong, S., Poulos, R., Olivier, J., Watson, W.L., Grzebieta, R.: Relative injury severity among vulnerable non-motorised road users: compara-tive analysis of injury arising from bicycle-motor vehicle and bicycle-pedestrian collisions. Accident Analysis & Prevention 42(1), 290–296 (2010)
3. Eydeland, A., Wolyniec, K.: Energy and power risk management, new developments in modeling, pricing, and hedging, John Wiley and Sons (2003)
4. Britain, G.: Management of risk: guidance for practitioners. Office of Government Commerce (2007)
5. Holzmann, R., Jorgensen, S.: Social protection as social risk man-agement: conceptual underpinnings for the social protection sector strategy paper. Journal of International Development 11(7), 1005–1027 (1999)
6. Weber, O., Scholz, R.W., Michalik, G.: Incorporating sustainability criteria into credit risk management. Business Strategy and the Environment 19(1), 39–50 (2010)
7. Zhang, J., Lu, J., Zhang, G.: A hybrid knowledge-based risk prediction method using fuzzy logic and CBR for avian influenza early warning. Accepted by Journal of Multi-Valued Logic and Soft Computing 17(4), 363–386 (2011)
8. Zhang, G., Ma, J., Lu, J.: Emergency management evaluation by a fuzzy multi-criteria group decision support system. Stochastic Environmental Research and Risk Assessment 23(4), 517–527 (2009)

Chapter 2
Computational Intelligence Techniques for Risk Management in Decision Making

İhsan Kaya[1], Cengiz Kahraman[2], and Selçuk Çebi[3]

[1] Department of Industrial Engineering, Yıldız Technical University,
Yıldız 34349, Istanbul, Turkey
ikayaem@yahoo.com
[2] Department of Industrial Engineering, Istanbul Technical University,
Maçka 34367, Istanbul, Turkey
kahramanc@itu.edu.tr
[3] Department of Industrial Engineering, Karadeniz Technical University,
61080, Trabzon, Turkey
cebiselcuk@gmail.com

Abstract. Risk management involves assessing risks, evaluating alternatives and implementing solutions and it is a problem of multicriteria decision making (MCDM), where retrofit alternatives are predefined and the decisionmaker(s) (DMs) evaluate them based on multiple criteria. Risk managers choose from a variety of methods to minimize the effects of accidental loss upon their organizations. In the literature it is possible to meet many techniques used for risk management. Besides of these techniques computational intelligence techniques have been used for risk management in decision making process in a wide area. The intelligence can be defined as the capability of a system to adapt its behavior to meet its goals in a range of environments and the life process itself provides the most common form of intelligence. Computational intelligence (CI) can be defined as the study of the design of intelligent agents. In this chapter the CI techniques for risk management in decision making is given with a wide literature review and a detailed classification of the existing methodologies is made. The direction of CI for risk management in the future is evaluated.

1 Introduction

Risk has different meanings to different people; that is, the concept of risk varies according to viewpoint, attitudes and experience. Engineers, designers and contractors view risk from the technological perspective; lenders and developers tend to view it from the economic and financial side; health professionals, environmentalists, chemical engineers take a safety and environmental perspective. Risk is therefore generally seen as an abstract concept whose measurement is very difficult. The Oxford Advanced Learner's Dictionary defines risk as the: ''chance of failure or the possibility of meeting danger or of suffering harm or loss'' (Baloi and Price, 2003). Risk is defined in ISO 31000 which is a family of standards relating to risk management codified by the International Organization for Standardization as the effect of uncertainty

J. Lu, L.C. Jain, and G. Zhang: Handbook on Decision Making, ISRL 33, pp. 9–38.
springerlink.com

on objectives, whether positive or negative. Risks can come from uncertainty in financial markets, project failures, legal liabilities, credit risk, accidents, natural causes and disasters as well as deliberate attacks from an adversary. Risk management is the identification, assessment, and prioritization of risks followed by coordinated and economical application of resources to minimize, monitor, and control the probability and/or impact of unfortunate events or to maximize the realization of opportunities (Hubbard, 2009). Risk analysis generally consists of risk assessment and risk management, where the former generally involves objective elements and the latter involves both objective and subjective elements Risk management is performed to ensure that risk is maintained within an acceptable (regulatory) level to avoid any serious adverse effect to the public and environment by selecting the most suitable alternative (e.g., least cost, most effective) (Tesfamariam et al., 2010). The risk assessment phase consists of risk identification, risk analysis, and risk evaluation. Risk identification is an important stage to identify all of the risks that require management, including those that might not be influenced by decision or recommendation. In identifying the risks it will be necessary to consider: What can happen? When and where? How and why? The next phase for risk assessment is the risk analysis whose objectives are the separation of the minor acceptable risks from the major risks, and the provision of data to assist in the evaluation and treatment of the risks. Risk analysis may employ qualitative, semi-quantitative, or quantitative methodologies. The risk evaluation phase is a relatively simple comparison between the risk criteria and the level of risk found during the risk analysis phase of the risk assessment. The fundamental question being asked here is whether the identified risk is acceptable, given that the risk criteria provide the cut-off between acceptable and unacceptable risk (Watson, 2005).

Fogel (1995) summarized Computational intelligence (CI) as "… These technologies of neural, fuzzy and evolutionary systems were brought together under the rubric of Computational Intelligence, a relatively new term offered to generally describe methods of computation that can be used to adapt solutions to new problems and do not rely on explicit human knowledge." According to Marks (1993), CI consists of neural networks, genetic algorithms, fuzzy systems, evolutionary programming, and artificial life. CI techniques have been used for risk management in decision making process successfully. The techniques are excellent tool to improve decision making process by decreasing uncertainty or improving calculations. For example, risk assessment is often done by DMs, because there is no exact and mathematical solution to the problem. Usually the human reasoning and perception process cannot be expressed precisely. Different DMs have different opinions about risk and its effects. Therefore the association of its dependent variables and the fuzzy set theory which is a CI technique provides an excellent framework to model risk assessment problem. CI is defined as the study of the design of intelligent agents. Since its relation to other branches of computer science is not well-defined, CI means different things to different people. In this chapter CI techniques for risk management in decision making are briefly explained. Then a special literature review on CI for risk management in decision making is summarized. The rest of this chapter is organized as follows: Risk management and its application in decision making are explained in Section 2. CI techniques which are used for risk management are briefly explained in Section 3. A literature survey on CI for risk management in decision making and the results of this survey are summarized in Section 4. The conclusions and further research directions are indicated in Section 5.

2 Risk Management in Decision Making

Risk may be described as the chance of something happening that will have an impact on objectives. Risk is usually considered in terms of consequence (outcome) and likelihood (probability or frequency). Risk management can be defined as the culture, processes, and structures that are directed toward the effective management of potential opportunities and adverse effects. Risk management processes involve first establishing the risk context, which includes the level of risk that is acceptable. Then risks are identified, assessed (analyzed and evaluated against the acceptable level of risk), and where appropriate treated to reduce the likelihood and/or the consequence. The components of such a risk management process, integrated with the essential communication/consultation and monitoring/reviewing processes, are shown schematically in Fig. 1 (Watson, 2005).

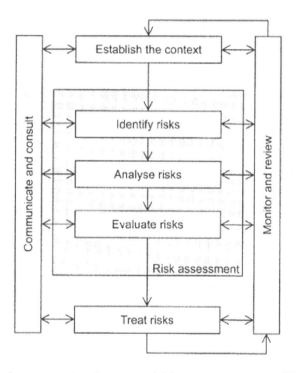

Fig. 1. Schematic representation of a structured risk management process (Watson, 2005)

There are many different techniques in use today for the identification and evaluation of risks. Multicriteria decision making (MCDM) techniques are used to select the best alternatives under multiple and often conflicting criteria. The rapid growth of MCDM over the past three decades is due to a number of factors, including dissatisfaction with conventional "single criterion" methods and the emergence of software and algorithms for solving risk management problems which includes assessment of risks and evaluation of solutions. MCDM can help decision makers in risk management problem to formulate their values and preferences, to quantify these priorities, and to apply them to a particular decision context.

Risk management is a problem of MCDM, where retrofit alternatives A_i ($i = 1, 2, \ldots, m$) are predefined and the decisionmakers (DMs) evaluate them based on multiple criteria C_j ($j = 1, 2, \ldots, n$). A typical MCDM problem with m alternatives and n criteria can be described by the following equation (Tesfamariam et al., 2010):

$$
\begin{array}{c}
\begin{array}{cccc} C_1 & C_2 & \ldots & C_n \end{array} \\
\begin{array}{c} A_1 \\ A_2 \\ \vdots \\ A_m \end{array}
\begin{array}{cccc}
x_{11} & x_{12} & \ldots & x_{1n} \\
x_{21} & x_{22} & \ldots & x_{2n} \\
\vdots & \vdots & \vdots & \vdots \\
x_{m1} & x_{m2} & \ldots & x_{mn}
\end{array}
\end{array}
\tag{1}
$$

where x_{ij} is an exogenous variable, called a performance (payoff or utility) rating of an alternative A_i for a given criterion C_j.

Decision-making problems are broadly categorized into deterministic, stochastic/risk and uncertain. Deterministic problems are those in which data are known with certainty; whereas stochastic problems are those in which data are not known with certainty but can be represented by a probability distribution; and uncertain refers to those problems in which data are not known. Most risk management problems fall into the last two categories because they are poorly structured problems for which few algorithms or mechanical methods exist (Baloi and Price, 2003). In this chapter, CI techniques for risk management in decision making are studied.

3 Computational Intelligence Techniques in Risk Management

A definition of intelligence which is needed to be defined to research in a field termed artificial intelligence and CI has only rarely been provided, and the definitions in the literature have often been of little operational value. The intelligence was defined as the ability of solving the hard problems. However there are basic questions in this definition, such as how hard the problem is or who decides which problem is hard (Chellapilla and Fogel, 1999). Chellapilla and Fogel (1999) defined the intelligence as "the capability of a system to adapt its behavior to meet its goals in a range of environments and the life process itself provides the most common form of intelligence".

The birth of CI is attributed to the IEEE World Congress on Computational Intelligence in 1994 Orlando, Florida. This term combines elements of learning, adaptation, evolution and fuzzy logic (rough sets) to create systems that is, in some sense, intelligent has been investigated by researchers until now. It is a necessity to answer the question what the "computational intelligence" is. CI can be broadly defined as the ability of a machine to react to an environment in new ways, making useful decisions in light of current and previous information. CI is generally accepted to include evolutionary computation, fuzzy systems, neural networks, and combinations thereof. CI, which consists of neural networks, fuzzy logic and evolutionary computing, and so on, is a novel technology to bring intelligence into computation. Compared with the traditional artificial intelligence, a significant characteristic of CI is that the precise model needs not to be established when dealing with imprecise, uncertain, and incomplete information (Kahraman et al., 2010). The term CI was first introduced by Bezdek (1994). Bezdek says "... A system is computationally intelligent when it: deals with only numerical (low-level) data, has a pattern recognition component, does not use knowledge in the AI sense; and additionally when it (begins to) exhibit (i) computational adaptivity; (ii) computational fault tolerance; (iii) speed approaching human-like turnaround, and (iv) error rates that approximate human performance". In the face of these difficulties fuzzy logic (FL), neural networks (NNs) and evolutionary computation (EC) were integrated under the name of CI as a hybrid system (Rudas and Fodor, 2008). While some techniques within CI are often counted as artificial intelligence techniques (e.g., genetic algorithms, or neural networks) there is a clear difference between these techniques and traditional logic based artificial intelligence techniques. In general, typical artificial intelligence techniques are top-to-bottom, where the structure of models, solutions, etc. is imposed. CI techniques are generally bottom-up, where order and structure emerges from an unstructured beginning. An artificial neural network is a branch of CI that is closely related to machine learning. CI is further closely associated with soft computing, connectionist systems and cybernetics. Fogel (1995) summarized CI as "... These technologies of neural, fuzzy and evolutionary systems were brought together under the rubric of Computational Intelligence, a relatively new term offered to generally describe methods of computation that can be used to adapt solutions to new problems and do not rely on explicit human knowledge." According to Marks (1993), CI consists of neural nets, genetic algorithms, fuzzy systems, evolutionary programming, and artificial life.

The CI techniques for risk management in decision making are counted as the fuzzy set theory, neural networks, and evolutionary computation includes genetic algorithms, ant colony optimization, classifier systems, and hybrid systems when the literature review is made. These techniques are briefly summarized as below (Kahraman et al., 2010):

3.1 The Fuzzy Set Theory

Zadeh published the first paper, called "fuzzy sets", on the theory of fuzzy logic (1965). He characterized non-probabilistic uncertainties and provided a methodology, the fuzzy set theory, for representing and computing data and information that are uncertain and imprecise. Zadeh (1996) defined the main contribution of fuzzy logic as "a methodology for computing with words" and pointed out two major necessities for computing with words: "First, computing with words is a necessity when the available information is too imprecise to justify the use of numbers, and second, when there is a tolerance for imprecision which can be exploited to achieve tractability, robustness, low solution cost, and better rapport with reality." While the complexity arise and the precision is receded, linguistic variables has been used to modeling. Linguistic variables are variables which are defined by words or sentences instead of numbers (Zadeh, 1975).

Many problems in real world deal with uncertain and imprecise data so conventional approaches cannot be effective to find the best solution. To cope with this uncertainty, the fuzzy set theory has been developed as an effective mathematical algebra under vague environment. Although humans have comparatively efficient in qualitative forecasting, they are unsuccessful in making quantitative predictions (Karwowski and Evans, 1986). Since fuzzy linguistic models permit the translation of verbal expressions into numerical ones, thereby dealing quantitatively with imprecision in the expression of the importance of each criterion, some methods based on fuzzy relations are used. When the system involves human subjectivity, fuzzy algebra provides a mathematical framework for integrating imprecision and vagueness into the models (Kaya and Çınar, 2008).

Uncertainties can be reflected in mathematical background by fuzzy sets. Fuzzy sets have reasonable differences with crisp (classical) sets. Crisp set A in a universe U can be defined by listing all of its elements denoted x. Alternatively, zero-one membership function, $\mu_A(x)$, which is given below can be used to define x.

$$\mu_A(x) = \begin{cases} 1, & x \in A \\ 0, & x \notin A \end{cases} \tag{2}$$

Unlike crisp sets, a fuzzy set $\tilde{A}$ in the universe of U is defined by a membership function $\mu_{\tilde{A}}(x)$ which takes on values in the interval [0, 1]. The definition of a fuzzy set is the extended version of a crisp set. While the membership function can take the value of 0 or 1 in crisp sets, it takes a value in interval [0, 1] in fuzzy sets. A fuzzy set, $\tilde{A}$, is completely characterized by the set of ordered pairs (Zimmermann, 1987):

$$\tilde{A} = \left\{ \left(x, \mu_{\tilde{A}}(x)\right) \middle| x \in X \right\} \tag{3}$$

A fuzzy set $\tilde{A}$ in X is convex if and only if for every pair of point x^1 and x^2 in X, the membership function of $\tilde{A}$ satisfies the inequality

$$\mu_{\tilde{A}}\left(\delta x^1 + (1-\delta)x^2\right) \geq \min\left(\mu_{\tilde{A}}\left(x^1\right), \mu_{\tilde{A}}\left(x^2\right)\right) \tag{4}$$

where $\delta \in [0, 1]$ (Jahanshahloo et al., 2006).

In fuzzy logic, basic sets operations, union, intersection and complement, are defined in terms of their membership functions. Let $\mu_{\tilde{A}}(x)$ and $\mu_{\tilde{B}}(x)$ be the membership functions of fuzzy sets $\tilde{A}$ and $\tilde{B}$. Fuzzy union has the membership function of

$$\mu_{\tilde{A} \cup \tilde{B}}(x) = \max\left[\mu_{\tilde{A}}(x), \mu_{\tilde{B}}(x)\right] \tag{5}$$

and fuzzy intersection has the membership function of

$$\mu_{\tilde{A}\cap\tilde{B}}(x) = \min\left[\mu_{\tilde{A}}(x), \mu_{\tilde{B}}(x)\right] \tag{6}$$

and fuzzy complement has the membership function of

$$\mu_{\tilde{\bar{A}}}(x) = 1 - \mu_{\tilde{A}}(x) \tag{7}$$

Most real world applications, especially in engineering, consist of real and nonnegative elements such as the Richter magnitudes of an earthquake (Ross, 1995). However there are some applications that some strict numbers cannot be defined for given situation on account of uncertainty and ambiguity. The fuzzy set theory provides a representing of uncertainties. There is a huge amount of literature on fuzzy logic and the fuzzy set theory. In recent studies, the fuzzy set theory has been concerned with engineering applications. Certain types of uncertainties are encountered in a variety of areas and the fuzzy set theory has pointed out to be very efficient to consider these (Kaya and Çınar, 2008). Automatic control, consumer electronics, signal processing, time-series prediction, information retrieval, database management, computer vision, data classification, and decision-making are some areas in which fuzzy logic and the fuzzy set theory are applied (Jang and Sun, 1995).

3.2 Neural Networks

Artificial neural networks (ANNs) are optimization algorithms which are recently being used in variety of applications with great success. ANNs models take inspiration from the basic framework of the brain. Brain is formed by neurons and Figure 2 contains a schematic diagram of real neuron. The main parts of a neuron are cell body, axon and dendrites. A signal transport from axon to dendrites and passing through other neurons by synapses (Fu, 1994). ANNs consist of many nodes and connecting synapses. Nodes operate in parallel and communicate with each other through connecting synapses. ANNs are used effectively for pattern recognition and regression. In recent years, experts prefer ANNs over classical statistical methods as a forecasting model. This increasing interest can be explained by some basic properties of ANNs. Extrapolating from historical data to generate forecasts, solving the complex nonlinear problems successively and high computation rate are features that are the reasons for many experts to prefer ANNs. Moreover, there is no requirement for any assumptions in ANNs (Pal and Mitra, 1992).

Neural networks can be classified according to their architectures as single layer feed-forward networks, multilayer feed-forward networks and recurrent neural network (Haykin, 1999). In feed-forward neural networks signals flow from the input layer to the output layer. Each layer has a certain number of neurons which are the basic processing elements of ANNs. Neurons are connected with the other neurons in further layers and each connection has an associated weight. The neurons connect from one layer to the next layer, not to the previous layer or within the same layer. Multilayer feed-forward neural network is composed of an input layer, some hidden layers and an output layer. Each neuron, except the neurons in input layer, is an operation unit in ANNs. The process starts with summation of weighted activation of

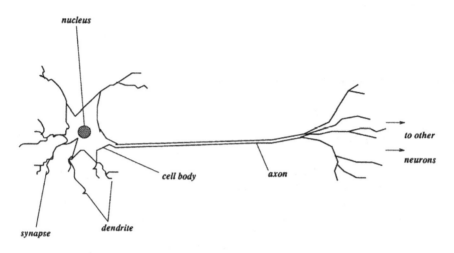

Fig. 2. A schematic diagram of real neuron (Cheng and Titterington, 1994)

other neurons through its incoming connections. Then the weighted sum is passed through a function which is called activation function and this activated value is the output of the neuron. Activation function defines the output of a neuron. There are three basic activation function types (Haykin, 1999):   (i) threshold function, (ii) piecewise-linear function and   (iii) sigmoid function.

The most important section of ANNs is training. Training is another name of seeking the correct weight values. While classical statistical techniques only estimate the coefficient of independent variables, ANNs select proper weights during training and keeps them for further use to predict the output (Cheng and Titterington, 1994). Three types of learning processes are supervised, unsupervised and reinforcement learning.

3.3 Evolutionary Computation

Evolutionary computation concept based on Darwin's evolution theory by applying the biological principle of natural evolution to artificial systems for the solution of optimization problems has received significant attention during the last two decade, although the origins were introduced the late 1950's (Back et al., 1997). The domain of evolutionary computation involves the study of the foundations and the applications of computational techniques based on the principles of natural evolution. Evolutionary algorithms employ this powerful design philosophy to find solutions to hard problems from different domains, including optimization, automatic programming, circuit design, machine learning, economics, ecology, and population genetics, to mention but a few.

There are several techniques of evolutionary computations, among which the best known ones are genetic algorithms, genetic programming, evolution strategies, classifier systems and evolutionary programming; though different in the specifics they are all based on the same general principles (Back et al., 1997; Dimopoulos and Zalzala, 2000; Pena-Reyes and Sipper, 2000).

3.3.1 Genetic Algorithms

Genetic algorithms (GAs) which was first developed by Holland in 1975 are based on mechanics of natural selection and genetics to search through decision space for optimal solutions. The metaphor underlying GAs is natural selection. Genetic algorithm is a search technique used in computing to find exact or approximate solutions to optimization and search problems. GAs can be categorized as global search heuristics. GAs are a particular class of evolutionary algorithms (also known as evolutionary computation) that use techniques inspired by evolutionary biology such as inheritance, mutation, selection, and crossover (also called recombination). GAs are stochastic search methods based on the genetic process of biological organisms. Unlike conventional optimization methods, GAs maintain a set of potential solutions (populations) in each generation. A GA is encoding the factors of a problem by chromosomes, where each gene represents a feature of problem. In evolution, the problem that each species faces is to search for beneficial adaptations to the complicated and changing environment. In other words, each species has to change its chromosome combination to survive in the living world. In GA, a string represents a set of decisions (chromosome combination), that is a potential solution to a problem. Each string is evaluated on its performance with respect to the fitness function (objective function). The ones with better performance (fitness value) are more likely to survive than the ones with worse performance. Then the genetic information is exchanged between strings by crossover and perturbed by mutation. The result is a new generation with (usually) better survival abilities. This process is repeated until the strings in the new generation are identical, or certain termination conditions are met. GAs consist of four main sections that are explained in the following subsections: Encoding, Selection, Reproduction, and Termination (Kaya and Engin, 2007; Engin et al., 2008; Kaya, 2009a; Kaya, 2009b).

3.3.1.1 Encoding. While using GAs, encoding a solution of a problem into a chromosome is very important. Various encoding methods have been created for particular problems to provide effective implementation of Gas for the last 10 years. According to what kind of symbol is used as the alleles of a gene, the encoding methods can be classified as follows (Gen and Cheng, 2000): Binary encoding, real number encoding, integer or literal permutation encoding, and general data structure encoding.

3.3.1.2 Selection. During each successive generation, a proportion of the existing population is selected to breed a new generation. Individual solutions are selected through a fitness-based process, where fitter solutions (as measured by a fitness function) are typically more likely to be selected. Certain selection methods rate the fitness of each solution and preferentially select the best solutions. Other methods rate only a random sample of the population, as this process may be very time-consuming. Most functions are stochastic and designed so that a small proportion of less fit solutions are selected. This helps keep the diversity of the population large, preventing premature convergence on poor solutions. Popular and well-studied selection methods include roulette wheel selection and tournament selection. The fitness function is the important section of selection. It is defined over the genetic representation and measures the quality of the represented solution. It is always problem dependent.

3.3.1.3 Reproduction. The next step of GAs is to generate a second generation population of solutions from those selected through genetic operators: crossover (also called recombination), and/or mutation. For each new solution to be produced, a pair of "parent" solutions is selected for breeding from the pool selected previously. By producing a "child" solution using the crossover and mutation, a new solution is created which typically shares many of the characteristics of its "parents". New parents are selected for each child, and the process continues until a new population of solutions of appropriate size is generated. These processes ultimately result in the next generation population of chromosomes that is different from the initial generation. Generally, the average fitness will have increased by this procedure for the population, since only the best organisms from the first generation are selected for breeding, along with a small proportion of less fit solutions, for reasons already mentioned above.

3.3.2 Genetic Programming

Genetic programming (GP) developed by Koza (1990, 1992) is an evolutionary algorithm-based methodology inspired by biological evolution to find computer programs that perform a user-defined task. It is defined by Koza (1990) as "…genetic programming paradigm, populations of computer programs are genetically bred using the Darwinian principle of survival of the fittest and using a genetic crossover (recombination) operator appropriate for genetically mating computer programs." In GP, instead of encoding possible solutions to a problem as a fixed-length character string, they are encoded as computer programs. The individuals in the population are programs that — when executed — are the candidate solutions to the problem. It is a specialization of genetic algorithms where each individual is a computer program. Therefore, it is a machine learning technique used to optimize a population of computer programs according to a fitness landscape determined by a program's ability to perform a given computational task.

3.3.3 Classifier Systems

Classifier systems (CSs), presented by Holland in 1970's, are evolution-based learning systems, rather than a 'pure' evolutionary algorithm. They can be thought of as restricted versions of classical rule-based systems, with the addition of input and output interfaces. A classifier system consists of three main components: (1) the rule and message system, which performs the inference and defines the behavior of the whole system, (2) the apportionment of credit system, which adapts the behavior by credit assignment, and (3) the GAs, which adapt the system's knowledge by rule discovery (Pena-Reyes and Siper, 2000). CSs exploit evolutionary computation and reinforcement learning to develop a set of condition-action rules (i.e., the classifiers) which represent a target task that the system has learned from on-line experience. There are many models of CSs and therefore also many ways of defining what a learning classifier system is. Nevertheless, all CS models, more or less, comprise four main components: (i) a finite population of condition action rules, called classifiers, that represents the current knowledge of the system; (ii) the performance component, which governs the interaction with the environment; (iii) the reinforcement component, which distributes the reward received from the environment to the classifiers accountable for the rewards obtained; (iv) the discovery component, which is responsible for discovering better rules and improving existing ones through a GA (Holmes et al., 2002).

3.3.4 Ant Colony Optimization

In the early 1990s, ant colony optimization (ACO) which is a class of optimization algorithms modeled on the actions of a real ant colony was introduced by Dorigo and colleagues as a novel nature-inspired metaheuristic for the solution of hard combinatorial optimization problems. The working principles of ACO based on a real ant colony can be explained as follows (Dorigo and Gambardella, 1997):

Real ants are capable of finding the shortest path from a food source to the nest without using visual cues. Also, they are capable of adapting to changes in the environment, e.g. finding a new shortest path once the old one is no longer feasible due to a new obstacle. Consider Figure 3-A: ants are moving on a straight line that connects a food source to their nest. It is well known that the primary means for ants to form and maintain the line is a pheromone trail. Ants deposit a certain amount of pheromone while walking, and each ant probabilistically prefers to follow a direction rich in pheromone. This elementary behaviour of real ants can be used to explain how they can find the shortest path that reconnects a broken line after the sudden appearance of an unexpected obstacle has interrupted the initial path (Figure 3-B). In fact, once the obstacle has appeared, those ants which are just in front of the obstacle cannot continue to follow the pheromone trail and therefore they have to choose between turning right or left. In this situation we can expect half the ants to choose to turn right and the other half to turn left. A very similar situation can be found on the other side of the obstacle (Figure 3-C). It is interesting to note that those ants which choose, by chance, the shorter path around the obstacle will more rapidly reconstitute the interrupted pheromone trail compared to those who choose the longer path. Thus, the shorter path will receive a greater amount of pheromone per time unit and in turn a larger number of ants will choose the shorter path. Due to this positive feedback (autocatalytic) process, all the ants will rapidly choose the shorter path (Figure 3-D). The most interesting aspect of this autocatalytic process is that finding the shortest path around the obstacle seems to be an emergent property of the interaction between the obstacle shape and ants distributed behaviour: although all ants move at approximately the same speed and deposit a pheromone trail at approximately the same rate, it is a fact that it takes longer to contour obstacles on their longer side than on their shorter side which makes the pheromone trail accumulate quicker on the shorter side. It is the ants' preference for higher pheromone trail levels which makes this accumulation still quicker on the shortest path.

Dorigo and Gambardella (1997) applied this philosophy to solve travelling salesman problem. After the initial proof-of-concept application to the traveling salesman problem (TSP), ACO has been applied to many other optimization problems such as assignment problems, scheduling problems, and vehicle routing problems. Recently, researchers have been dealing with the relation of ACO algorithms to other methods for learning and optimization.

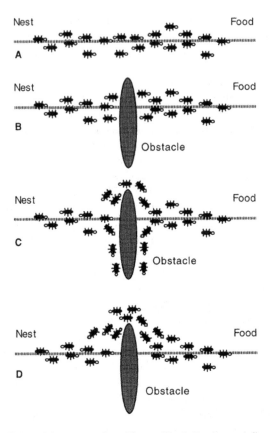

Fig. 3. Real Ant Colony Movements from Nest to Food (Dorigo and Gambardella, 1997)

3.3.5 Particle Swarm Optimization

Particle swarm optimization (PSO) which is population based stochastic optimization technique developed by Kennedy and Eberhart (1995), inspired by social behavior of bird flocking or fish schooling is a global optimization algorithm for dealing with problems in which a best solution can be represented as a point or surface in an n-dimensional space. PSO shares many similarities with evolutionary computation techniques such as GA. The system is initialized with a population of random solutions and searches for optima by updating generations. However, unlike GA, PSO has no evolution operators such as crossover and mutation. In PSO, the potential solutions, called particles, fly through the problem space by following the current optimum particles. Each individual of the population, called a 'particle', flies around in a multidimensional search space looking for the optimal solution. Particles, then, may adjust their position according to their own and their neighboring-particles experience, moving toward their best position or their neighbor's best position. In order to achieve this, a particle keeps previously reached 'best' positions in a cognitive memory. PSO performance is measured according to a predefined fitness function. Balancing between global and local exploration abilities of the flying particles could be achieved through user-defined parameters. PSO has many advantages over other heuristic techniques such that it can be implemented in a few lines of computer code, it requires only primitive mathematical operators, and it has great capability of escaping local optima. In a PSO system, multiple candidate solutions coexist and collaborate simultaneously. Each solution candidate, called a 'particle', flies in the problem search space (similar to the search process for food of a bird swarm) looking for the optimal position to land. A particle, as time passes through his quest, adjusts its position according to its own 'experience', as well as according to the experience of neighboring particles. Tracking and memorizing the best position encountered build particle's experience. For that reason, the PSO algorithm possesses a memory (i.e. every particle remembers the best position it reached during the past). PSO system combines local search methods (through self experience) with global search methods (through neighboring experience), attempting to balance exploration and exploitation. Two factors characterize a particle status on the search space: its position and velocity (Salman et al., 2002).

3.3.6 Hybrid Systems

Solving complex problems, such as financial investment planning, foreign exchange trading, and knowledge discovery from large/multiple databases, involves many different components or sub-tasks, each of which requires different types of processing. To solve such complex problems, a great diversity of intelligent techniques, including traditional hard computing techniques (e.g., expert systems) and CI techniques (e.g., fuzzy logic, NNs, and GAs), are required. For example, in financial investment planning, NNs can be used as a pattern watcher for the stock market; GAs can be used to predict interest rates; and approximate reasoning based on fuzzy logic can be used to evaluate financial risk tolerance ability of clients. These techniques are complementary rather than competitive, and thus must be used in combination and not exclusively. These systems are called hybrid intelligent systems (Zhang and Zhang, 2004). Although the sufficient results found by CI techniques, more effective solutions can be obtained when used combination of these techniques. Each combination has an aim to decrease the limitation of one method. For example, genetic algorithm has been used to improve the performance of ANNs in literature. Usage of genetic algorithms in ANNs for training or finding the appropriate architecture can keep from getting trapped at local minima. Another example is using fuzzy inference with other CI techniques. A fuzzy inference system can take linguistic information (linguistic rules) from human experts and also adapt itself using numerical data to achieve better performance. Generally, using hybrid systems provides synergy to the resulting system in the advantages of the constituent techniques and avoids their shortcomings (Jang and Sun, 1995).

4 Literature Survey

In this section a literature survey on CI for risk management in decision making has been achieved and the results of this review are summarized in Table 1. In the first column, these papers are classified as roughly based on their CI technique(s) in risk management. Then these papers are classified based on their main topics.

Table 1. CI Papers for risk management in decision making

CI Technique(s)	Author(s)	Main Topic(s)
Fuzzy logic	Fanghua and Guanchun (2009), Qin et al. (2008), Chowdhury and Husain (2006), Kangas and Kangas (2004), Prodanovic and Simonovic (2002), Ducey and Larson (1999), Stansbury et al. (1999), Donald and Ross (1996), Jablonowski (1995)	Ecological and Environmental Risk Management
	Tesfamariam et al. (2010)	Seismic Risk Management
	Takács (2009), Gao et al. (2008), Jasinevicius and Petrauskas (2008), Xie et al. (2008), Yang et al. (2007), Cameron and Peloso (2005), Simonovic (2002), Ghyym (1999), Donald and Ross (1995)	Construction of the Risk Management System
	Yu et al. (2009), Liang et al. (2009), Wei (2008), Dia and Zéghal (2008)	Financial Risk Management
	Hong-Jun (2009), Li and Zhang (2008), Pan et al. (2008), Popovic and Popovic (2004)	Supplier Risk Management
	Shan et al. (2009)	Risk Management for Floating Ship-Repair System
	Yuansheng et al. (2008), Au et al. (2006), Shamsuzzaman et al. (2003), Karsak and Tolga (2001),	Risk Management for Investments
	Nasirzadeh et al. (2008), Alvanchi and AbouRizk (2008), Zeng et al. (2007), Zheng and Ng (2005), Lin and Chen (2004), Knight, and Fayek (2002)	Risk Management for Construction Projects
	Guangyu et al. (2008)	Risk Management For Human Resource Management
	Fukayama et al. (2008), Zhou et al. (2003),	Risk Management For Aeronautical Industry
	Zhuang et al. (2008)	Enterprise Risk Management
	Kyoomarsi et al. (2008), Chen 2001,	Software Risk Management

Table 1. (*continued*)

CI Technique(s)	Author(s)	Main Topic(s)
	Jasinevicius and Petrauskas (2008)	Risk Management for Port Security System
	Kleiner et al. (2006)	Risk Management for Buried Infrastructure Assets
	Arenas Parra et al. (2001)	Risk Management for Portfolio Selection
	Baloi and Price (2003), Kuchta (2001), Chen and Chang (2001)	Risk Management for Project Management
	Schmutz et al. (2002)	Risk Management for Electricity Market Environment
	Wu et al. (2010), Lee and Wong (2007)	Financial Risk Management
	Au et al. (2006),	Risk Management for Investments
Fuzzy, NNs	Haslum et al. (2008)	Online Risk Assessment
	Christodoulou et al. (2009)	Ecological and Environmental Risk Management
	Hunter and Serguieva (2000), Zhengyuan and Lihua (2008)	Risk Management for Project Management
Fuzzy, NNs, PSO	Huang et al. (2006)	Financial Risk Management
Fuzzy, GAs	Leu et al. (2000), Ebert and Baisch (1998)	Risk Management for Project Management
	Durga Rao et al. (2007)	Risk Management for Nuclear Power Plant
	Pan et al. (2008)	Risk Management for Cascade Hydropower Stations
GAs	Lu et al. (2009), Sun et al. (2008)	Enterprise Risk Management
	Cao and Gao (2006)	Supplier Risk Management
	Jain et al. (2009), Lu et al. (2008), Sepúlveda et al. (2007)	Risk Management for Electricity Market Environment

Table 1. (*continued*)

CI Technique(s)	Author(s)	Main Topic(s)
	Reneses et al. (2004)	Financial Risk Management
	Kelly et al. (2007), Hegazy et al. (2004)	Ecological and Environmental Risk Management
	Li et al. (2006), Vinod et al. (2004)	Risk Management For Power Plants
GAs, ACO	Marinakis et al. (2008)	Financial Risk Management
GAs, PSO	Sun et al. (2006)	Financial Risk Management
	Cao et al. (2009), Zhao and Chen (2009), Chen et al. (2007), Jangmin et al. (2004), Campi and Pullo (1999)	Financial Risk Management
	Wu et al. (2009), Ge et al. (2008), Cullmann et al. (2007), Alonso-Betanzos et al. (2003), Chan-Yan and Lence (2000)	Ecological and Environmental Risk Management
	Yang and Cui (2009)	Risk Management System for Tourism Sector
NNs	Okoroh et al. (2007), Atoui et al. (2006)	Risk Management for Healthcare
	Wang et al.(2004)	Maritime Risk Assessment
	Tian et al. (2008)	Enterprise Risk Management
	Jin et al. (2008)	Risk Management for Civil Aviation
	Wang and Hua (2008)	Risk Management for Construction Projects
	Ju et al. (2009), Venables et al. (2008), Ross (2002)	Risk Management for Investments
NNs, GAs	Al-Sobiei et al. (2005)	Risk Management for Construction Projects
	Lu et al. (2008a, 2009, 2009a, 2009b)	Enterprise Risk Management
PSO	Azevedo and Vale (2008)	Risk Management for Electricity Market Environment
	Lee et al. (2009)	Risk Management for Project Management

From Table 1, we conclude that the papers using CI techniques for risk management (RM) in decision making have gained more attention in recent years. It seems that CI techniques are generally used for "Ecological and Environmental Risk Management", "Construction of the Risk Management System" and "Financial Risk Management".

In Table 2, the literature review results for the frequency of papers whose titles (T) include "computational intelligence" "and decision making (DM)" are given for the years from 1995 to 2010. When a search was made with respect to paper's title including "computational intelligence" and "decision making", totally 52 articles were viewed. When a search was made with respect to paper's keywords (K) including "computational intelligence" and "decision making", totally 4263 articles were viewed. At the same time, these papers were classified based on their keywords. For example, the number of the papers using *the fuzzy set theory* (FST) is 154.

A pie chart which summarizes the number of the papers on risk management of decision making processes is given in Figure 4. According to Figure 4, the papers

Table 2. Search Result Based on the Publication Years

Year	T: RM T: DM	T: RM K: DM	K: RM K: DM	K: RM K: DM K: FST	K: RM K: DM K: NNs	K: RM K: DM K: GAs	K: RM K: DM K: PSO
2010	3	22	91	6	2		
2009	6	138	621	36	9	4	3
2008	4	151	612	44	8	5	2
2007	7	72	374	11	4	3	2
2006	5	51	346	11	3	3	
2005	4	43	367	5	2	1	
2004	5	46	402	9	4	3	
2003	0	24	349	6	1	1	
2002	4	25	290	7	2		
2001	6	17	160	5	1		
2000	1	12	169	4	2	1	
1999	2	17	169	4	1		
1998	1	9	133	2		1	
1997	0	14	63	1			
1996	3	19	64	1			
1995	1	9	53	2			
TOTAL	52	669	4263	154	39	22	7

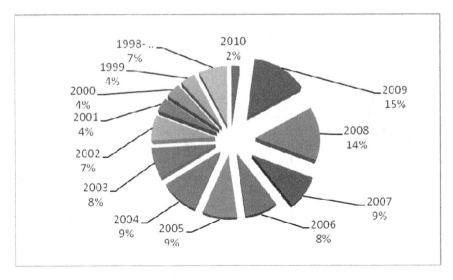

Fig. 4. Distribution of articles on risk management in decision making

related to risk management in decision making have gained more attention year by year. A result that 48% of these papers have been published in the last 5 years can be observed. It is easy to obtain a result that risk management in decision making processes will be gain more attention in the next years.

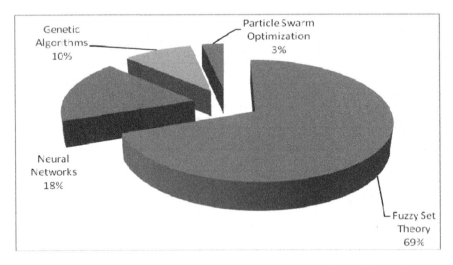

Fig. 5. Percentages of using CI techniques for risk management in decision making

In the other half of Table 2, the literature review has been extended to the papers whose keywords include "risk management", "decision making" and "related CI technique" for the years from 1995 to 2010. The papers were classified based on their keywords. In this case, total number of papers whose keywords include FST, NNs,

GAs, and PSO were determined as 154, 39, 22 and 7, respectively. The percentages of CIs techniques for risk management in decision making are shown in Figure 5. As a result from this pie chart, the FST is the most used CI technique for the risk management in decision making with 69% whereas the NNs and GAs are the followings with 18% and 10%, respectively. PSO is also the least CI technique in this area with 3% since it is a new technique used in the literature.

The papers related to the FST for risk management in decision making are also classified based on their subject areas and the frequency distribution of these papers is illustrated in Figure 6. As it is seen from Figure 6, the most of fuzzy papers are related to "engineering" and totally 92 papers have been published in this area. The next areas are "computer science" and "environmental sciences" with 56 and 22 papers, respectively. A pie chart is drawn to analyze the percentage of these papers with respect to their publication years and the chart is shown in Figure 7.

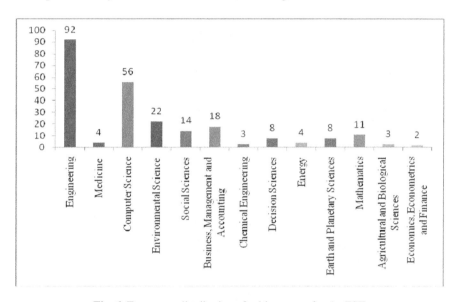

Fig. 6. Frequency distribution of subject areas for the FST

According to Figure 7, 56% of the papers have been published in the last three years and the usage of FST for risk management in decision making is being more popular day by day. According to Figure 7, the papers using FST have the highest percentage in 2008 with 29%. Also the largest increase in percentage is observed from 2007 (7%) to 2008 (29%).

NNs are the second most used CI technique for the risk management in decision making. The papers related to NNs for risk management in decision making are classified with respect to their publication years and a pie chart for this classification is shown in Figure 8. As it can be seen from Figure 8, 66% of the papers have been published for the last five years. Also the usage of NNs for risk management in decision making has an increasing trend.

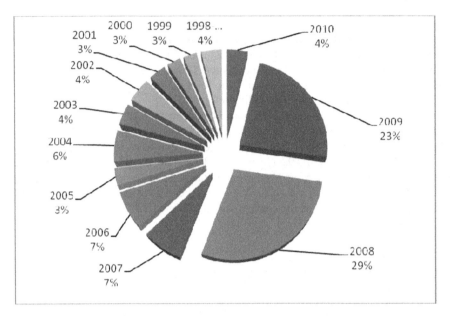

Fig. 7. Distribution of articles based on their publication years on risk management in decision making by using FST

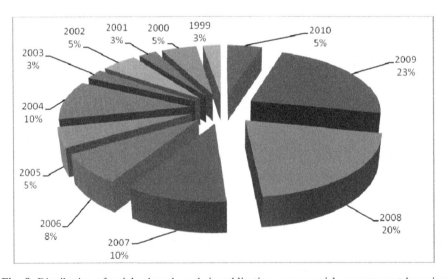

Fig. 8. Distribution of articles based on their publication years on risk management by using NNs

The published papers are also classified based on their subject areas and a pie chart for this analysis is illustrated in Figure 9. As it is seen from Figure 9, the area "Computer Science" is the most popular area for the application of NNs with 32% whereas the areas, "Engineering" and "Medicine", are the following areas with the percentages 28 and 10, respectively.

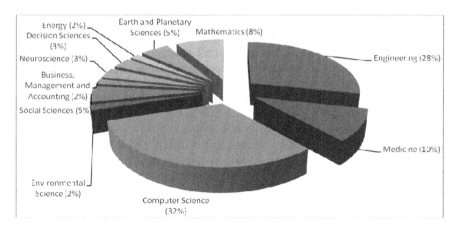

Fig. 9. A pie chart for subject area classification of NNs

The other CI technique which is used for risk management in decision making is GAs. As it is seen from Figure 10, GAs have generally been used in the area "Engineering" for risk management in decision making problems and the percentage of this area is 43. The second most popular area for the using of GAs is the area "Computer Science" with 15%.

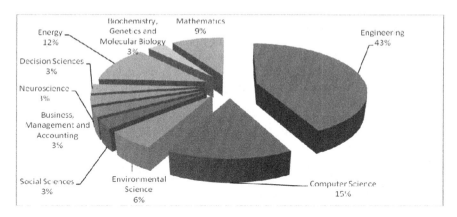

Fig. 10. A pie chart for subject area classification of GAs

5   Conclusions

Risk management is performed to ensure that risk is maintained within an acceptable level to avoid any serious adverse effect to the public and environment by selecting the most suitable alternative. Risk managers make decisions to minimize the effects of risks. They do so using a variety of risk management techniques. These techniques are selected based on problem's characteristics. Multicriteria decision-making analysis (MCDA), as a hybrid methodology of a selected combination of qualitative and quantitative methods, has been broadly used in risk decision-making process. Over the past quarter century the need for risk based decision-making has been increasingly recognized. This chapter surveys the using of CI techniques for risk management in decision making. The CI techniques have been successfully used for modeling and evaluating of risk management problem in decision making. Most of the existing literatures focus on identifying risk, providing risk evaluation method and developing risk management models.

While the DMs make a risk evaluation they should always consider whether their decisions are optimal, whether they use the best evidence available, and whether they help to manage risk effectively. They need risk management to decrease the negative effects of their decisions. Effective risk management often requires identifying the "optimal" risk control option based on multiple uncertain attributes. While traditional utility theory based techniques have been generated to deal with the "multiplicity" of the attributes, many problems regarding their uncertainty are observed, but not well addressed. The main objective of risk management is to improve decision-making by reducing uncertainty which has many different types of sources. The main types of uncertainty include error, imprecision, variability, vagueness, ambiguity and ignorance. The diversity in terms of different types of uncertainty makes the modelling process a very difficult task because the information concerning each specific uncertainty is scarce. Several formal techniques for managing the different types of uncertainty for risk management have been developed and the fuzzy set theory seems as the best technique to manage uncertainty and therefore the fuzzy set theory is the most appropriate CI technique for risk management in decision making. Risk management has traditionally been based on experience and subjective judgment; that is, it features humanistic systems that are not characterized by precision. Since the fuzzy set theory involves human intuitive thinking, it seems a suitable tool for risk management. Also fuzzy logic modeling techniques can be used as well in risk management systems to assess risks level in the cases where the experts have not enough reliable data to apply the statistical approaches.

The results of the literature survey show that such CI techniques can be successfully applied for risk management in decision making. CI includes a highly interdisciplinary framework useful for supporting the design and development of intelligent systems. CI involves innovative models that often come with machine learning, but the researchers do not have any consensus on these models. Neural Networks, genetic algorithms, and fuzzy systems are common methods encountered for risk management in decision making. Clusterization, classification, and approximation capabilities of CI systems are improving day by day and so many methods have been already developed that it is always possible to find alternative solutions with these CI techniques.

To obtain a better performance two or more CI techniques can be integrated for risk management. In the recent years, evolutionary computation methods PSO and GAs and NNs have been widely used together to solve decision-making problems due to their excellent performances in treating non-linear data with self-learning capability. Although the fuzzy set theory is the most used CI technique for risk management in decision making it has a disadvantage that is the lack of effective learning capability. The other CI technique, NNs, can be used to eliminate this disadvantage. The neural learning technique can improve the performance model and make the risk assessment model more robust.

In the future research, the usage of intelligent systems which involves one or more CI techniques or hybrid systems which is a combination of CI techniques to integrate the advantages of these techniques can be analyzed for risk management problems.

References

Alonso-Betanzos, A., Fontenla-Romero, O., Guijarro-Berdiñas, B., Hernández-Pereira, E., Paz Andrade, M.I., Jiménez, E., Soto, J.L.L., Carballas, T.: An intelligent system for forest fire risk prediction and fire fighting management in Galicia. Expert Systems with Applications 25(4), 545–554 (2003)

Al-Sobiei, O.S., Arditi, D., Polat, G.: Managing owner's risk of contractor default. Journal of Construction Engineering and Management 131(9), 973–978 (2005)

Alvanchi, A., AbouRizk, S.: Applying fuzzy set theory for the project risk management using MATLAB fuzzy tool box. In: Proceedings, Annual Conference - Canadian Society for Civil Engineering, vol. 1, pp. 368–378 (2008)

Arenas Parra, M., Bilbao Terol, A., Rodríguez Uría, M.V.: A fuzzy goal programming approach to portfolio selection. European Journal of Operational Research 133(2), 287–297 (2001)

Atoui, H., Fayn, J., Gueyffier, F., Rubel, P.: Cardiovascular risk stratification in decision support systems: A probabilistic approach. Application to pHealth, Computers in Cardiology 33, 281–284 (2006)

Au, K.F., Wong, W.K., Zeng, X.H.: Decision model for country site selection of overseas clothing plants. International Journal of Advanced Manufacturing Technology 29(3-4), 408–417 (2006)

Azevedo, F., Vale, Z.A.: A short-term risk management tool applied to OMEL electricity market using Particle Swarm Optimization. In: 5th International Conference on the European Electricity Market, pp. 1–6 (2008)

Back, T., Hammel, U., Schwefel, H.P.: Evolutionary computation: comments on the history and current state. IEEE Transactions on Evolutionary Computation 1(1), 3–17 (1997)

Baloi, D., Price, A.D.F.: Modelling global risk factors affecting construction cost performance. International Journal of Project Management 21(4), 261–269 (2003)

Bezdek, J.C.: What is computational intelligence? In: Zurada, J.M., Marks, R.J., Robinson, C.J. (eds.) Computational Intelligence, Imitating Life, pp. 1–12. IEEE Computer Society Press (1994)

Cameron, E., Peloso, G.F.: Risk management and the precautionary principle: A fuzzy logic model. Risk Analysis 25(4), 901–911 (2005)

Campi, C., Pullo, M.: Project of a new products line: An application of neural networks. Production Planning and Control 10(4), 317–323 (1999)

Cao, H., Gao, Y.: Penalty guided genetic algorithm for partner selection problem in agile manufacturing environment. In: Proceedings of the World Congress on Intelligent Control and Automation (WCICA), vol. 1, pp. 3276–3280 (2006)

Cao, L.-J., Liang, L.J., Li, Z.-X.: The research on the early-warning system model of operational risk for commercial banks based on BP neural network analysis. In: Proceedings of the 2009 International Conference on Machine Learning and Cybernetics, vol. 5, pp. 2739–2744 (2009)

Chan-Yan, D.A., Lence, B.J.: Turbidity modelling for reservoirs: The development of risk-based performance indicators. In: Annual Conference Abstracts- Canadian Society for Civil Engineering, p. 40 (2000)

Chellapilla, K., Fogel, D.B.: Evolution, Neural Networks, Games, and Intelligence. Proceedings of the IEEE 87(9), 1471–1496 (1999)

Chen, Q.S., Zhang, D.F., Wei, L.J., Chen, H.W.: A modified genetic programming for behavior scoring problem. In: Proceedings of the 2007 IEEE Symposium on Computational Intelligence and Data Mining, CIDM 2007, pp. 535–539 (2007)

Chen, S.M.: Fuzzy group decision making for evaluating the rate of aggregative risk in software development. Fuzzy Sets and Systems 118(1), 75–88 (2001)

Chen, S.M., Chang, T.H.: Finding multiple possible critical paths using fuzzy PERT. IEEE Transactions on Systems, Man, and Cybernetics, Part B: Cybernetics 31(6), 930–937 (2001)

Cheng, B., Titterington, D.M.: Neural Networks: A Review from a Statistical Perspective. Statistical Science 9(1), 2–54 (1994)

Chowdhury, S., Husain, T.: Evaluation of drinking water treatment technology: An entropy-based fuzzy application. Journal of Environmental Engineering 132(10), 1264–1271 (2006)

Christodoulou, S., Deligianni, A., Aslani, P., Agathokleous, A.: Risk-based asset management of water piping networks using neurofuzzy systems. Computers, Environment and Urban Systems 33(2), 138–149 (2009)

Cullmann, J., Schmitz, G.H., Gorner, W.: A new system for online flood forecasting - Performance and implications, vol. (317), pp. 330–336. IAHS-AISH Publication (2007)

Dia, M., Zéghal, D.: Fuzzy evaluation of risk management profiles disclosed in corporate annual reports, Canadian. Journal of Administrative Sciences 25(3), 237–254 (2008)

Dimopoulos, C., Zalzala, A.M.S.: Recent developments in evolutionary computation for manufacturing optimization: problems, solutions, and comparisons. IEEE Transactions on Evolutionary Computation 4(2), 93–113 (2000)

Donald, S., Ross, T.J.: Fuzzy multi-objective approach to risk management. Computing in Civil Engineering 2, 1400–1403 (1995)

Donald, S., Ross, T.J.: Use of fuzzy logic and similarity measures in the risk management of hazardous waste sites. Computing in Civil Engineering, 376–382 (1996)

Dorigo, M., Gambardella, L.M.: Ant colonies for the travelling salesman problem. BioSystems 43, 73–81 (1997)

Ducey, M.J., Larson, B.C.: A fuzzy set approach to the problem of sustainability. Forest Ecology and Management 115(1), 29–40 (1999)

Durga Rao, K., Gopika, V., Kushwaha, H.S., Verma, A.K., Srividya, A.: Test interval optimization of safety systems of nuclear power plant using fuzzy-genetic approach. Reliability Engineering and System Safety 92(7), 895–901 (2007)

Ebert, C., Baisch, E.: Industrial application of criticality predictions in software development. In: Proceedings of the International Symposium on Software Reliability Engineering, ISSRE, pp. 80–89 (1998)

Engin, O., Çelik, A., Kaya, İ.: A fuzzy approach to define sample size for attributes control chart in multistage processes: An application in engine valve manufacturing process. Applied Soft Computing 8(4), 1654–1663 (2008)

Fanghua, H., Guanchun, C.: A fuzzy multi-criteria group decision-making model based on weighted bord a scoring method for watershed ecological risk management: a case study of three gorges reservoir area of China. Water Resources Management, 1–27 (2009)

Fogel, D.: Review of Computational intelligence: imitating life. IEEE Transactions Neural Networks 6, 1562–1565 (1995)

Fu, L.: Neural Networks in Computer Intelligence. McGraw-Hill, United States of America (1994)

Fukayama, H., Fernandes, E., Ebecken, N.F.F.: Risk management in the aeronautical industry: Results of an application of two methods. WIT Transactions on Information and Communication Technologies 39, 195–204 (2008)

Gao, Y., Zhang, G., Lu, J., Dillon, T., Zeng, X.: A λ-cut approximate algorithm for goal-based bilevel risk management systems. International Journal of Information Technology and Decision Making 7(4), 589–610 (2008)

Ge, Y., Liu, J., Li, F., Shi, P.: Quantifying social vulnerability for flood disasters of insurance company. Journal of Southeast University, 147–150 (2008)

Gen, M., Cheng, R.: Genetic Algorithms and Engineering Optimization. John Wiley and Sons, New York (2000)

Ghyym, S.H.: Semi-linguistic fuzzy approach to multi-actor decision-making: Application to aggregation of experts' judgments. Annals of Nuclear Energy 26(12), 1097–1112 (1999)

Guangyu, Z., Huajun, L., Depeng, Z.: Risk management on knowledge employees turnover in high-tech firms - Based on fuzzy comprehensive evaluation. In: Proceedings of the International Conference on Information Management, Innovation Management and Industrial Engineering, ICIII 2008, vol. 3, pp. 399–402 (2008)

Haslum, K., Abraham, A., Knapskog, S.: HiNFRA: Hierarchical neuro-fuzzy learning for on-line risk assessment. In: Proceedings - 2nd Asia International Conference on Modelling and Simulation, AMS 2008, pp. 631–636 (2008)

Haykin, S.: Neural Networks: A Comprehensive Foundation. Prentice Hall (1999)

Hegazy, T., Elbeltagi, E., El-Behairy, H.: Bridge deck management system with integrated life-cycle cost optimization. Transportation Research Record 1866, 44–50 (2004)

Holmes, J.H., Lanzi, P.L., Stolzmann, W., Wilson, S.W.: Learning classifier systems: New models, successful applications. Information Processing Letters 82, 23–30 (2002)

Hong-Jun, G.: The application research of rough set on risk management of suppliers. In: Proceedings - 2009 International Conference on Electronic Computer Technology, ICECT 2009, pp. 682–685 (2009)

Huang, F.Y., Li, R.J., Liu, H.X., Li, R.: A modified Particle Swarm Algorithm combined with Fuzzy Neural Network with application to financial risk early warning. In: Proceedings of 2006 IEEE Asia-Pacific Conference on Services Computing, pp. 168–173 (2006)

Hubbard, D.: The failure of risk management: Why it's broken and how to fix it. John Wiley & Sons (2009)

Hunter, J., Serguieva, A.: Project risk evaluation using an alternative to the standard present value criteria. Neural Network World 10(1), 157–172 (2000)

Jablonowski, M.: Recognizing knowledge imperfection in the risk management process. In: Annual Conference of the North American Fuzzy Information Processing Society - NAFIPS, pp. 1–4 (1995)

Jahanshahloo, G.R., Lotfi, F.H., Izadikhah, M.: Extension of the TOPSIS method for decision-making problems with fuzzy data. Applied Mathematics and Computation 181, 1544–1551 (2006)

Jain, A.K., Srivastava, S.C.: Strategic bidding and risk assessment using genetic algorithm in electricity markets. International Journal of Emerging Electric Power Systems 10(5), 1–20 (2009)

Jang, J.S.R., Sun, C.T.: Neuro-Fuzzy modeling and control. Proceedings of the IEEE 83(3), 378–406 (1995)

Jang, J.S.R., Sun, C.T.: Neuro-Fuzzy modeling and control. Proceedings of the IEEE 83(3), 378–406 (1995)

Jangmin, O., Lee, J. W., Lee, J., Zhang, B.-T.: Dynamic Asset Allocation Exploiting Predictors in Reinforcement Learning Framework. In: Boulicaut, J.-F., Esposito, F., Giannotti, F., Pedreschi, D. (eds.) ECML 2004. LNCS (LNAI), vol. 3201, pp. 298–309. Springer, Heidelberg (2004)

Jasinevicius, R., Petrauskas, V.: Fuzzy expert maps for risk management systems. In: US/EU-Baltic International Symposium: Ocean Observations, Ecosystem-Based Management and Forecasting - Provisional Symposium Proceedings, Baltic, pp. 1–4 (2008)

Jin, L., Yan-Qi, C., Pei, H.: The risk monitoring system of China's civil aviation industry. In: Proceedings of the Flight Safety Foundation Annual International Air Safety Seminar, International Federation of Airworthiness International Conference, and the International Air Transport Association, vol. 2, pp. 669–708 (2008)

Ju, Y.-J., Qiang, M., Qian, Z.: A study on risk evaluation of real estate project based on BP neural Networks. In: International Conference on E-Business and Information System Security, EBISS 2009, pp. 1–4 (2009)

Kahraman, C., Kaya, İ., Çınar, D.: Computational Intelligence: Past, Today, and Future. In: Ruan, D. (ed.) Computational Intelligence in Complex Decision Systems, pp. 1–46. Atlantis Press (2010)

Kangas, A.S., Kangas, J.: Probability, possibility and evidence: Approaches to consider risk and uncertainty in forestry decision analysis. Forest Policy and Economics 6(2), 169–188 (2004)

Karsak, E., Tolga, E.: Fuzzy multi-criteria decision-making procedure for evaluating advanced manufacturing system investments. International Journal of Production Economics 69(1), 49–64 (2001)

Karwowski, W., Evans, G.W.: Fuzzy Concepts in Production Management Research – A Review. International Journal of Production Research 24(1), 129–147 (1986)

Kaya, İ.: A genetic algorithm approach to determine the sample size for attribute control charts. Information Sciences 179(10), 1552–1566 (2009a)

Kaya, İ.: A genetic algorithm approach to determine the sample size for control charts with variables and attributes. Expert Systems with Applications 36(5), 8719–8734 (2009b)

Kaya, İ., Çınar, D.: Facility Location Selection Using A Fuzzy Outranking Method. Journal of Multiple-Valued Logic and Soft Computing 14, 251–263 (2008)

Kaya, İ., Engin, O.: A new approach to define sample size at attributes control chart in multistage processes: An application in engine piston manufacturing process. Journal of Materials Processing Technology 183(1), 38–48 (2007)

Kelly, M., Guo, Q., Liu, D., Shaari, D.: Modeling the risk for a new invasive forest disease in the United States: An evaluation of five environmental niche models. Computers, Environment and Urban Systems 31(6), 689–710 (2007)

Kennedy, J., Eberhart, R.C.: Particle swarm optimization. In: Proceedings of the IEEE International Conference on Neural Networks, pp. 1942–1948 (December 1995)

Kleiner, Y., Rajani, B., Sadiq, R.: Failure risk management of buried infrastructure using fuzzy-based techniques. Journal of Water Supply: Research and Technology - AQUA 55(2), 81–94 (2006)

Knight, K., Fayek, A.R.: Use of fuzzy logic for predicting design cost overruns on building projects. Journal of Construction Engineering and Management 128(6), 503–512 (2002)

Koza, J.R.: Genetic programming: a paradigm for genetically breeding populations of computer programs to solve problems, Technical Report STANCS-90-1314, Department of Computer Science, Stanford University (1990)

Koza, J.R.: Genetic Programming. MIT Press, Cambridge (1992)

Kuchta, D.: Use of fuzzy numbers in project risk (criticality) assessment. International Journal of Project Management 19(5), 305–310 (2001)

Kyoomarsi, F., Dehkordy, P.K., Peiravy, M.H., Heidary, E.: Using UML and fuzzy logic in optimizing risk management modeling. In: MCCSIS 2008 - IADIS Multi Conference on Computer Science and Information Systems, Proceedings of Informatics 2008 and Data Mining, pp. 177–182 (2008)

Lee, K.C., Lee, N., Li, H.: A particle swarm optimization-driven cognitive map approach to analyzing information systems project risk. Journal of the American Society for Information Science and Technology 60(6), 1208–1221 (2009)

Lee, V.C.S., Wong, H.T.: A multivariate neuro-fuzzy system for foreign currency risk management decision making. Neurocomputing 70(4-6), 942–951 (2007)

Leu, S.-S., Chen, A.-T., Yang, C.-H.: A GA-based fuzzy optimal model for construction time-cost trade-off. International Journal of Project Management 19(1), 47–58 (2000)

Li, T., Zhang, Q.: Interaction analysis among the risk management in the supplier network. In: Proceedings of the 8th International Conference of Chinese Logistics and Transportation Professionals - Logistics: The Emerging Frontiers of Transportation and Development in China, pp. 4768–4773 (2008)

Li, Y.G., Shen, J., Liu, X.C.: New bidding strategy for power plants based on chance-constrained programming. Proceedings of the Chinese Society of Electrical Engineering 26(10), 120–123 (2006)

Liang, X., Chen, D., Ruan, D., Tang, B.: Evaluation models of insurers' risk management based on large system theory. Stochastic Environmental Research and Risk Assessment 23(4), 415–423 (2009)

Lin, C.T., Chen, Y.T.: Bid/no-bid decision-making - A fuzzy linguistic approach. International Journal of Project Management 22(7), 585–593 (2004)

Lu, F., Huang, M., Wang, X.: PSO based stochastic programming model for risk management in Virtual Enterprise. In: 2008 IEEE Congress on Evolutionary Computation, 682–687 (2008a)

Lu, F.Q., Huang, M., Ching, W.K., Wang, X.W., Sun, X.L.: Multi-swarm particle swarm optimization based risk management model for virtual enterprise. In: World Summit on Genetic and Evolutionary Computation, 2009 GEC Summit - Proceedings of the 1st ACM/SIGEVO Summit on Genetic and Evolutionary Computation, GEC 2009, pp. 387–392 (2009)

Lu, F.Q., Huang, M., Wang, X.-W.: Two levels distributed decision making model for risk management of virtual enterprise. Computer Integrated Manufacturing Systems 15(10), 1930–1937 (2009a)

Lu, F.Q., Huang, M., Wang, X.-W.: Risk management of virtual enterprise based on stochastic programming and GA. Journal of Northeastern University 30(9), 1241–1244 (2009b)

Lu, G., Wen, F., Chung, C.Y., Wong, K.P.: Mid-term operation planning with risk management of generation company in electricity market environment. Electric Power Automation Equipment 28(7), 1–6 (2008)

Marinakis, Y., Marinaki, M., Doumpos, M., Matsatsinis, N., Zopounidis, C.: Optimization of nearest neighbor classifiers via metaheuristic algorithms for credit risk assessment. Journal of Global Optimization 42(2), 279–293 (2008)

Marks, R.: Computational versus artificial. IEEE Transactions on Neural Networks 4, 737–739 (1993)

Nasirzadeh, F., Afshar, A., Khanzadi, M., Howick, S.: Integrating system dynamics and fuzzy logic modelling for construction risk management. Construction Management and Economics 26(11), 1197–1212 (2008)

Okoroh, M.I., Ilozor, B.D., Gombera, P.: An artificial neural networks approach in managing healthcare. International Journal of Electronic Healthcare 3(3), 279–292 (2007)

Pal, S.K., Mitra, S.: Multilayer perceptron, fuzzy sets, and classification. IEEE Transactions on Neural Networks 3(5), 683–697 (1992)

Pan, B., Chi, H., Xu, J., Qi, M.: Decision-making model for risk management of cascade hydropower stations. Journal of Southeast University 24, 22–26 (2008)

Pan, W., Wang, S., Zhang, J., Hua, G., Fang, Y.: Fuzzy multi-objective order allocation model for risk management in a supply chain. In: Proceedings - 2nd Asia International Conference on Modelling and Simulation, AMS 2008, pp. 771–776 (2008)

Pena-Reyes, C.A., Siper, M.: Evolutionary computation in medicine: an overview. Artificial Intelligence in Medicine 19, 1–23 (2000)

Popovic, D.S., Popovic, Z.N.: A Risk management procedure for supply restoration in distribution networks. IEEE Transactions on Power Systems 19(1), 221–228 (2004)

Prodanovic, P., Simonovic, S.P.: Comparison of fuzzy set ranking methods for implementation in water resources decision-making. Canadian Journal of Civil Engineering 29(5), 692–701 (2002)

Qin, X.S., Huang, G.H., Li, Y.P.: Risk management of BTEX contamination in ground water - An integrated fuzzy approach. Ground Water 46(5), 755–767 (2008)

Reneses, J., Centeno, E., Barquín, J.: Medium-term marginal costs in competitive generation power markets. IEE Proceedings: Generation, Transmission and Distribution 151(5), 604–610 (2004)

Ross, C.P.: Refining AVO interpretation for reservoir characterization with neural networks. In: Proceedings of the Annual Offshore Technology Conference, pp. 1319–1325 (2002)

Ross, T.: Fuzzy Logic with Engineering Applications. McGraw-Hill, USA (1995)

Rudas, I.J., Fodor, J.: Intelligent Systems. International Journal of Computers Communications & Control III, 132–138 (2008)

Salman, A., Ahmad, I., Al-Madani, S.: Particle swarm optimization for task assignment problem. Microprocessors and Microsystems 26, 363–371 (2002)

Schmutz, A., Gnansounou, E., Sarlos, G.: Economic performance of contracts in electricity markets: A fuzzy and multiple criteria approach. IEEE Transactions on Power Systems 17(4), 966–973 (2002)

Sepúlveda, M., Onetto, E., Palma-Behnke, R.: Iterative heuristic response surface method for transmission expansion planning. Journal of Energy Engineering 133(2), 69–77 (2007)

Shamsuzzaman, M., Sharif Ullah, A.M.M., Bohez, E.L.J.: Applying linguistic criteria in FMS selection: Fuzzy-set-AHP approach. Integrated Manufacturing Systems 14(3), 247–254 (2003)

Shan, X.-L., Yu, Q., Tian, J.: Risk management of mooring operation of floating dock. Journal of Tianjin University Science and Technology 42(4), 335–339 (2009)

Simonovic, S.P.: Understanding risk management. In: Annual Conference of the Canadian Society for Civil Engineering, pp. 1–12 (2002)

Stansbury, J., Bogardi, I., Stakhiv, E.Z.: Risk-cost optimization under uncertainty for dredged material disposal. Journal of Water Resources Planning and Management 125(6), 342–351 (1999)

Sun, J., Xu, W., Fang, W.: Solving Multi-period fINANCIAL Planning Problem Via Quantum-behaved Particle Swarm Algorithm. In: Huang, D.-S., Li, K., Irwin, G.W. (eds.) ICIC 2006. LNCS (LNAI), vol. 4114, pp. 1158–1169. Springer, Heidelberg (2006)

Sun, X., Huang, M., Wang, X.: Organizational-DDM risk management model and algorithm for virtual enterprise. In: Proceedings of the 27th Chinese Control Conference, pp. 552–556 (2008)

Takács, M.: Multilevel fuzzy model for risk management. In: ICCC 2009 - IEEE 7th International Conference on Computational Cybernetics, pp. 153–156 (2009)

Tesfamariam, S., Sadiq, R., Najjaran, H.: Decision making under uncertainty - An example for seismic risk management. Risk Analysis 30(1), 78–94 (2010)

Tian, Y., Chen, J., Wang, X., Shi, C.: The research on B to C e-commerce trade risk identification system based on SOM neural network. In: Proceedings of International Conference on Risk Management and Engineering Management, pp. 133–136 (2008)

Venables, A., Dickson, M., Boon, P.: Use of neural networks for integrated assessments of wetlands in the Gippsland region of south-east Australia. In: Proceedings of the 2008 International Conference on Data Mining, DMIN 2008, pp. 185–190 (2008)

Vinod, G., Kushwaha, H.S., Verma, A.K., Srividya, A.: Optimisation of ISI interval using genetic algorithms for risk informed in-service inspection. Reliability Engineering and System Safety 86(3), 307–316 (2004)

Wang, J., Sii, H.S., Yang, J.B., Pillay, A., Yu, D., Liu, J., Maistralis, E., Saajedi, A.: Use of advances in technology for maritime risk assessment. Risk Analysis 24(4), 1041–1063 (2004)

Wang, T., Hua, Z.: Research on construction projects decision based on Life cycle engineering model. In: Proceedings of International Conference on Risk Management and Engineering Management, pp. 132–135 (2008)

Watson, D.B.: Aeromedical Decision-Making: An Evidence-Based Risk Management Paradigm. Aviation, Space, and Environmental Medicine 76(1), 58–62 (2005)

Wei, R.: Development of credit risk model based on fuzzy theory and its application for credit risk management of commercial banks in China. In: International Conference on Wireless Communications, Networking and Mobile Computing, WiCOM 2008, pp. 1–4 (2008)

Wu, C., Guo, Y., Zhang, X., Xia, H.: Study of personal credit risk assessment based on support vector machine ensemble. International Journal of Innovative Computing, Information and Control 6(5), 2353–2360 (2010)

Wu, Y., Yin, K., Jiang, W.: Study on risk assessment and management of landslide hazard in New Bodong County, Three Gorge Reservoir. In: Proceedings - International Conference on Management and Service Science, MASS 2009, pp. 1–4 (2009)

Xie, G., Zhang, J., Lai, K.K., Yu, L.: Variable precision rough set for group decision-making: An application. International Journal of Approximate Reasoning 49(2), 331–343 (2008)

Yang, J., Cui, Z.: BP neural network for the risk management of MICE. In: Proceedings of the 1st International Workshop on Education Technology and Computer Science, ETCS 2009, vol. 1, pp. 119–122 (2009)

Yang, Z.L., Bonsall, S., Wang, J., Wong, S.: Risk management with multiple uncertain decision making attributes. In: Proceedings of the European Safety and Reliability Conference, ESREL 2007 - Risk, Reliability and Societal Safety, vol. 1, pp. 763–772 (2007)

Yu, L., Wang, S., Lai, K.K.: An intelligent-agent-based fuzzy group decision making model for financial multicriteria decision support: The case of credit scoring. European Journal of Operational Research 195(3), 942–959 (2009)

Yuansheng, H., Yufang, L., Jiajia, D.: Method of risk management in realty item based on decomposition of process and variable weight fuzzy evaluation. In: International Conference on Wireless Communications, Networking and Mobile Computing, WiCOM 2008, pp. 1–4 (2008)

Zadeh, L.A.: Fuzzy Sets. Information and Control 8, 338–353 (1965)

Zadeh, L.A.: The concept of a linguistic variable and its application to approximate reasoning. Information Sciences 8, 199–249 (1975)

Zadeh, L.A.: Fuzzy Logic Equals Computing with Words. IEEE Transactions on Fuzzy Systems 4(2), 103–111 (1996)

Zeng, J., An, M., Smith, N.J.: Application of a fuzzy based decision making methodology to construction project risk assessment. International Journal of Project Management 25(6), 589–600 (2007)

Zhang, Z., Zhang, C. (eds.): Agent-Based Hybrid Intelligent Systems. LNCS (LNAI), vol. 2938. Springer, Heidelberg (2004)

Zhao, S.-F., Chen, L.C.: The BP neural networks applications in bank credit risk management system. In: Proceedings of the 8th IEEE International Conference on Cognitive Informatics, ICCI 2009, pp. 527–532 (2009)

Zheng, D.X.M., Ng, S.T.: Stochastic time-cost optimization model incorporating fuzzy sets theory and nonreplaceable front. Journal of Construction Engineering and Management 131(2), 176–186 (2005)

Zhengyuan, J., Lihua, G.: The project risk assessment based on rough sets and neural network (RS-RBF). In: International Conference on Wireless Communications, Networking and Mobile Computing, WiCOM 2008, pp. 1–4 (2008)

Zhou, P., Zhu, S.L., Jiang, S.S.: Risk management of aeronautic projects with fuzzy analytic hierarchy process. Computer Integrated Manufacturing Systems, CIMS 9(12), 1062–1066 (2003)

Zhuang, P., Li, Y., Hou, H., Qu, T.: Visualization of current core research groups of enterprise risk management. In: Proceedings - 5th International Conference on Fuzzy Systems and Knowledge Discovery, FSKD 2008, vol. 4, pp. 310–314 (2008)

Zimmermann, H.J.: Fuzzy sets, decision making, and expert systems, Boston (1987)

Chapter 3
Using Belief Degree Distributed Fuzzy Cognitive Maps for Energy Policy Evaluation

Lusine Mkrtchyan[1] and Da Ruan[2]

[1] IMT Lucca Institute for Advanced Studies,
P.zza San Ponziano 6, 55100 Lucca, Italy
l.mkrtchyan@imtlucca.it
[2] Belgian Nuclear Research Centre (SCK•CEN), 2400 Mol, Belgium
druan@sckcen.be

Abstract. *Cognitive maps* (CMs) were initially for graphical representation of uncertain causal reasoning. Later Kosko suggested *Fuzzy Cognitive Maps* (FCMs) in which users freely express their opinions in linguistic terms instead of crisp numbers. However, it is not always easy to assign some linguistic term to a causal link. In this paper we suggest a new type of CMs namely, *Belief Degree-Distributed FCMs* (BDD-FCMs) in which causal links are expressed by belief structures which enable getting the links' evaluations with distributions over the linguistic terms. We propose a general framework to construct BDD-FCMs by directly using belief structures or other types of structures such as interval values, linguistic terms, or crisp numbers. The proposed framework provides a more flexible tool for causal reasoning as it handles any kind of structures to evaluate causal links. We propose an algorithm to find a similarity between experts judgments by BDD-FCMs for a case study in Energy Policy evaluation.

1 Introduction

Decision making processes usually involve several experts and the final decision is based on the opinions of the experts as a group. Decision tasks are often characterized by very interrelated features making the decision process to be a non trivial task. The decision making becomes even more difficult due to the conflicting interests and different backgrounds of the experts. Using quantitative mathematical models may be very difficult as usually the experts cannot express their opinions in exact mathematical terms. To this end, the application of *fuzzy set theory* (FST) seems more appropriate, as such analysis can handle subjectivity as well as inexact and vague information.

However, experts are not always available, therefore we may need to reduce the number of experts or in certain situations even to replace one expert with another. To reduce the number of experts, we need to know which experts are *similar*; thus we can take those similar experts into one group. This study aims to find a similarity between various experts based on their judgments by using *Fuzzy Cognitive Maps* (FCMs). FCMs

J. Lu, L.C. Jain, and G. Zhang: Handbook on Decision Making, ISRL 33, pp. 39–67.
springerlink.com   © Springer-Verlag Berlin Heidelberg 2012

are descriptive tools to describe and model dynamic systems, analyze the causal relationships between different concepts inside a given dynamic system. FCMs model the system using concepts and cause-effects relationships, which link concepts and describe how they affect each other.

However, for complex problems it is not easy to assign linguistic terms for relationships, it is not convenient to use FCMs. Particularly, if the judgment has to be done for the future prediction, experts have some difficulty to decide the exact linguistic term as they may be not sure about their opinion. To overcome this problem, we propose to use FCMs in which the relationships between concepts (nodes) are expressed not by a single linguistic term, but by a *belief structure*. In this paper we propose a new type of FCMs mainly *Belief Degree-Distributed FCMs* (BDD-FCMs), in which experts have more freedom to evaluate causal links by different structures such as crisp numbers, linguistic terms, interval values, or belief structures.

We apply the proposed approach to find a similarity between experts using BDD-FCMs to a case study of *energy policy evaluation*. Note that we do not provide the answer which expert is better or more knowledgeable than another but find if there exists a similarity between the experts, we do not give a preference of one expert over another.

The rest of this paper is organized as follows: in Section 2 we give theoretical background of CMs and FCMs. In Section 3 we introduce Belief Degree-Distributed FCMs (BDD-FCMs). In Section 4 we show the use of BDD-FCMs for merging opinions of several experts as a group. In Section 5 we detail the algorithm to compare FCMs. In Section 6 we explain the proposed algorithm to compare BDD-FCMs. In Section 7 we give an illustrative example of BDD-FCMs comparison. In Section 8 we apply our approach for *Energy Policy Evaluation*. In Section 9 we explain the proposed algorithm in real-life case study. Finally, we conclude this study in Section 11.

2   Fuzzy Cognitive Maps

In this section we outline basic concepts of CMs, and FCMs, and their related drawbacks. CMs were first introduced by Axelroid [1] who focused on the policy domain. Since then many researchers have applied CMs in various ill-structured or not well-defined problems. The application areas of CMs or FCMs are very broad and diverse, and, in general CMs and FCMs can be used in any kind of decision making problems as a decision supporting tool. Table 1 summarizes the most application fields. As far as we found, there is no research that uses FCMs for an energy policy domain.

A CM has two types of elements: *Concepts* and *Causal Beliefs*. The former are variables while the latter are relationships between variables. Causal relationships can be either positive or negative, as specified by a '+', respectively a '-', sign on the arrow connecting two variables. The variables that cause a change are called *cause variables* and the ones that undergo the effect of the change are called *effect variables*. If the relationship is positive, increase or decrease of the cause variable causes the effect variable to change in the same direction (e.g., increase in the cause variable causes increase of the effect variable). In the case of negative relationship, the change of effect variable is in the opposite direction (e.g., increase in the cause variable causes decrease of the effect variable).

Table 1. Application fields of CMs and FCMs

Research Areas	Descriptions	References
Engineering		
Failure System modeling	Analyses failure modes and effects using FCMs	[2]
Network Security	Describes a decision engine for an intelligent intrusion detection system using FCMs	[3]
Electrical Circuits		[4]
Ecology		
Environmental Management	Gives a modeling of environmental and ecological problems using FCMs	[5], [6]
Sport Fisheries Management	Uses FCMs in decision making about sport fisheries	[7]
Information Technology		
Information Management	Uses FCMs for strategic planning of information systems	[8]
Business Process Management	Models and analyses the business performance indicators using FCMs	[9]
Electronic Data Management	Provides a design of electronic data interchange controls using FCMs	[10]
Data Mining	Uses CMs to retrieve news information from Internet and based on that predicts interest rates	[11]
Risk Analysis and Management		
Software Project Management	Uses extended FCMs for risk analysis in software project management	[12]
Health Care	Assesses information security issues using FCMs	[13]
Security Analysis	Uses FCMs discussing port security system	[14]
Social Sciences		
Politics	Uses CMs and FCMs for political decisions	[1]
Others		
Virtual Words	Applies FCMs to undersea virtual world of dolphins	[15]
Chemical Industry	Applies a weight adaptation method for FCM to a process control problem in chemical industry	[16]
Food Industry	Provides a new algorithm to compare FCMs discussing an application example of food industry	[17]

In graph theory terms, a CM is a directed graph with concepts as *nodes* and causal believes as *edges*.

Figure 1 (a) shows a simple cognitive map in a public health domain [18]. If we consider the relationship between C_5 and C_6, C_5 is a *cause variable* whereas C_6 is an *effect variable*, so that the increase/decrease in C_5 will cause the decrease/increase in C_6. Indeed, the *increase* of *Sanitation facilities decreases Number of diseases per 100 residents*, and visa versa. On the other hand, if we consider the relationship between C_4 and C_7, in this case the increase/decrease of C_4 *cause variable* will cause the increase/decrease of C_7 *effect variable*. In fact, increasing *Garbage per area* causes *increase* of *Bacteria per area*.

In his work [1] Axelroid introduced also *Weighted CMs* and *Functional CMs*. In Weighted CMs the sign in the map is replaced by a positive or a negative number, which shows the direction of the effect as well as its magnitude. Functional CMs are CMs in which a function is associated with each causal relationship showing more precisely the direction and the magnitude of the effect. These two types of CMs give more flexibilities as they can handle and provide more detailed information.

However, CMs, whatever their type, are not easy to define and the magnitude of the effect is difficult to express in numbers. Usually CMs are constructed by gathering information from experts and generally experts are more likely to express themselves in qualitative rather than quantitative terms. To this end, it may be more appropriate to use Fuzzy CMs (FCMs), suggested by Kosko [19] to represent the concepts linguistically with an associated fuzzy set. Actually, FCMs are weighted cognitive maps with fuzzy weights.

The degree of a relationship between concepts in an FCM is either a number in $[-c, c]$ (c is a non-negative integer number), or a linguistic term, such as *often*, *extremely*, and *some*, etc. Figure 1 (b) shows a simple example of FCM where the causal relationships are expressed by using fuzzy linguistic terms. For example, if we again consider the relationship between C_5 and C_6, the increase/decrease of C_5 cause variable will cause *very high* decrease/increase in C_6.

To analyze and compare FCMs, we define the concept of *adjacency matrix*.

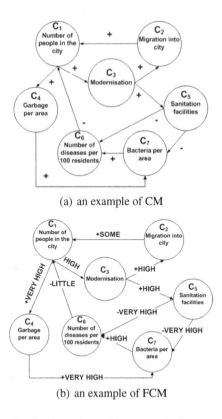

(a) an example of CM

(b) an example of FCM

Fig. 1. Illustrations of CM (a) and FCM (b)

Definition 1. *Consider the nodes/concepts C_1, C_2,..., C_n of an FCM. The matrix $E=(e_{ij})$, where e_{ij} is the weight of the directed edge C_iC_j, is called an* adjacency *or connection matrix of the FCM.*

The following matrix shows the adjacency matrix of the *public health* example with reference to Figure 1 (a).

$$
E = \begin{pmatrix}
 & c_1 & c_2 & c_3 & c_4 & c_5 & c_6 & c_7 \\
c_1 & 0 & 0 & +1 & +1 & 0 & 0 & 0 \\
c_2 & +1 & 0 & 0 & 0 & 0 & 0 & 0 \\
c_3 & 0 & +1 & 0 & 0 & +1 & 0 & 0 \\
c_4 & 0 & 0 & 0 & 0 & 0 & 0 & +1 \\
c_5 & 0 & 0 & 0 & 0 & 0 & -1 & -1 \\
c_6 & -1 & 0 & 0 & 0 & 0 & 0 & 0 \\
c_7 & 0 & 0 & 0 & 0 & 0 & +1 & 0
\end{pmatrix}
$$

Notice that all matrices associated with FCMs are always square matrices with diagonal entries as zero.

Definition 2. *For a given node C_j, the* centrality *value $CEN(C_i)$ is decided as follows: $CEN(C_i) = IN(C_i) + OUT(C_i)$, where $IN(C_i)$ is the column sum of absolute values of a variable in the adjacency matrix and shows the cumulative strength of variables entering the unit, and $OUT(C_i)$ is the row sum of absolute values of a variables and shows the cumulative strengths of connections exiting the variable.*

Definition 3. *A node C_i is called* transmitter *if $IN(C_i) = 0$ and $OUT(C_i) > 0$, and is called receiver if $IN(C_i) > 0$ and $OUT(C_i) = 0$.*

The total number of receiver nodes in a map is considered an index of its *complexity*. Large number of receiver variables indicates a complex map, while large number of transmitters shows a formal hierarchical system where causal arguments are not well elaborated [20].

Definition 4. *An FCM is called* dynamic *if the causal relations flow through a cycle.*

Definition 5. *Let C_1, C_2,..., C_n be the nodes of a cyclic FCM. When C_i is switched on and if the causality flows through the edges of a cycle and if it again causes C_i, then dynamical system goes round and round. This is true for any node C_i for $i = 1,...,n$. The equilibrium state for the dynamical system is called the hidden pattern.*

Definition 6. *If the equilibrium state of a dynamical state is a unique state vector, then it is called a fixed point.*

Definition 7. *Let C_1, C_2,..., C_n be the nodes of an FCM and $A = (a_1, a_2, ..., a_n)$ where $a_i \in \{0, \}$. A is called the* instantaneous state vector *and it denotes the on-off position of the node at an instant. $a_i = 0$ if a_i is off and $a_i = 1$ if a_i is on for all $i = 1, 2, ..., n$.*

Suppose we want to analyze the FCM in Figure 1 (a). As an instantaneous state vector we take $A_1 = [0100000]$, means we want to see if *Migration into city* is on, what does it direct or indirect affect. We multiply the state vector with an adjacency matrix, then after thresholding and updating the state vector repeat the multiplication, until the system settles down to a fixed point. For our example we will have the following steps.

1. $A_1 \cdot E = [1\,0\,0\,0\,0\,0\,0] = A_2$, $A_2 = [1\,1\,0\,0\,0\,0\,0]$
2. $A_2 \cdot E = [1\,0\,1\,1\,0\,0\,0] = A_3$, $A_3 = [1\,1\,1\,1\,0\,0\,0]$
3. $A_3 \cdot E = [1\,1\,1\,1\,1\,0\,1] = A_4$, $A_4 = [1\,1\,1\,1\,1\,0\,1]$
4. $A_4 \cdot E = [1\,1\,1\,1\,1\,0\,0] = A_5$, $A_5 = [1\,1\,1\,1\,1\,0\,0]$
5. $A_5 \cdot E = [1\,1\,1\,1\,1\,-1\,0] = A_6$, $A_6 = [1\,1\,1\,1\,1\,0\,0] = A_5$

where the second column is the thresholded and updated state vectors.

So the increase of *migration into city* results in the increase of *number of people, modernization, garbage per area* and *sanitation facilities*.

The FCMs drawn by different experts can be different. Each expert can be assigned a weight in the $[0, 1]$ range that shows the importance, experience or trustworthiness of the expert.

In general, if the CM matrices have different sizes, matrices with fewer concepts are augmented by including any missing concept(s) through the addition of extra rows and columns of all zeros. The final matrix, representing the group opinion, becomes

$$E = \sum_{i=1}^{m} w_i \cdot E_i \tag{1}$$

where m is the number of experts, $w_i \geq 0$ shows the i-th expert weight, and E_i is the adjacency matrix of a map provided by the i-th expert.

Suppose two experts evaluate the same problem and the maps provided by them differ by sizes, by nodes and by their structures as in Figure 2. Expert e_1 provided six nodes whereas expert e_2 evaluated the relationships only among five nodes. Table 2 shows the adjacency matrices of each expert according to their maps.

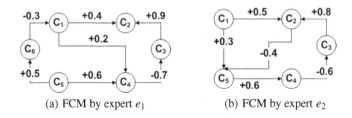

(a) FCM by expert e_1 (b) FCM by expert e_2

Fig. 2. Evaluations of the same problem provided by two different experts

Table 2. Adjacency matrices of two experts from Figure 2

(a) Expert e_1

	c_1	c_2	c_3	c_4	c_5	c_6
c_1	0	0.4	0	0.2	0	0
c_2	0	0	0	0	0	0
c_3	0	0.9	0	0	0	0
c_4	0	0	-0.7	0	0	0
c_5	0	0	0	0.6	0	0.5
c_6	-0.3	0	0	0	0	0

(b) Expert e_2

	c_1	c_2	c_3	c_4	c_5
c_1	0	0.5	0	0	0.3
c_2	0	0	0	0	-0.4
c_3	0	0.8	0	0	0
c_4	0	0	-0.6	0	0
c_5	0	0	0	0.6	0

Note that as expert e_2 did not consider node c_6, in the augmented matrix the sixth row and column will be filled by zeros.

Assume that two experts have different weights: $w_{e_1} = 0.4$ and $w_{e_1} = 0.6$. Table 3 the aggregated adjacency matrix by using (1) and Figure 3 shows the final map generated from Table 3.

Table 3. The aggregated results of two experts opinion as a group

	c_1	c_2	c_3	c_4	c_5	c_6
c_1	0	0.46	0	0.08	0.18	0
c_2	0	0	0	0	-0.24	0
c_3	0	0.84	0	0	0	0
c_4	0	0	-0.64	0	0	0
c_5	0	0	0	0.6	0	0.2
c_6	-0.12	0	0	0	0	0

Among several ways of developing CMs and FCMs, the most common methods are:

- extracting from questionnaires,
- extracting from written texts,

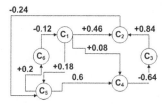

Fig. 3. Group map generated from Table 3

- conducting interviews,
- drawing maps from data.

To obtain a cognitive map from questionnaires requires first to identify the most important variables for the given problem, and then give experts ordered pairs of variables in a questionnaire format. Afterward, experts decide the strength of causal links relying on their knowledge and experience.

The second method is a type of content analysis in which the causal relationships are identified by analyzing texts. The main problem related to this method is that usually the relationships are not explicitly stated, and the language structure in which texts are written can vary from one language to another.

The detailed description how to construct a cognitive map through interviews is provided in [20]. The methodology is composed from the following steps: decide the most important variables, provide an expert completely unrelated map explaining how to draw a cognitive map, and finally, ask them to draw their own maps of the issue under investigation.

The automatic construction of cognitive maps (particularly FCMs) based on user provided data is discussed in [21]. This method first finds the degree of a similarity between any two variables, then decides whether the relation between the two variables is direct or inverse.

However, in this paper we will adopt the different approaches for different problems in real situations. For *energy policy* evaluation we construct maps from data, in which we compare experts after aggregating data about each scenario and each criterion. Therefore, for this problem experts data represent their evaluation of all scenarios.

As we already mentioned, for complex problems it is not easy to assign linguistic terms for relationships, it is not convenient to use FCMs. For that reason, we propose to use FCMs in which the relationships between concepts (nodes) are expressed not by a single linguistic term, but by a *belief structure*. In the next section, we will explain our approach in more details.

3 Belief Degree-Distributed FCMs

In this section we propose a new type of FCMs in which the relationships between nodes are expressed in terms of a *belief degree distribution* [22].

In BDD-FCM we use belief structures to represent the general belief of experts about the given relationship between two nodes. Suppose we have a set of linguistic terms

$S = \{s_i\}$ where i is the maximum number of linguistic terms and depends on the task under consideration. Here we assume $i = 7$, and assign the following meanings to the linguistic terms: s_1: *definite belief* of negative impact, s_2: *strong belief* of negative impact, s_3: *weak belief* of negative impact, s_4: *undetermined* of impact magnitude and direction, s_5: *weak belief* of positive impact, s_6: *strong belief* of positive impact, s_7: *definite belief* of positive impact. Notice that the number of linguistic terms and their meanings depend on the problem but, generally, we restrict the minimum number of linguistic terms of five and the maximum number seven. Note that positive/negative impact does not assume positive/negative consequences but the fact the increase or decrease of cause node changes the effect variable in the same/opposite direction.

For example, for the relationship between C_1 and C_2 an expert may express his/her opinion by the following statement: he/she has 60 % *definite belief* about the negative impact between cause and effect variables, 20 % is *undetermined* about the change magnitude and direction, and has 10 % is *weak belief* of the positive impact. Here the percentages are referred to as the belief degrees that indicate the extents that the corresponding grades are assessed. The belief structure in this case will be $\{(s_1, 0.6), (s_4, 0.2), (s_5, 0.1)\}$.

In general, the belief structure for the causal relationship L_{ij} between the i-th and j-th nodes can be defined as follows:

$$B^e(L_{ij}) = \left\{ (\beta^e_{ij,k}, s_k), k = 1, ..., m \right\}, \forall i, \forall j, \forall e \tag{2}$$

in which $\sum\limits_{i,j=1}^{l} \beta^e_{ij,k} \leq 1, \forall i, \forall j, \forall e$, where $\beta^e_{ij,k}$ is a belief degree of expert e at level s_k for the causal link between i-th and j-th nodes, l is the number of nodes used by expert e, and m is the number of linguistic terms to describe all belief degrees.

Notice that the sum of belief degrees for a given link may be also less than 1, which indicates the incompleteness of the judgments. The incompleteness can have several reasons, for example it can be due to lack of knowledge about the causal link, or lack of experience and expertise of the expert. Thus, we give flexibility for the experts to express their opinions in the case of partial information or unclear idea about the concepts and their relationships.

The experts can express their opinions not only by belief degrees but also by other types of structures like linguistic terms, crisp numbers, and interval values. In all cases, we can transfer all types of values to belief structures and then apply our algorithm to find the similarity between the experts.

If an expert e provides the judgment by a linguistic term s_k, then the corresponding belief structure will be:

$$B^e(L_{ij}) = \{(1, s_k)\}, k = 1, ..., m. \tag{3}$$

If an expert e provides the judgment by assigning a crisp number $n^e_{ij} \in [-c, c]$ (c is a scale decided upon on the problem) to the causal link from node C_i to node C_j, then the related belief degrees will be the membership values of the assigned value related to each linguistic term. If we have m linguistic terms, and μ_{s_k}, k=1,...,m, are membership functions of the linguistic terms s_k, then the belief structure for assignment n^e_{ij} will be as follows:

$$B^e(L_{ij}) = \{(\mu_{s_k}(n_{ij}^e), s_k)\}, \forall i, \forall j, \forall e. \tag{4}$$

Figure 4 is an example of transforming a crisp number assignment into a belief structure. Here we have seven linguistic terms with their membership functions. If the expert's judgment is ϑ, then the corresponding belief structure will be non-zero membership values of $\mu(\vartheta)$ in fuzzy sets; thus for this example we will have $\{(\delta, s_2), (\varepsilon, s_3)\}$.

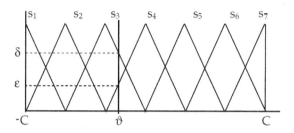

Fig. 4. Transforming crisp number assignment to belief structures

Another way for experts to make an assignment for a causal link is to use an interval value assignment, where they indicate the lower and upper bounds for the evaluation in the decided scale. If the lower and upper bounds of the evaluation of the link between the i-th and j-th nodes by the expert e are $n_{ij}^{Low_e}$ and $n_{ij}^{Up_e}$ respectively, then the related belief structure will be the normalized value of the area constrained by the interval values.

The area restricted by the upper and level bounds of the interval assignment will be:

$$A_{ij,k}^e = \left\{ \int_{n_{ij}^{Low_e}}^{n_{ij}^{Up_e}} \mu_{s_k}(x)\, dx \right\}, \forall i, \forall j, \forall e. \tag{5}$$

The related belief structure can be obtained by normalizing $A^e(ij,k)$ values as follows:

$$B^e(L_{ij}) = \left(\frac{A_{ij,k}^e}{\sum_{l=1}^{m} A_{ij,l}^e}, s_k \right), k = 1, ..., 1, \forall i, \forall j, \forall e. \tag{6}$$

Figure 5 is an example of transferring an interval value assignment into belief structures. In this example the value for s_3 is A_3 area constrained by $\vartheta^U - \vartheta^L$.

The main advantage of transforming all kinds of input values to belief structures, is that we do not loose any information of experts' judgments.

The *centrality* of a node C_i is decided as for FCMs, but for calculating $IN(C_i)$ and $OUT(C_i)$ we first sum all belief degrees of each linguistic term, then sum all belief degrees of the final result. In FCMs, a node can be more central although it has fewer connections if the connections carry larger weight [19]. Similarly, in BDD-FCMs, a node can be more central having fewer connections if the connections carry greater

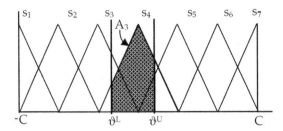

Fig. 5. Transforming an interval assignment into belief structures

beliefs. The centrality in BDD-FCMs shows the contribution of a node showing how the node is connected with other nodes and in which degrees are the beliefs for the connections.

To evaluate the system when one or more variables are on (the event associated with the variable occurs), we transfer belief degree structure into a crisp number, and follow the same structure as we discussed for FCMs. If for a task we used m linguistic terms, for the link between i-th and j-th nodes described by (2) will be:

$$B(L_{ij}) = \sum_{k=1}^{m} \lambda_k \cdot \beta_{ij,k}, \ k = 1, ..., m, \tag{7}$$

where

$$\lambda_k = \begin{cases} \frac{k-p}{p-1}, & \text{if } k < p \\ 0, & \text{if } k = p \\ \frac{p-k}{p-1}, & \text{otherwise} \end{cases} \tag{8}$$

where $p = \lceil \frac{m}{2} \rceil$. For example suppose we have seven linguistic terms, and link is described by the belief structure as $\{(s_1, 0.2), (s_2, 0.4), (s_5, 0.3), (s_7, 0.1)\}$, then applying (7) and (8), we will have $0.2 \cdot (-3/3) + 0.4 \cdot (-2/3) + 0.3 \cdot 1/3 + 0.1 \cdot 3/3 = -0.33$.

After having collected judgments from the experts and transferring them to belief structures, we apply our algorithm described in the next section to compare cognitive maps and find the similarity between the experts.

4   Application Example of BDD-FCMs Group Mapping

In this section we explain the BDD-FCM model by a numerical example. Suppose we have four experts: e_1, e_2, e_3, and e_4, who give their judgments about a transportation problem [23]. The problem is for a fixed source, a fixed destination and a unique route from the source to the destination, with the assumption that all the passengers travel in the same route. Only the peak-hour is considered since the passenger demand is very high only during this time period. Assume all experts have the same weights; thus $w_1 = w_2 = w_3 = w_4 = 1$. Initially, they are given the list of different nodes and they are asked to choose the most important nodes and draw the relationships between those

Table 4. List of nodes with their descriptions

Node	Node description
C_1	Frequency of the vehicles along the route (%)
C_2	In-vehicle travel time along the route
C_3	Travel fare along the route
C_4	Speed of the vehicles along the route
C_5	Number of intermediate points in the route
C_6	Waiting time
C_7	Number of transfers in the route
C_8	Congestion in the vehicle.

nodes. They expressed their opinions differently by crisp numbers, linguistic terms, interval values and belief structures as shown in Figure 6. Table 4 lists all used nodes and their descriptions.

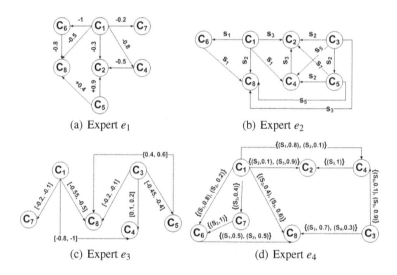

(a) Expert e_1 (b) Expert e_2

(c) Expert e_3 (d) Expert e_4

Fig. 6. Cognitive maps obtained from four different experts

We first transfer all values of the causal links to belief structures. Table 5 shows the transformation results.

To build the group map from the given maps, we augment all matrices to bring them into the same dimension. For each causal link between the i-th and j-th nodes at s_k level, we aggregate all m experts' judgments as follows:

$$\beta_{ij,k} = \frac{\sum_{i=1}^{m} w_i \cdot \beta_{ij,k}^{e_i}}{\sum_{i=1}^{m} w_i}, \forall i, \forall j \tag{9}$$

where w_i is the weight of the expert e_i.

Table 5. Adjacency matrices of four experts from Figure 6

(a) Expert e_1

	C_1	C_2	C_4	C_5	C_6	C_7	C_8
C_1	0	$\{(s_3,1)\}$	$\{(s_1,0.6),(s_2,0.4)\}$	0	$\{(s_1,1)\}$	$\{(s_3,0.65),(s_4,0.35)\}$	$\{(s_2,0.5),(s_3,0.5)\}$
C_2	0		0	0	0	0	0
C_4	0	$\{(s_3,1)\}$	0	0	0	0	0
C_5	0	$\{(s_6,0.3),(s_7,0.7)\}$	0	0	0	0	$\{(s_5,0.8),(s_6,0.2)\}$
C_6	0		0	0	0	0	$\{(s_1,0.6),(s_2,0.4)\}$
C_7	0		0	0	0	0	0
C_8	0		0	0	0	0	0

(b) Expert e_2

	C_1	C_2	C_3	C_4	C_5	C_6	C_8
C_1	0	$\{(s_3,1)\}$	0	$\{(s_1,1)\}$	0	$\{(s_1,1)\}$	$\{(s_2,1)\}$
C_2	0	0	0	0	0	0	0
C_3	0	$\{(s_2,1)\}$	0	$\{(s_5,1)\}$	$\{(s_2,1)\}$	0	$\{(s_3,1)\}$
C_4	0	$\{(s_3,1)\}$	0	0	0	0	0
C_5	0	$\{(s_7,1)\}$	0	$\{(s_2,1)\}$	0	0	$\{(s_5,1)\}$
C_6	0	0	0	0	0	0	$\{(s_1,1)\}$
C_8	0	0	0	0	0	0	0

(c) Expert e_3

	C_1	C_3	C_4	C_5	C_7	C_8
C_1	0	0	$\{(s_1,0.7),(s_2,0.3)\}$	0	$\{(s_3,0.55),(s_4,0.45)\}$	$\{(s_2,0.65),(s_3,0.35)\}$
C_3	0	0	$\{(s_4,0.45),(s_5,0.55)\}$	$\{(s_2,0.6),(s_3,0.4)\}$	$\{(s_2,0.6),(s_3,0.4)\}$	0 $\{(s_3,0.55),(s_4,0.45)\}$
C_4	0	0	0	0	0	0
C_5	0	0	0	0	0	$\{(s_5,0.5),(s_6,0.5)\}$
C_7	0	0	0	0	0	0
C_8	0	0	0	0	0	0

(d) Expert e_4

	C_1	C_2	C_3	C_4	C_6	C_7	C_8
C_1	0	$\{(s_2,0.1),(s_3,0.9)\}$	0	0	$\{(s_1,0.8),(s_2,0.2)\}$	$\{(s_3,0.4)\}$	$\{(s_2,0.4),(s_3,0.6)\}$
C_2	0	0	0	0	0	0	0
C_3	0	0	0	$\{(s_4,0.1),(s_5,0.9)\}$	0	0	$\{(s_3,0.7),(s_4,0.3)\}$
C_4	0	$\{(s_3,1)\}$	0	0	0	0	0
C_6	0	0	0	0	0	0	$\{(s_1,0.5),(s_2,0.5)\}$
C_7	0	0	0	0	$\{(s_2,1)\}$	0	0
C_8	0	0	0	0	0	0	0

Table 6 shows the aggregation results using (9) generated from Table 5. Figure 7 shows the final group map generated from Table 6 where n_{ij} corresponds the i-th row and the j-th column in Table 6.

Table 6. Group Aggregation Results

	C_1	C_2	C_3	C_4	C_5	C_6	C_7	C_8
C_1	0	$\{(s_2,0.025),(s_3,0.725)\}$	0	$\{(s_1,0.575),(s_2,0.175)\}$	0	$\{(s_1,0.7),(s_2,0.05)\}$	$\{(s_3,0.4),(s_4,0.2)\}$	$\{(s_2,0.637),(s_3,0.362)\}$
C_2	0		0	0	0	0	0	0
C_3	0	$\{(s_2,0.25)\}$	0	$\{(s_4,0.137),(s_6,0.612)\}$	$\{(s_2,0.4),(s_3,0.1)\}$	0	$\{(s_2,0.15),(s_3,0.1)\}$	$\{(s_3,0.562),(s_4,0.187)\}$
C_4	0	$\{(s_3,0.75)\}$	0	0	0	0	0	0
C_5	0	$\{(s_6,0.075),(s_7,0.425)\}$	0	$\{(s_2,0.25)\}$	0	0	0	$\{(s_5,0.575),(s_6,0.175)\}$
C_6	0	0	0	0	0	0	0	$\{(s_1,0.525),(s_2,0.225)\}$
C_7	0	0	0	0	0	$\{(s_2,0.25)\}$	0	0
C_8	0	0	0	0	0	0	0	0

From Figure 7 we observe some important facts. For example, notice that according to group decision, C_1 and C_3 are transmitters, consequently they have an effect on other nodes, but no other node has an effect on them. On the other hand, C_8 and C_2 are receivers, thus they are affected by other nodes and do not have an effect to any other node. The most central node is C_1 with the central value equal to 4.1. Nodes with high centrality values deserve special attention in any analysis for decision support.

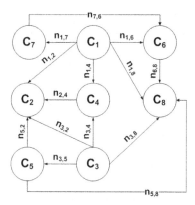

Fig. 7. Group map generated from Table 6

Suppose that *Number of transfers in the route* (C_7) has been increased. According, to group opinion there is a *strong belief* that it will decrease *Waiting time* (C_6) which in its turn will decrease *Congestion in the vehicle*. For the last statement experts have 52% *definite* believe and 22% of *strong belief*. Note that the sum of beliefs is less than 100% as not all experts share the same opinion about the link.

Note that the more the sum of belief degrees of a causal link is close to one, the more agreement exists among experts in judging the link.

For sensitivity analysis we will transform the belief degree distribution values in Table 6 into crisp numbers by using (7) and (8) then use different instantaneous state vectors depending on the task. Table 7 shows the transformed results by applying (7) and (8).

Table 7. Group Aggregation Results By Crisp Numbers

	C_1	C_2	C_3	C_4	C_5	C_6	C_7	C_8
C_1	0	-0.26	0	-0.69	0	-0.73	-0.13	-0.55
C_2	0	0	0	0	0	0	0	0
C_3	0	-0.17	0	0.3	-0.3	0	-0.13	-0.19
C_4	0	-0.25	0	0	0	0	0	0
C_5	0	0.475	0	-0.17	0	0	0	0.31
C_6	0	0	0	0	0	0	0	-675
C_7	0	0	0	0	0	-0.17	0	0
C_8	0	0	0	0	0	0	0	0

Now suppose the *number of intermediate points in the route* has been increased by 50 %. The state vector to analyse the impact of this increase will be $A = (0\,0\,0\,0\,0.5\,0\,0\,0)$. Multiplying it with the adjacency matrix, we will have the immediate results of this change. The resulted vector is $(0\,0.285\,0 - 0.102\,0\,0\,0\,0.186)$. So, the change in C_5 results of 28 % increase of *in vehicle travel time along the route*, 10 % decrease of *speed of the vehicles along the route*, and finally 18 % increase in the *vehicles congestion*.

Obviously, we can use different combinations of changes in different nodes, and see the most sensitive variables that are affected by that change the most.

In general, the obtained group map can be used for a static analysis of the domain for exploring nodes' relative importance, and indirect and total causal effects between concept nodes. Dynamic analysis of a group map is concerned with the evolution over time of a system.

5   Comparison of FCMs

While there are many cases to use CMs or FCMs in different application fields, there are quite few attempts to develop algorithms to compare cognitive maps.

CMs and FCMs can be compared in two dimensions: comparing the *content* and the *structure* of each map [24]. The Content difference is associated with the differences in elements in both maps and the differences of the relationships between those elements. The structural difference, on the other hand, is associated with the varying complexity degrees of the maps structure. In this paper, we do not need to perform structural comparison, but only focus on the content difference analysis.

In this study we will use the algorithm suggested by Langfield-Smith and Wirth [24] (from now on we will refer to the algorithm as LW algorithm), and we will modify the algorithm to fit our needs. In [17] the authors argue LW algorithm mentioning that it does not consider the missing values properly and it lacks of generalizibility. However, the modification provided by the authors is very complex to be applied. We will take into account only the comment about generalizibility, and will adjust LW algorithm to be applicable in all cases. Notice that by generalizibility, we mean that in LW algorithm the authors fix the number of linguistic terms, and the comparison formula cannot handle different number of linguistic terms. More specifically, they fixed the number of linguistic terms to 7, assigning maximum strength to 3, minimum strength to -3, and the provided formula fixes maximum strength as 3 and in other cases it cannot calculate the comparison.

Eliminating this drawback of LW algorithm, the *distance ratio* (DR) to compare two FCMs will be:

$$DR = \frac{\sum_{i=1}^{p} \sum_{1=0}^{p} (a_{ij}^* - b_{ij}^*)}{2\alpha p_c^2 + 2p_c(p_{u_A} + p_{u_B}) + p_{u_A}^2 + p_{u_B}^2 - (2\alpha p_c + p_{u_A} + p_{u_B})} \tag{10}$$

where

$$m_{ij}^* = \begin{cases} 1(i), & \text{if } m_{ij} \neq 0 \text{ and } i \text{ or } j \notin P_c \\ m_{ij}(ii), & \text{otherwise} \end{cases} \tag{11}$$

In (10), a_{ij} and b_{ij} are the adjacency matrices of the first and second map respectively, p is the total number of possible nodes, P_c is the set of nodes common to both maps, p_c is the number of such nodes, p_{u_A} is the number of nodes unique to A and p_{u_B} is the number of nodes unique to B, m_{ij} is the value of the i-th row and j-th column in the augmented adjacency matrix, α is maximum strength decided upon on the problem. One would argue that only the numerator is enough to calculate the difference between two maps.

However, it is important also to consider how much in common two maps have, as well as, in which extend maps have unique elements. Thus, the more the common nodes, and the less the unique nodes, the more similar are the maps. In this regard, (10) represents a complete comparison between two maps, consequently two experts.

Below we explain (10)and (11) in more details. Consider again Figure 2 that gives evaluation of two different experts about the same problem. The set of common elements between two maps is: $P_c = (c_1, c_2, c_3, c_4, c_5)$; thus number of common elements is $p_c = 5$. The first map has only one unique element: c_6 and the second map does not have unique element. Thus, $p_{u_A} = 1$ and $p_{u_B} = 0$. Maximum strength used for this example is one, thus $\alpha = 1$. Therefore, the denumerator in (10) will be: $2 \cdot 5^2 + 2 \cdot 5(1+0) + 1^2 + 0^2 - (2 \cdot 5 + 1 + 0) = 50$.

(14) is applied only for the cases where one of the nodes of the relationship belongs to the set of common elements, and the value of the causal link is not zero. Table 8 shows augmented matrices of two experts from Figure 2. When we compare the link between c_5 and c_6, the difference is not $(0.5 - 0)$ but 1 by (14) as c_5 belongs to common nodes' set, while c_6 does not belong. The same observation holds also for the causal link between c_6 and c_1.

Table 8. Augmented matrices of two experts from Figure 2

(a) Expert e_1

	C_1	C_2	C_3	C_4	C_5	C_6
C_1	0	0.4	0	0.2	0	0
C_2	0	0	0	0	0	0
C_3	0	0.9	0	0	0	0
C_4	0	0	-0.7	0	0	0
C_5	0	0	0	0.6	0	0.5
C_6	-0.3	0	0	0	0	0

(b) Expert e_2

	C_1	C_2	C_3	C_4	C_5	C_6
C_1	0	0.5	0	0	0.3	0
C_2	0	0	0	0	-0.4	0
C_3	0	0.8	0	0	0	0
C_4	0	0	-0.6	0	0	0
C_5	0	0	0	0.6	0	0
C_6	0	0	0	0	0	0

Langfield-Smith and Wirth [24] proposed three types of differences between two individuals in their research:

1. *Existence or non-existence of elements*: thus one expert considers certain element as important for the given domain, the other has the opposite opinion. In this case the adjacency matrix for the cognitive map of the first expert contains the element/elements while the other matrix does not contain.
2. *Existence or non-existence of beliefs*: thus one expert considers that there is a casual relationship between two concepts, while the other has the opposite opinion. In this case two experts should agree upon the fact that the nodes are important for the given domain, but have opposite opinions towards the causal link.
3. *Different values for identical beliefs*: thus two experts agree that there is a relationship between two nodes, but one expert holds the belief more strongly than the other. In adjacency matrices this difference is expressed by non-identical non-zero values for the cell showing the causal link between two nodes.

Note that the above mentioned differences are related only to the *content* of FCMs and not the *structure*. We will focus only on content difference and not the complexity and the structure of FCMs. Since our analysis is based on adjacency matrices, we do not need to go further in structure analysis of FCMs as adjacency matrices catch the most important properties of FCMs.

In the next section we present the modification of LW algorithm to compare BDD-FCMs.

6   Comparison of BDD-FCMs

Our modified algorithm considers that the causal links are expressed by belief structures, and we do not restrict and fix the number of linguistic terms. Thus, the similarity between two experts e_A and e_B is computed as follows:

$$SIM(e_A, e_B) = \frac{1}{1 + DR(e_A, e_B)} \qquad (12)$$

and

$$DR(e_A, e_B) = \frac{\sum\limits_{i=1}^{p} \sum\limits_{j=1}^{p} \sum\limits_{k=1}^{m} \left| (\beta_{ij,k}^{e_A})^* - (\beta_{ij,k}^{e_B})^* \right|}{2\alpha p_c^2 + 2p_c(p_{u_A} + p_{u_B}) + p_{u_A}^2 + p_{u_B}^2 - (2\alpha p_c + p_{u_A} + p_{u_B})} \qquad (13)$$

where

$$(\beta_{ij,k}^{e})^* = \begin{cases} 2(i), & \text{if } \beta_{ij,k}^{e_A} \neq 0 \text{ and } C_i \text{ or } C_j \notin P_c \\ \beta_{ij,k}^{e}(ii), & \text{otherwise} \end{cases} \qquad (14)$$

In (13), $\beta_{ij,k}^{e_A}/\beta_{ij,k}^{e_B}$ are belief structures of e_A/e_B experts, and the other variables have the same meaning as in (10). As we use belief structures, the maximum strength is one, therefore, $\alpha = 1$. Note that, in (14), (i) clause with respect to (11) the difference between a causal link in two maps in case when one or two nodes of the link do not belong to the common nodes, will be two. Indeed, in any number of linguistic terms, maximum difference of two opinions occurs when the beliefs of two experts are completely different; thus there is no intersection between the provided belief degrees. For example, if we have $i = 5$ linguistic terms, one combination of the maximum difference between the two opinions would be the following case: $\{(s_1, x_1), (s_2, x_2), (s_3, x_3)\}$ and $\{(s_4, x_4), (s_5, x_5)\}$, where x_i (i=1, 2, ..., 5) are believe degrees at s_i level, therefore, $x_1 + x_2 + x_3 \leq 1$ and $x_4 + x_5 \leq 1$. In other words, for the given causal link maximum possible difference occurs when both experts have $non - zero$ evaluation and in their belief structures they do not use the same linguistic term. Obviously, the maximum belief degree difference is two.

However, we cannot apply (13), as it does not differentiate the difference of the strengths of beliefs. For example, if we have two experts' evaluations $\{(s_1, 1)\}$ and $\{(s_2, 1)\}$ respectively, the difference of belief degrees is 2, and if we have $\{(s_1, 1)\}$ and $\{(s_7, 1)\}$ the difference is *again* 2. However, if we associate linguistic term for each belief degree, obviously the difference for the two cases should be not equal. Indeed, the difference between *definite positive impact* and *strong positive impact* should be not equal to the difference between *definite positive impact* and *definite negative impact*. To solve this problem, we suggest to induce a *belief-degree coefficient* that will consider the *indexes* of all non-zero belief degrees. The extended version of (13) with the described *belief-degree coefficient* is introduced as follows:

$$DR(e_A, e_B) = \frac{\sum_{i=1}^{p} \sum_{j=1}^{p} \sum_{k=1}^{m} \left| (\beta_{ij,k}^{e_A})^* \cdot \gamma_{ij}^{e_A} - (\beta_{ij,k}^{e_B})^* \cdot \gamma_{ij}^{e_B} \right|}{2p_c^2 + 2p_c(p_{u_A} + p_{u_B}) + p_{u_A}^2 + p_{u_B}^2 - (2\alpha p_c + p_{u_A} + p_{u_B})} \tag{15}$$

where

$$\gamma_{ij}^e = \frac{\sum_{k=1}^{m} k^e}{\sum_{k=1}^{m} k} \tag{16}$$

In (16) the numerator is the sum of the *indexes* of all *non-zero* belief degrees and the denominator is the sum of the indexes of all belief degrees. For instance, if expert e evaluated causal link as $\{(s_1, 0.4), (s_2, 0.5), (s_4, 0.1)\}$, and we have five linguistic terms, $\gamma^e = \frac{1+2+4}{1+2+3+4+5} = \frac{7}{15}$.

In the next section we illustrate a numerical example for explaining the algorithm in more details.

7   An Example of Comparing Experts Using BDD-FCMs

In this section we explain our algorithm by a numerical example. Suppose that we have four experts: e_1, e_2, e_3, and e_4, who give their judgments about the same problem. Initially, they are given the list of 12 different nodes and they are asked to choose the most important nodes and draw the relationships between those nodes. They expressed their opinions differently by crisp numbers, linguistic terms, interval values and belief structure as shown in Figure 8.

To apply the (13), we first transfer all values of the causal links to belief structures. Table 9 shows the transformation results.

Suppose we want to compare the maps from experts e_1 and e_2. The set of all common nodes used by these two experts is the following: $P_c = \{n_1, n_2, n_4, n_8, n_9, n_{10}\}$. The set of unique elements of the first map is $P_{U_{exp1}} = \{n3, n5\}$ and the second map does not have any unique element. Consequently, the number of the common and unique elements will be: $p_c = 6$, $p_{U_{exp1}} = 2$ and $p_{U_{exp1}} = 0$.

For simplicity we do not use (16). Applying (13) for the first and second experts we will have:

DR(e_1, e_2)= $\frac{14.2}{86} = 0.15$

SIM(e_1, e_2)= $\frac{1}{1+0.15} = 0.87$

Table 10 shows the results showing the distance ratio between experts according to their BDD-FCMs.

Figure 9 shows the comparison graph of the experts. The most similar experts are e_2 and e_4, while e_3 is the most different expert with respect to the others.

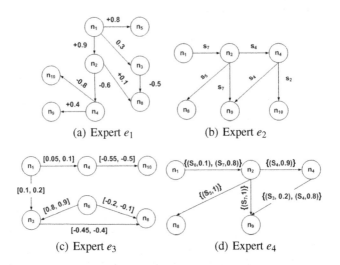

(a) Expert e_1

(b) Expert e_2

(c) Expert e_3

(d) Expert e_4

Fig. 8. Cognitive maps obtained from four different experts

Table 9. Adjacency matrices of four experts from the maps in Figure 8

(a) Expert e_1

	n_1	n_2	n_3	n_4	n_5	n_8	n_9	n_{10}
n_1	0	$\{(s_6,0.3),(s_7,0.7)\}$	$\{(s_4,0.1),(s_5,0.9)\}$	0	$\{(s_6,0.6),(s_7,0.4)\}$	0	0	0
n_2	0	0	-	$\{(s_2,0.8),(s_3,0.2)\}$	0	$\{(s_4,0.7),(s_5,0.3)\}$	0	0
n_3	0	0	0	0	0	$\{(s_2,0.8),(s_3,0.2)\}$	0	0
n_4	0	0	0	0	0	0	$\{(s_5,0.8),(s_6,0.2)\}$	$\{(s_1,0.6),(s_2,0.4)\}$
n_5	0	0	0	0	0	0	0	0
n_8	0	0	0	0	0	0	0	0
n_9	0	0	0	0	0	0	0	0
n_{10}	0	0	0	0	0	0	0	0

(b) Expert e_2

	n_1	n_2	n_4	n_8	n_9	n_{10}
n_1	0	$\{(s_7,1)\}$	0	0	0	0
n_2	0	0	$\{(s_4,1)\}$	$\{(s_5,1)\}$	$\{(s_7,1)\}$	0
n_4	0	0	0	0	$\{(s_4,1)\}$	$\{(s_2,1)\}$
n_8	0	0	0	0	0	0
n_9	0	0	0	0	0	0
n_{10}	0	0	0	0	0	0

(c) Expert e_3

	n_1	n_3	n_4	n_6	n_8	n_{10}
n_1	0	$\{(s_4,0.45),(s_5,0.55)\}$	$\{(s_4,0.3),(s_5,0.7)\}$	0	0	0
n_3	0	0	0	0	$\{(s_2,0.65),(s_3,0.35)\}$	0
n_4	0	0	0	0	0	$\{(s_2,0.65),(s_3,0.35)\}$
n_6	0	$\{(s_6,0.45),(s_7,0.55)\}$	0	0	$\{(s_3,0.55),(s_4,0.45)\}$	0
n_8	0	0	0	0	0	0
n_{10}	0	00	0	0	0	0
n_{11}	0	0	0	0	0	0

(d) Expert e_4

	n_1	n_2	n_4	n_8	n_9
n_1	0	$\{(s_6,0.1),(s_7,0.8)\}$	0	0	0
n_2	0	0	$\{(s_4,0.9)\}$	$\{(s_5,1)\}$	$\{(s_7,1)\}$
n_4	0	0	0	0	$\{(s_3,0.2),(s_4,0.8)\}$
n_8	0	0	0	0	0
n_9	0	0	0	0	0

Table 10. The DR and SIM values of two experts BDD-FCMs

(a) The DR values

	e_1	e_2	e_3	e_4
e_1	-	0.15	0.16	0.18
e_2	0.15	-	0.42	0.11
e_3	0.16	0.42	-	0.37
e_4	0.18	0.11	0.37	-

(b) The SIM values

	e_1	e_2	e_3	e_4
e_1	-	0.87	0.86	0.85
e_2	0.87	-	0.70	0.90
e_3	0.86	0.70	-	0.73
e_4	0.85	0.90	0.73	-

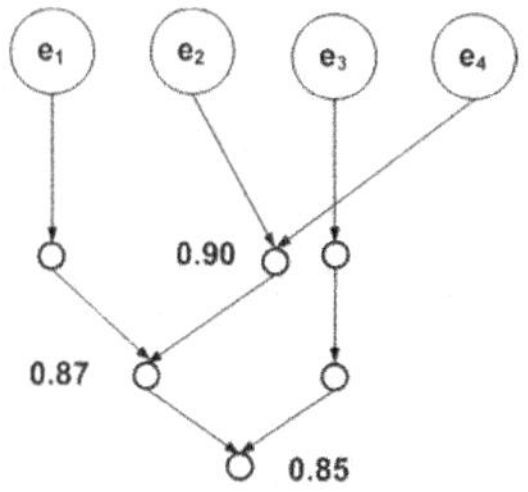

Fig. 9. Comparison of four experts' similarity from the maps in Figure 8

8   Application Examples: Energy Policy Evaluation

We will consider three main issues of energy policy: *Environment and human health and safety*, *Economic welfare* and *Social, political, cultural and ethical needs*. Each issue in its turn depends on many attributes. Here we mention only some of them. Table 11 shows each attribute, its description and the corresponding node name.

Table 11. List of criteria and subcriteria

Criteria	Subcriteria	Node Label
Environment and health safety n_1	Air pollution	n_4
	Land use	n_5
	Water use	n_6
	Catastrophic risk	n_7
	Geographical distribution risks	n_8
Economic welfare n_2	Intensity of energy use	n_9
	Security of energy supply	n_{10}
	Economic risks	n_{11}
	Compatibility with international RD agenda	n_{12}
Social, political and cultural needs n_3	Control and concentration of power	n_{13}
	Socio-political stability	n_{14}
	Job opportunities	n_{15}

Figure 10 is a kind of BDD-FCM presenting the relationships between the criteria and subcriteria of the energy policy evaluation.

Suppose we have six experts e_1,..., e_6 that evaluate seven different energy policy scenarios a_1,..., a_7. After aggregating the results we have evaluations of all experts for each scenario as in Table 12. Our task is to find similarities between the six experts. In Table 12 we took values from [25], as we also wanted validate our algorithm by this example. The results of our algorithm show that we get similar results as in [25], and our algorithm is more efficient in terms of computational effort.

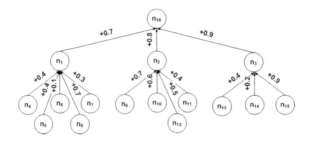

Fig. 10. A possible representation of the Energy Policy evaluation

As already mentioned in Section 6, the number and values of linguistic terms depend on the task under considerations. For the energy policy evaluation, we consider the following set of seven linguistic terms: s_1: Very Bad, s_2: Bad, s_3: Fairly Bad, s_4: Moderate, s_5: Fairly Good, s_6: Good, and s_7: Very Good.

Table 12. The evaluation of each scenario after aggregating all criteria for each expert

Experts	Energy policy scenarios						
	a_1	a_2	a_3	a_4	a_5	a_6	a_7
e_1	Very good	Good	Moderate	Bad	Fairly Good	Fairly Bad	Very Bad
e_2	Very good	Good	Moderate	Bad	Fairly Good	Very Bad	Very Bad
e_3	Very bad	Fairly Bad	Moderate	Good	Very Good	Good	Moderate
e_4	Very bad	Fairly Bad	Fairly Bad	Good	Fairly Good	Good	Very Good
e_5	Very bad	Bad	Fairly Bad	Moderate	Fairly Good	Good	Very Good
e_6	Very bad	Good	Bad	Good	Good	Good	Very Good

In this application, we do not need to compare experts for the evaluation of each criterion. We compare the experts' opinions *only after aggregating* data by all criteria. Therefore, the adjacency matrices of all experts' BDD-FCMs will be vectors representing the the expert' belief degrees for each energy policy scenario. Notice that all vectors have the same number of p elements (thus there are not unique elements for neither of expert), therefore (13) for energy policy evaluation becomes as follows:

$$DR(e_A, e_B) = \frac{\sum_{k=0}^{m}\sum_{i=1}^{p} \left| \beta_{i,k}^{e_A} - \beta_{i,k}^{e_B} \right|}{2p^2 - 2p} \tag{17}$$

Table 13 shows the belief structure of each expert generated from Table 12.

Table 13. Experts' belief structures generated from Table 12

	a_1	a_2	a_3	a_4	a_5	a_6	a_7
e_1	$\{(s_7,1)\}$	$\{(s_6,1)\}$	$\{(s_4,1)\}$	$\{(s_2,1)\}$	$\{(s_5,1)\}$	$\{(s_3,1)\}$	$\{(s_1,1)\}$
e_2	$\{(s_7,1)\}$	$\{(s_6,1)\}$	$\{(s_4,1)\}$	$\{(s_2,1)\}$	$\{(s_5,1)\}$	$\{(s_1,1)\}$	$\{(s_1,1)\}$
e_3	$\{(s_1,1)\}$	$\{(s_3,1)\}$	$\{(s_4,1)\}$	$\{(s_6,1)\}$	$\{(s_7,1)\}$	$\{(s_6,1)\}$	$\{(s_4,1)\}$
e_4	$\{(s_1,1)\}$	$\{(s_3,1)\}$	$\{(s_3,1)\}$	$\{(s_6,1)\}$	$\{(s_5,1)\}$	$\{(s_6,1)\}$	$\{(s_7,1)\}$
e_5	$\{(s_1,1)\}$	$\{(s_2,1)\}$	$\{(s_3,1)\}$	$\{(s_4,1)\}$	$\{(s_5,1)\}$	$\{(s_6,1)\}$	$\{(s_7,1)\}$
e_6	$\{(s_1,1)\}$	$\{(s_6,1)\}$	$\{(s_2,1)\}$	$\{(s_6,1)\}$	$\{(s_6,1)\}$	$\{(s_6,1)\}$	$\{(s_7,1)\}$

Table 14 shows the DR and SIM values of six experts with the reference to the data in Table 13.

Table 14. The DR and SIM values of experts for Energy Policy Evaluation

(a) DR

	e_1	e_2	e_3	e_4	e_5	e_6
e_1	-	0.033	0.2	0.2	0.2	0.2
e_2	0.033	-	0.2	0.2	0.2	0.2
e_3	0.2	0.2	-	0.1	0.16	0.13
e_4	0.2	0.2	0.1	-	0.066	0.1
e_5	0.2	0.2	0.16	0.066	-	0.13
e_6	0.2	0.2	0.13	0.1	0.13	-

(b) SIM

	e_1	e_2	e_3	e_4	e_5	e_6
e_1	-	0.97	0.83	0.83	0.83	0.83
e_2	-	-	0.83	0.83	0.83	0.83
e_3	-	-	-	0.91	0.86	0.88
e_4	-	-	-	-	0.94	0.91
e_5	-	-	-	-	-	0.88
e_6	-	-	-	-	-	-

Figure 11 is a graphical representation of the similarity between the experts. It includes more details as the similar graph in [25].

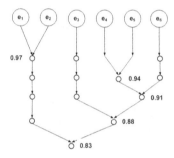

Fig. 11. Graphical representation of the clustering of experts by their similarity

9 Real-Life Case Study: Energy Policy Evaluation

In this section we show the proposed methodology to real-life case study of energy policy evaluation. The Belgian parliament in 2003 enacted a law to progressively phase out existing nuclear power plants [26]. To evaluate the possible future of Belgian Energy policies, 8 scenarios, 44 criteria are developed and 10 experts are asked to evaluate each scenario. Table 15 lists of all criteria.

The experts are asked to choose among all criteria the ones that they consider as more important (or they have knowledge to evaluate) and evaluate each scenario against all the chosen criteria. The maximum number of criteria chosen by an expert was 20, and the minimal number of the evaluated criteria given by another expert was 5. All experts chose different nodes, thus none of any two experts chose the exact same set of nodes.

As already discussed, for multiple scenarios we first aggregate data and then apply our algorithm to find the similarity between experts. We are not interested in how much experts are different by chosen criteria and given values to the common criteria, but more interested in how much they are different in choosing the energy policy scenarios. As an aggregation algorithm we used the one suggested in [22]. This algorithm is more

Table 15. Criteria List for Energy Policy Domain

Criteria	Node	Criteria	Node
Impacts of air pollution on human health: midterm	n_1	Impacts of air pollution on human health: long term	n_2
Impacts on occupational health (gas+coal)	n_3	Radiobiological health impacts (nuclear)	n_4
Need for long-term management of HLW	n_5	Visual impact on landscape	n_6
Noise amenity	n_7	Impact on natural ecosystems (air pollution): midterm	n_8
Impact on natural ecosystems (air pollution): longterm	n_9	Environmental impact from solid waste (coal)	n_{10}
(Land use) Occupational Risk(nuclear)	n_{11}	Water Use	n_{12}
Catastrophic risk: nuclear	n_{13}	Geographical distribution risks/benefits	n_{14}
Intensity of energy use	n_{15}	Security of energy supply	n_{16}
Distribution of economic benefits/burdens	n_{17}	Economic risks	n_{18}
Overall cost energy system: 2010	n_{19}	Overall cost energy system: 2030	n_{20}
Overall cost energy system: 2050	n_{21}	Ability to provide specialist market	n_{22}
Marginal cost electricity: midterm	n_{23}	Marginal cost electricity: longterm	n_{24}
Strategic factors for export	n_{25}	Compatibility with international RD agenda	n_{26}
Amount of direct or indirect subsidies needed	n_{27}	Consumer choice	n_{28}
Citizen participation	n_{29}	Contribution to rational energy use	n_{30}
Degree of decentralization	n_{31}	Need for intermediary storage of spent fuel	n_{32}
Control and concentration of power	n_{33}	Influence on political decision-making	n_{34}
Need for socio-political stability	n_{35}	Need for direct political intervention	n_{36}
Reversibility of technology choice	n_{37}	Knowledge specialization	n_{38}
Need for institutional non-prolefiration measures	n_{39}	Potential for technology transfer	n_{40}
Leaving resources for development	n_{41}	Equity(general)	n_{42}
Job opportunities	n_{43}	Diversification	n_{44}

convenient for our task as it used belief structures to compute the *attractiveness* of each scenario for each expert. We used computer program described in [22] to compute the evaluation of each scenario by each expert. Initially, all data are given as interval values, and we use (5) to convert them to belief structures. Table 16 shows the aggregation results that we got running the program [22].

Table 16. The aggregation results of 10 experts' evaluation for 8 scenarios

	MLCS	MPCS	MPLCS	MPLCSI	RLCS	RPCS	RPLCS	RPLCCSI
e_1	0.5043	0.5549	0.5873	0.6438	0.6604	0.6697	0.6894	0.7538
e_2	0.7208	0.5872	0.6617	0.7222	0.8227	0.728	0.7945	0.8154
e_3	0.6622	0.568	0.6312	0.6317	0.8142	0.761	0.8077	0.7622
e_4	0.6451	0.6127	0.7024	0.7602	0.7339	0.5856	0.739	0.758
e_5	0.6478	0.5602	0.5595	0.552	0.735	0.6637	0.6678	0.6445
e_6	0.8613	0.617	0.7887	0.8865	0.881	0.6775	0.8725	0.892
e_7	0.2925	0.6777	0.7175	0.5573	0.4715	0.7125	0.741	0.7203
e_8	0.8092	0.5447	0.6808	0.7272	0.8257	0.6937	0.7563	0.771
e_9	0.531	0.524	0.618	0.6187	0.5798	0.5673	0.7177	0.7283
e_{10}	0.6401	0.4796	0.588	0.5693	0.7237	0.6078	0.6774	0.6449

Table 17 shows the same results expressed by belief structures. Our aim to convert Table 16 to belief structures is being able to compare our algorithm with LW algorithm as LW method is applied only when the links are expressed by numbers or somehow are converted to numbers. Thus, we applied (11) for Table 16, and (13) for Table 17.

Table 18 shows the distance ratio of all experts applying (11) for Table 16 and Table 19 shows the same result when we apply (13) for Table 17.

Figure 12 shows the final result of comparison of 10 experts according to their evaluations of 8 scenarios. From Figure 12, experts e_2 and e_8 are the most similar, whereas expert e_7 is the most dissimilar comparing with all other experts. Depending on the threshold we can combine the experts into groups based on their similarities. For example, if we fix the threshold 0.996, experts e_2, e_3 and e_8 will make one group, and experts e_1, e_5, e_9 will combine another group. On the other hand, if we fix greater threshold, like 0.995, the group will have more members, like in this case, experts e_1, e_2, e_3, e_4, e_5, e_8, e_9 and e_{10} will belong to the same group.

Table 17. The aggregation results with belief structures generated from Table 16

(a)

	MLCS	MPCS	MPLCS	MPLCSI
e_1	$\{(s_1,0.02),(s_2,0.98)\}$	$\{(s_1,0.22),(s_2,0.78)\}$	$\{(s_1,0.35),(s_2,0.65)\}$	$\{(s_1,0.58),(s_2,0.42)\}$
e_2	$\{(s_1,0.88),(s_2,0.12)\}$	$\{(s_1,0.35),(s_2,0.65)\}$	$\{(s_1,0.65),(s_2,0.35)\}$	$\{(s_1,0.89),(s_2,0.11)\}$
e_3	$\{(s_1,0.65),(s_2,0.35)\}$	$\{(s_1,0.27),(s_2,0.73)\}$	$\{(s_1,0.52),(s_2,0.48)\}$	$\{(s_1,0.53),(s_2,0.47)\}$
e_4	$\{(s_1,0.58),(s_2,0.42)\}$	$\{(s_1,0.45),(s_2,0.55)\}$	$\{(s_1,0.81),(s_2,0.19)\}$	$\{(s_0,0.04),(s_1,0.96)\}$
e_5	$\{(s_1,0.59),(s_2,0.41)\}$	$\{(s_1,0.24),(s_2,0.76)\}$	$\{(s_1,0.24),(s_2,0.76)\}$	$\{(s_1,0.21),(s_2,0.79)\}$
e_6	$\{(s_0,0.45),(s_1,0.55)\}$	$\{(s_1,0.47),(s_2,0.53)\}$	$\{(s_0,0.15),(s_1,0.85)\}$	$\{(s_0,0.55),(s_1,0.45)\}$
e_7	$\{(s_2,0.17),(s_3,0.83)\}$	$\{(s_1,0.71),(s_2,0.29)\}$	$\{(s_1,0.13),(s_2,0.87)\}$	$\{(s_1,0.23),(s_2,0.77)\}$
e_8	$\{(s_0,0.24),(s_1,0.76)\}$	$\{(s_1,0.18),(s_2,0.82)\}$	$\{(s_1,0.72),(s_2,0.28)\}$	$\{(s_1,0.91),(s_2,0.09)\}$
e_9	$\{(s_1,0.12),(s_2,0.88)\}$	$\{(s_1,0.1),(s_2,0.9)\}$	$\{(s_1,0.47),(s_2,0.53)\}$	$\{(s_1,0.47),(s_2,0.53)\}$
e_{10}	$\{(s_1,0.56),(s_2,0.44)\}$	$\{(s_2,0.92),(s_3,0.08)\}$	$\{(s_1,0.35),(s_2,0.65)\}$	$\{(s_1,0.28),(s_2,0.72)\}$

(b)

	RLCS	RPCS	RPLCS	RPLCCSI
e_1	$\{(s_1,0.64),(s_2,0.36)\}$	$\{(s_1,0.68),(s_2,0.32)\}$	$\{(s_1,0.76),(s_2,0.24)\}$	$\{(s_0,0.02),(s_1,0.98)\}$
e_2	$\{(s_0,0.29),(s_1,0.71)\}$	$\{(s_1,0.91),(s_2,0.09)\}$	$\{(s_0,0.18),(s_1,0.82)\}$	$\{(s_0,0.26),(s_1,0.74)\}$
e_3	$\{(s_0,0.26),(s_1,0.74)\}$	$\{(s_0,0.04),(s_1,0.96)\}$	$\{(s_0,0.23),(s_1,0.77)\}$	$\{(s_0,0.05),(s_1,0.95)\}$
e_4	$\{(s_1,0.94),(s_2,0.06)\}$	$\{(s_1,0.34),(s_2,0.66)\}$	$\{(s_1,0.96),(s_2,0.04)\}$	$\{(s_0,0.03),(s_1,0.97)\}$
e_5	$\{(s_1,0.94),(s_2,0.06)\}$	$\{(s_1,0.65),(s_2,0.35)\}$	$\{(s_1,0.67),(s_2,0.33)\}$	$\{(s_1,0.58),(s_2,0.42)\}$
e_6	$\{(s_0,0.52),(s_1,0.48)\}$	$\{(s_1,0.71),(s_2,0.29)\}$	$\{(s_0,0.49),(s_1,0.51)\}$	$\{(s_0,0.57),(s_1,0.43)\}$
e_7	$\{(s_2,0.89),(s_3,0.11)\}$	$\{(s_1,0.85),(s_2,0.15)\}$	$\{(s_1,0.96),(s_2,0.04)\}$	$\{(s_1,0.88),(s_2,0.12)\}$
e_8	$\{(s_0,0.3),(s_1,0.7)\}$	$\{(s_1,0.77),(s_2,0.23)\}$	$\{(s_0,0.03),(s_1,0.97)\}$	$\{(s_0,0.08),(s_1,0.92)\}$
e_9	$\{(s_1,0.32),(s_2,0.68)\}$	$\{(s_1,0.27),(s_2,0.73)\}$	$\{(s_1,0.87),(s_2,0.13)\}$	$\{(s_1,0.91),(s_2,0.09)\}$
e_{10}	$\{(s_1,0.89),(s_2,0.11)\}$	$\{(s_1,0.43),(s_2,0.57)\}$	$\{(s_1,0.71),(s_2,0.29)\}$	$\{(s_1,0.58),(s_2,0.42)\}$

Table 18. The DR value applying our algorithm generated from Table 17

	e_1	e_2	e_3	e_4	e_5	e_6	e_7	e_8	e_9	e_{10}
e_1	-	0.043	0.033	0.045	0.036	0.072	0.062	0.044	0.039	0.037
e_2		-	0.021	0.035	0.037	0.045	0.069	0.02	0.067	0.05
e_3		-	-	0.04	0.037	0.055	0.062	0.027	0.0045	0.041
e_4		-	-	-	0.046	0.052	0.073	0.030	0.045	0.043
e_5		-	-	-	-	0.077	0.056	0.050	0.045	0.014
e_6		-	-	-	-	-	0.078	0.048	0.080	0.058
e_7		-	-	-	-	-	-	0.072	0.060	0.065
e_8		-	-	-	-	-	-	-	0.054	0.063
e_9		-	-	-	-	-	-	-	-	0.037
e_{10}		-	-	-	-	-	-	-	-	-

Table 19. The SIM value applying our algorithm generated from Table 17

	e_1	e_2	e_3	e_4	e_5	e_6	e_7	e_8	e_9	e_{10}
e_1	-	0.96	0.97	0.96	0.96	0.93	0.94	0.96	0.96	0.96
e_2		-	0.98	0.97	0.96	0.96	0.93	0.98	0.94	0.95
e_3		-	-	0.96	0.96	0.95	0.94	0.97	0.96	0.96
e_4		-	-	-	0.96	0.95	0.93	0.97	0.96	0.96
e_5		-	-	-	-	0.93	0.95	0.95	0.96	0.99
e_6		-	-	-	-	-	0.93	0.95	0.92	0.94
e_7		-	-	-	-	-	-	0.93	0.94	0.94
e_8		-	-	-	-	-	-	-	0.95	0.94
e_9		-	-	-	-	-	-	-	-	0.96
e_{10}		-	-	-	-	-	-	-	-	-

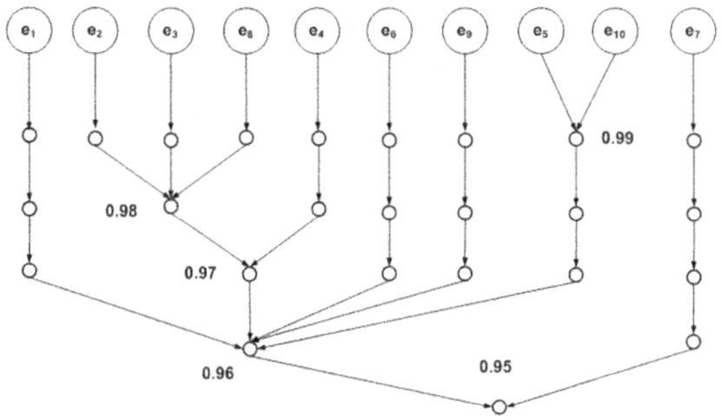

Fig. 12. Final Result of Similarity among 10 experts

10   Adding Confidence Levels(CVs) for Criteria: GBDD-FCMs and Experts' Comparison Considering CL Values

In the previous sections we discussed GBDD-FCM and the algorithm to compare experts. In both cases we assumed that experts are confident about their judgment; thus we assumed that $CV = 1$. However, in real life applications, often for a given criterion experts give an evaluation with not fully confidence. In this section we discuss GBDD-FCM and the algorithm of the experts' comparison, considering the fact that $CV \leq 1$.

Table 20 shows an example in which five experts provided judgments for three different scenarios and for each criterion they provided also the confident values of the judgments.

Table 20. Adjacency matrices of five experts' FCMs

(a) Expert e_1

Criteria	Scenario 1		Scenario 2		Scenario 3	
C_1	24	29	27	30	23	25
C_2	79	79	63	63	78	78
C_3	0	0	0	0	0	0
C_4	0	43	92	95	88	93
C_5	18	59	24	63	53	91
C_6	55	55	24	24	69	69
C_7	75	88	79	92	86	93
C_8	87	87	87	87	87	87
C_9	0	38	6	40	48	81

(b) Expert e_2

Criteria	Scenario 1	CV	Scenario 2	CV	Scenario 3	CV
C_1	3	90%	3	100%	2	70%
C_2	8	80%	6	90%	8	90%
C_4	4	100%	10	100%	9	100%
C_5	4	90%	4	90%	7	90%
C_7	8	90%	9	90%	9	100%
C_8	9	100%	9	100%	9	80%
C_9	2	100%	2	90%	6	90%

(c) Expert e_3

Criteria	Scenario 1	CV	Scenario 2	CV	Scenario 3	CV
C_1	L	100%	L	90%	VL	80%
C_2	VH	100%	H	90%	VH	80%
C_3	N	100%	N	90%	N	80%
C_5	L	100%	L	90%	H	80%
C_7	VH	100%	M	90%	VH	80%
C_8	VH	100%	VH	90%	VH	80%

(d) Expert e_4

Criteria	Scenario 1	CV	Scenario 2	CV	Scenario 3	CV
C_1	$(s_2,0.3)$	100%	$(s_2,-0.4)$	80%	$(s_1,0)$	80%
C_2	$(s_5,-0.1)$	900%	$(s_4,0.1)$	100%	$(s_5,0)$	100%
C_3	$(s_0,0)$	100%	$(s_0,0.3)$	80%	$(s_0,0.2)$	90%
C_4	$(s_3,0)$	90%	$(s_3,0.3)$	100%	$(s_5,0.2)$	90%
C_7	$(s_5,0.2)$	90%	$(s_5,-0.1)$	100%	$(s_5,-0.3)$	100%
C_8	$(s_5,0.2)$	100%	$(s_5,-0.3)$	100%	$(s_5,0.1)$	90%
C_9	$(s_1,-0.4)$	80%	$(s_1,0.2)$	90%	$(s_4,0.5)$	100%

(e) Expert e_5

Criteria	Scenario 1								Scenario 2								Scenario 3							
	s_0	s_1	s_2	s_3	s_4	s_5	s_6	CV	s_0	s_1	s_2	s_3	s_4	s_5	s_6	CV	s_0	s_1	s_2	s_3	s_4	s_5	s_6	CV
C_1	0	0.4	0.6	0	0	0	0	60%	0	0.3	0.7	0	0	0	0	80%	0	0.6	0.4	0	0	0	0	100%
C_2	0	0	0	0.3	0.7	0	0	100%	0	0	0	0.2	0.8	0	0	100%	0	0	0	0	0.3	0.7	0	60%
C_4	0.2	0.4	0.4	0	0	0	0	90%	0	0	0	0	0	0.4	0.6	80%	0	0	0	0	0	0.6	0.4	80%
C_5	0	0.2	0.4	0.3	0.1	0	0	60%	0	0	0.4	0.4	0.2	0	0	100%	0	0	0	0.1	0.4	0.4	0.1	60%
C_8	0	0	0	0	0	0.8	0.2	90%	0	0	0	0	0	0.8	0.2	90%	0	0	0	0	0	0.7	0.3	90%
C_9	0.2	0.4	0.4	0	0	0	0	100%	0.1	0.5	0.4	0	0	0	0	100%	0	0	0	0.3	0.5	0.2	0	100%

As can be seen from Table 20, experts have evaluated each criterion with different formats. The expert e_1 has used interval values in $[0, 100]$ range. The expert e_2 has used a numerical format in $[0, 10]$ range. The expert e_3 has used a linguistic format. The expert e_4 has done the evaluation by 2-tuple format, and finally the expert e_5 has used belief degree distribution format.

First we transfer the evaluations of the criteria of each scenario for each expert to belief structures according with the provided data format. Table 21 shows the transformation results.

Table 21. Belief structures of the adjacency matrices of five experts' FCMs

(a) Expert e_1

Criteria	Scenario 1							Scenario 2							Scenario 3						
	s_0	s_1	s_2	s_3	s_4	s_5	s_6	s_0	s_1	s_2	s_3	s_4	s_5	s_6	s_0	s_1	s_2	s_3	s_4	s_5	s_6
C_1	0	0.41	0.59	0	0	0	0	0	0.29	0.71	0	0	0	0	0	0.44	0.56	0	0	0	0%
C_2	0	0	0	0	0.26	0.74	0	0	0	0	0.22	0.78	0	0	0	0	0	0	0.32	0.68	0%
C_3	1	0	0	0	0	0	0	1	0	0	0	0	0	0	1	0	0	0	0	0	0%
C_4	0.19	0.39	0.35	0.07	0	0	0	0	0	0	0	0	0.39	0.61	0	0	0	0	0	0.57	0.43%
C_5	0	0.17	0.41	0.36	0.06	0	0	0	0.06	0.39	0.42	0.13	0	0	0	0.15	0.43	0.37	0.05%
C_6	0	0	0	0.7	0.3	0	0	0	0.56	0.44	0	0	0	0	0	0	0	0.86	0.14	0%
C_7	0	0	0	0.16	0.74	0.05	0	0	0	0	0	0.044	0.79	0.17	0	0	0	0	0	0.63	0.37%
C_8	0	0	0	0	0	0.78	0.22	0	0	0	0	0	0.78	0.22	0	0	0	0	0	0.78	0.22%
C_9	0.22	0.44	0.32	0.02	0	0	0	0.1	0.46	0.41	0.03	0	0	0	0	0	0	0.31	0.5	0.19	0

(b) Expert e_2

Criteria	Scenario 1							CV	Scenario 2							CV	Scenario 3							CV
	s_0	s_1	s_2	s_3	s_4	s_5	s_6		s_0	s_1	s_2	s_3	s_4	s_5	s_6		s_0	s_1	s_2	s_3	s_4	s_5	s_6	
C_1	0	0.2	0.8	0	0	0	0	90%	0	0.3	0.7	0	0	0	0	80%	0	0.8	0.2	0	0	0	0	70%
C_2	0	0	0	0	0	0.2	0.8	80%	0	0	0	0.4	0.6	0	0	90%	0	0	0	0	0	0.2	0.8	90%
C_4	0	0	0.6	0.4	0	0	0	100%	0	0	0	0	0	0	1	100%	0	0	0	0	0	0.6	0.4	100%
C_5	0	0	0.6	0.4	0	0	0	90%	0	0	0.6	0.4	0	0	0	90%	0	0	0	0	0.8	0.2	0	90%
C_7	0	0	0	0	0.2	0.8	0	90%	0	0	0	0	0	0.6	0.4	90%	0	0	0	0	0	0.6	0.4	100%
C_8	0	0	0	0	0	0.6	0.4	100%	0	0	0	0	0	0.6	0.4	100%	0	0	0	0	0	0.6	0.4	80%
C_9	0	0.8	0.2	0	0	0	0	100%	0	0.8	0.2	0	0	0	0	90%	0	0	0	0.4	0.6	0	0	90%

(c) Expert e_3

Criteria	Scenario 1							CV	Scenario 2							CV	Scenario 3							CV
	s_0	s_1	s_2	s_3	s_4	s_5	s_6		s_0	s_1	s_2	s_3	s_4	s_5	s_6		s_0	s_1	s_2	s_3	s_4	s_5	s_6	
C_1	0	0	1	0	0	0	0	100%	0	0	1	0	0	0	0	90%	0	0	1	0	0	0	0	80%
C_2	0	0	0	0	0	1	0	100%	0	0	0	0	1	0	0	90%	0	0	0	0	0	1	0	80%
C_3	1	0	0	0	0	0	0	100%	1	0	0	0	0	0	0	90%	1	0	0	0	0	0	0	80%
C_5	0	0	1	0	0	0	0	100%	0	0	1	0	0	0	0	90%	0	0	0	0	1	0	0	80%
C_7	0	0	0	0	0	1	0	100%	0	0	0	1	0	0	0	90%	0	0	0	0	0	1	0	80%
C_8	0	0	0	0	0	1	0	100%	0	0	0	0	0	1	0	90%	0	0	0	0	0	1	0	80%

(d) Expert e_4

Criteria	Scenario 1							CV	Scenario 2							CV	Scenario 3							CV
	s_0	s_1	s_2	s_3	s_4	s_5	s_6		s_0	s_1	s_2	s_3	s_4	s_5	s_6		s_0	s_1	s_2	s_3	s_4	s_5	s_6	
C_1	0	0	0.7	0.3	0	0	0	100%	0	0.4	0.6	0	0	0	0	80%	0	1	0	0	0	0	0	80%
C_2	0	0	0	0	0	0.1	0.9	90%	0	0	0	0.9	0.1	0	100%	0	0	0	0	0	1	0	60%
C_3	1	0	0	0	0	0	0	100%	0.7	0.3	0	0	0	0	0	80%	0.8	0.2	0	0	0	0	0	90%
C_4	0	0	0	1	0	0	0	90%	0	0	0	0.7	0.3	0	0	100%	0	0	0	0	0	0.8	0.2	90%
C_7	0	0	0	0	0.8	0.2	0	90%	0	0	0	0.1	0.9	0	100%	0	0	0	0	0.3	0.7	0	100%
C_8	0	0	0	0	0.8	0.2	0	100%	0	0	0	0.3	0.7	0	100%	0	0	0	0	0	0.9	0.1	90%
C_9	0.4	0.6	0	0	0	0	0	80%	0	0.8	0.2	0	0	0	0	90%	0	0	0	0	0.5	0.5	0	100%

(e) Expert e_5

Criteria	Scenario 1							CV	Scenario 2							CV	Scenario 3							CV
	s_0	s_1	s_2	s_3	s_4	s_5	s_6		s_0	s_1	s_2	s_3	s_4	s_5	s_6		s_0	s_1	s_2	s_3	s_4	s_5	s_6	
C_1	0	0.4	0.6	0	0	0	0	60%	0	0.3	0.7	0	0	0	0	80%	0	0.6	0.4	0	0	0	0	100%
C_2	0	0	0	0	0.3	0.7	0	100%	0	0	0	0.2	0.8	0	0	100%	0	0	0	0	0.3	0.7	0	60%
C_4	0.2	0.4	0.4	0	0	0	0	90%	0	0	0	0	0.4	0.6	0	80%	0	0	0	0	0	0.6	0.4	80%
C_5	0	0.2	0.4	0.3	0.1	0	0	60%	0	0.4	0.4	0.2	0	0	0	100%	0	0	0	0.1	0.4	0.4	0.1	60%
C_8	0	0	0	0	0	0.8	0.2	90%	0	0	0	0	0	0.8	0.2	90%	0	0	0	0	0	0.7	0.3	90%
C_9	0.2	0.4	0.4	0	0	0	0	100%	0.1	0.5	0.4	0	0	0	0	100%	0	0	0	0.3	0.5	0.2	0	100%

To conduct a group matrix from Table 21 we first multiply belief values with the corresponding CV (Table 22), then apply (9). Thus, the belief values that include also CVs will be decided as follows:

$$B^e(L_{ij}) = \left\{ (\beta_{ij,k}^e \times CV_{ij}^e, s_k), k = 0, ..., m \right\}, \forall i, \forall j, \forall e \tag{18}$$

where CV_{ij} is the confidence of the expert e for the causal link between the i-th and the j-th nodes

Table 23 shows the group decision about each scenario applying (9) from Table 22.

Table 22. Belief structures (with CV) of the adjacency matrices of five experts' FCMs

(a) Expert e_1

Criteria	Scenario 1							Scenario 2							Scenario 3						
	s_0	s_1	s_2	s_3	s_4	s_5	s_6	s_0	s_1	s_2	s_3	s_4	s_5	s_6	s_0	s_1	s_2	s_3	s_4	s_5	s_6
C_1	0	0.41	0.59	0	0	0	0	0	0.29	0.71	0	0	0	0	0	0.44	0.56	0	0	0	0
C_2	0	0	0	0	0.26	0.74	0	0	0	0	0.22	0.78	0	0	0	0	0	0	0.32	0.68	0
C_3	1	0	0	0	0	0	0	1	0	0	0	0	0	0	1	0	0	0	0	0	0
C_4	0.19	0.39	0.35	0.07	0	0	0	0	0	0	0	0	0.39	0.61	0	0	0	0	0	0.57	0.43
C_5	0	0.17	0.41	0.36	0.06	0	0	0	0.06	0.39	0.42	0.13	0	0	0	0	0	0.15	0.43	0.37	0.05
C_6	0	0	0	0.7	0.3	0	0	0	0.56	0.44	0	0	0	0	0	0	0	0	0.86	0.14	0
C_7	0	0	0	0	0.16	0.79	0.05	0	0	0	0	0.044	0.79	0.17	0	0	0	0	0	0.63	0.37
C_8	0	0	0	0	0	0.78	0.22	0	0	0	0	0	0.78	0.22	0	0	0	0	0	0.78	0.22
C_9	0.22	0.44	0.32	0.02	0	0	0	0.1	0.46	0.41	0.03	0	0	0	0	0	0	0.31	0.5	0.19	0

(b) Expert e_2

Criteria	Scenario 1							Scenario 2							Scenario 3						
	s_0	s_1	s_2	s_3	s_4	s_5	s_6	s_0	s_1	s_2	s_3	s_4	s_5	s_6	s_0	s_1	s_2	s_3	s_4	s_5	s_6
C_1	0	0.12	0.72	0	0	0	0	0	0.24	0.56	0	0	0	0	0	0.56	0.14	0	0	0	0
C_2	0	0	0	0	0.16	0.64	0	0	0	0	0.36	0.54	0	0	0	0	0	0	0	0.18	0.72
C_4	0	0	0.6	0.4	0	0	0	0	0	0	0	0	0	1	0	0	0	0	0	0.6	0.4
C_5	0	0	0.54	0.36	0	0	0	0	0	0.54	0.36	0	0	0	0	0	0	0	0.72	0.18	0
C_7	0	0	0	0	0.18	0.72	0	0	0	0	0	0	0.54	0.36	0	0	0	0	0	0.6	0.4
C_8	0	0	0	0	0	0.6	0.4	0	0	0	0	0	0.6	0.4	0	0	0	0	0	0.48	0.32
C_9	0	0.8	0.2	0	0	0	0	0	0.72	0.18	0	0	0	0	0	0	0	0.36	0.54	0	0

(c) Expert e_3

Criteria	Scenario 1							Scenario 2							Scenario 3						
	s_0	s_1	s_2	s_3	s_4	s_5	s_6	s_0	s_1	s_2	s_3	s_4	s_5	s_6	s_0	s_1	s_2	s_3	s_4	s_5	s_6
C_1	0	0	1	0	0	0	0	0	0	0.9	0	0	0	0	0	0	0.8	0	0	0	0
C_2	0	0	0	0	0	1	0	0	0	0	0	0.9	0	0	0	0	0	0	0	0.8	0
C_3	1	0	0	0	0	0	0	0.9	0	0	0	0	0	0	0.8	0	0	0	0	0	0
C_5	0	0	1	0	0	0	0	0	0	0.9	0	0	0	0	0	0	0	0	0.8	0	0
C_7	0	0	0	0	0	1	0	0	0	0	0	0.9	0	0	0	0	0	0	0	0.8	0
C_8	0	0	0	0	0	1	0	0	0	0	0	0	0.9	0	0	0	0	0	0	0.8	0

(d) Expert e_4

Criteria	Scenario 1							Scenario 2							Scenario 3						
	s_0	s_1	s_2	s_3	s_4	s_5	s_6	s_0	s_1	s_2	s_3	s_4	s_5	s_6	s_0	s_1	s_2	s_3	s_4	s_5	s_6
C_1	0	0	0.7	0.3	0	0	0	0	0.32	0.48	0	0	0	0	0	0.8	0	0	0	0	0
C_2	0	0	0	0	0	0.09	0.81	0	0	0	0	0.9	0.1	0	0	0	0	0	0	0.6	0
C_3	1	0	0	0	0	0	0	0.56	0.24	0	0	0	0	0	0	0.72	0.18	0	0	0	0
C_4	0	0	0	0.9	0	0	0	0	0	0	0.7	0.3	0	0	0	0	0	0	0	0.72	0.18
C_7	0	0	0	0	0	0.72	0.18	0	0	0	0	0.1	0.9	0	0	0	0	0	0	0.3	0.7
C_8	0	0	0	0	0	0.8	0.2	0	0	0	0	0.3	0.7	0	0	0	0	0	0	0.81	0.09
C_9	0.32	0.48	0	0	0	0	0	0	0.72	0.18	0	0	0	0	0	0	0	0	0.5	0.5	0

(e) Expert e_5

Criteria	Scenario 1							Scenario 2							Scenario 3						
	s_0	s_1	s_2	s_3	s_4	s_5	s_6	s_0	s_1	s_2	s_3	s_4	s_5	s_6	s_0	s_1	s_2	s_3	s_4	s_5	s_6
C_1	0	0.24	0.36	0	0	0	0	0	0.24	0.56	0	0	0	0	0	0.6	0.4	0	0	0	0
C_2	0	0	0	0	0.3	0.7	0	0	0	0	0.2	0.8	0	0	0	0	0	0	0.18	0.42	0
C_4	0.18	0.36	0.36	0	0	0	0	0	0	0	0	0	0.32	0.48	0	0	0	0	0	0.48	0.32
C_5	0	0.12	0.24	0.18	0.06	0	0	0	0	0.4	0.4	0.2	0	0	0	0	0	0.06	0.24	0.24	0.06
C_8	0	0	0	0	0	0.72	0.18	0	0	0	0	0	0.72	0.18	0	0	0	0	0	0.63	0.27
C_9	0.2	0.4	0.4	0	0	0	0	0.1	0.5	0.4	0	0	0	0	0	0	0	0.3	0.5	0.2	0

Table 23. Belief Structured(with CV) adjacency matrix of GBDD-FCM from Table 22

Criteria	Scenario 1							Scenario 2							Scenario 3						
	s_0	s_1	s_2	s_3	s_4	s_5	s_6	s_0	s_1	s_2	s_3	s_4	s_5	s_6	s_0	s_1	s_2	s_3	s_4	s_5	s_6
C_1	0	0.15	0.67	0.06	0	0	0	0	0.22	0.64	0	0	0	0	0	0.48	0.38	0	0	0	0
C_2	0	0	0	0	0.11	0.54	0.29	0	0	0.16	0.78	0.02	0	0	0	0	0	0	0.1	0.54	0.14
C_3	0.6	0	0	0	0	0	0	0.49	0.05	0	0	0	0	0	0.51	0.04	0	0	0	0	0
C_4	0.07	0.15	0.26	0.55	0	0	0	0	0	0.14	0.06	0.14	0.42	0	0	0	0	0	0	0.47	0.27
C_5	0	0.14	0.44	0.18	0.02	0	0	0	0.01	0.47	0.24	0.17	0	0	0	0	0	0.05	0.44	0.16	0.02
C_6	0	0	0	0.14	0.06	0	0	0	0.11	0.09	0	0	0	0	0	0	0	0	0.17	0.29	0
C_7	0	0	0	0	0.07	0.67	0.05	0	0	0	0.18	0.03	0.45	0.21	0	0	0	0	0.6	0.55	0.31
C_8	0	0	0	0	0	0.78	0.2	0	0	0	0	0.06	0.74	0.16	0	0	0	0	0	0.7	0.18
C_9	0.15	0.42	0.18	0	0	0	0	0.04	0.48	0.23	0	0	0	0	0	0	0	0.19	0.41	0.18	0

Table 24. DR values of five experts for three scenarios

Criteria	Scenario 1					Scenario 2					Scenario 3				
	e_1	e_2	e_3	e_4	e_5	e_1	e_2	e_3	e_4	e_5	e_1	e_2	e_3	e_4	e_5
e_1	-	0.078	0.092	0.080	0.070	-	0.056	0.1	0.072	0.066	-	0.066	0.089	0.067	0.074
e_2	-	-	0.15	0.087	0.079	-	-	0.14	0.10	0.063	-	-	0.14	0.096	0.067
e_3	-	-	-	0.13	00.15	-	-	-	0.13	0.16	-	-	-	0.12	0.15
e_4	-	-	-	-	0.15	-	-	-	-	0.13	-	-	-	-	0.12
e_5	-	-	-	-	-	-	-	-	-	-	-	-	-	-	-

Table 25. Similarity measures of five experts for three scenarios

Criteria	Scenario 1					Scenario 2					Scenario 3				
	e_1	e_2	e_3	e_4	e_5	e_1	e_2	e_3	e_4	e_5	e_1	e_2	e_3	e_4	e_5
e_1	-	0.93	0.91	0.93	0.93	-	0.95	0.90	0.93	0.94	-	0.94	0.92	0.94	0.93
e_2	-	-	0.87	0.92	0.93	-	-	0.88	0.90	94	-	-	0.88	0.91	0.94
e_3	-	-	-	0.88	0.87	-	-	-	0.88	0.86	-	-	-	0.89	0.87
e_4	-	-	-	-	0.87	-	-	-	-	0.88	-	-	-	-	0.89
e_5	-	-	-	-	-	-	-	-	-	-	-	-	-	-	-

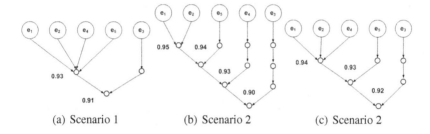

(a) Scenario 1   (b) Scenario 2   (c) Scenario 2

Fig. 13. Comparison of five experts from Table 25

In the previous section we showed the similarity of 10 experts over 8 different scenarios aggregating the results of all scenarios. However, it may happen that in certain case we are interested to see the similarity and conflict of experts for only some scenarios. In this section we do comparison without aggregating all criteria for each scenario, but we will aggregate after getting the similarity between experts for each scenario. It depends on the task under consideration, whether the experts' judgment will be first aggregated then applied the comparison algorithm, or first similarities for each scenario will be found, and then aggregation of the results will be done to find the final similarity.

Table 24 and Table 25 show the distance ratios and similarities of experts for each scenario, respectively. Figure 13 shows the comparison of the five experts for each scenario.

We can make some observations: for all scenarios e_1 and e_2 have the most similar opinions, whereas expert e_3 is the most different from all the other experts.

Note that it depends on the application domain if we aggregate for all scenarios then find similarity measures or we find for each scenario.

11 Conclusions

CMs have been used for analyzing decision-making by exploring causal links among relevant domain concepts. FCM was introduced as an extension of CMs with the additional capability of representing feedback through weighted causal links. However, in real life situations it is not always easy to give weights or linguistic terms to causal links.

In this paper we proposed to use BDD-FCMs as a tool to model dynamic systems and control the causal relationships inside the systems. The main advantage of BDD-FCMs is the flexibility and freedom of users to express their judgments by using different structures.

We focused our attention on similarity and conflict analysis between different experts. We suggested an algorithm to find a similarity between experts based on BDD-FCMs.

For our future study, we will consider not only the weights of the experts but also the confidence levels of their judgment for each causal relationship. Moreover, we will use BDD-FCMs in *multi criteria decision making* and investigate sensitivity analysis with BDD-FCMs applied to energy policy [26] and safeguard evaluation problems [22]. A further software tool for BDD-FCMs will be also considered in the future need. Moreover, we will use the proposed approach for decision masking applied in real life case study.

References

1. Axelrod, R.: Structure of Decision: the Cognitive Maps of Political Elites, Princeton (1976)
2. Pelaez, C.E., Bowles, J.B.: Using fuzzy cognitive maps as a system model for failure modes and effects analysis. Information Sciences 88, 177–199 (1996)
3. Siraj, A., et al.: Fuzzy cognitive maps for decision support in an iintelligent intrusion detection system. In: Proceedings of IFSA/NAFIPS Conference on Soft Computing, pp. 173–189. MIT Press (2001)
4. Styblinski, M.A., Meyer, B.D.: Fuzzy cognitive maps, signal flow graphs, and qualitative circuit analysis. In: Proceedings of the 2nd IEEE Int. Conf. on Neural Networks, pp. 549–556 (1988)
5. Ozesmi, U., Ozesmi, S.L.: Ecological model based on people's knowledge: a multi-step cognitive mapping approach. Ecological Modelling 176, 43–64 (2004)
6. Hobbsand, B.F., et al.: Fuzzy cognitive mapping as a tool to define mnagement objectives for complex ecosystems. Ecological Applications 12, 1548–1565 (2002)
7. Radomski, P.J., Goeman, P.J.: Decision making and modeling in freshwater sport-fisheries management. Fisheries 21, 14–21 (1996)
8. Kardaras, D., Karakostas, B.: The use of fuzzy cognitive maps to simulate the information systems strategic planning process. Information and Software Technology 41, 197–210 (1999)
9. Kardaras, D., Mentzas, G.: Using fuzzy cognitive maps to model and analyse business performance assessment. Advances in Industrial Engineering Applications and Practice 2, 63–68 (1997)
10. Lee, S., Han, I.: Fuzzy cognitive map for the design of edi controls. Information and Management 37, 37–50 (2000)
11. Hong, T., Han, I.: Knowledge-based data mining of news information on the internet using cognitive maps and neural networks. Expert Systems with Applications 23, 1–8 (2002)

12. Lazzerini, B., Mkrtchyan, L.: Risk analysis using extended fuzzy cognitive maps. In: 2010 International Conference on Intelligent Computing and Cognitive Informatics (ICICCI), Kuala Lumpur, June 22-23, pp. 179–182 (2010),
http://dx.doi.org/10.1109/ICICCI.2010.105,
doi:10.1109/ICICCI.2010.105, ISBN: 978-1-4244-6640-5
13. Smith, E., Eloff, J.: Cognitive fuzzy modeling for enhanced risk assessment in a health care institution. IEEE Intelligent Systems, 69–75 (2002)
14. Jasinevicius, R., Petrauskas, V.: Fuzzy expert maps for risk management systems. In: IEEE/OES US/EU-Baltic International Symposium, pp. 1–4 (2008)
15. Dickerson, J.A., Kosko, B.: Virtual worlds as fuzzy cognitive maps. Presence, 173–189 (1994)
16. Papageorgiou, E., Groumpos, P.: A Weight Adaptation Method for Fuzzy Cognitive Maps to a Process Control Problem. In: Bubak, M., van Albada, G.D., Sloot, P.M.A., Dongarra, J. (eds.) ICCS 2004. LNCS, vol. 3037, pp. 515–522. Springer, Heidelberg (2004)
17. Markoczy, L., Goldberg, J.: A method for eliciting and comparing causal maps. Journal of Management (2), 305–333 (1995)
18. Zhang, W., Chen, S.: A logical architecture for cognitive maps. In: Proceedings of the 2nd Int. Conf. on Neural Networks, pp. 231–238 (1988)
19. Kosko, B.: Fuzzy cognitive maps. International Journal on Man-Machine 24(1), 65–75 (1996)
20. Ozesmi, U., Ozesmi, S.L.: Automatic construction of fcms. Ecological Modelling 176, 43–64 (2004)
21. Schneider, M., et al.: Automatic construction of fcms. Fuzzy Sets Syst. 93, 161–172 (1998)
22. Kabak, O., Ruan, D.: A cumulative belief-degree approach for nuclear safeguards evaluation. IEEE Transactions on Knowledge and Data Management (in Press) (2010)
23. Kandasamy, W.B.V., Smarandache, F.: Fuzzy cognitive maps and neutrosophic cognitive maps. Phoenix (2003)
24. Langfield-Smith, K., Wirth, A.: Measuring differences between cognitive maps. The Journal of the Operational Research Society 43(12), 1135–1150 (1992)
25. Munda, G.: A conflict analysis approach for ulluminating distributional issues in sustainability policy. European Journal of Operational Research (1), 307–322 (2009)
26. Ruan, D., et al.: Multi-criteria group decision support with linguistic variables in long-term scenarios for belgian energy policy. Journal of Universal Computer Science 15(1), 103–120 (2010)

Chapter 4
The Risk of Comparative Effectiveness Analysis for Decision Making Purposes

Patricia Cerrito

University of Louisville, Department of Mathematics, Louisville, KY 40292

Abstract. The purpose of comparative effectiveness analysis is ordinarily defined as a means to compare the benefits of drug A versus drug B. However, particularly in relation to cancer drugs, there is only drug A, and comparative effectiveness analysis tends to compare drug A to a quality adjusted threshold value, with a frequent conclusion that the cost of the drug is not worth the additional life given to the patient. Ordinarily, a societal perspective is used to deny the drugs, since the additional life may be worth the drug cost for the patient, although not to the payer. The British organization, the National Institute for Clinical Excellence (NICE) has denied many cancer drugs to their patients because the cost exceeds a threshold value. The Centers for Medicaid and Medicare are examining a similar process to deny treatments that exceed a quality adjusted price of $50,000. There are similar provisions in the Healthcare Reform Act. With the emphasis upon medications, medical procedures are not as subject to this comparative effectiveness scrutiny; procedures can frequently exceed the cost of medication treatments. However, each medication is considered separately; no analysis examines the total contribution of the treatment to the overall cost of healthcare. We examine different aspects of comparative analysis using techniques of data mining.

1   Introduction

In many ways, comparative effectiveness analysis is used to inflate the actual cost of treatment based upon the perceived quality of life of patients generally. Quality of life is defined lower for patients who are older or disabled, or both compared to younger, healthier patients. The adjusted cost, then, does not necessarily reflect the actual required reimbursement for treatment. In addition, the payer does not pay the adjusted cost, only the actual cost. However, if the adjusted cost is above a threshold value, that threshold is justified by the payer as a reason to deny the treatment rather than to use the actual cost that would actually be paid.

We will examine different aspects of comparative effective analysis and some of the problems that are not considered generally when defining the models but where data mining techniques can greatly enhance the process. Otherwise, there are some missing pieces that result in a risk that inferior drugs are deemed as the most cost effective, or that medications are denied because of a poor understanding of quality of life.

J. Lu, L.C. Jain, and G. Zhang: Handbook on Decision Making, ISRL 33, pp. 69–89.

Currently, the focus on the definition of quality of life is on functioning; relationships are ignored. That there are differing perspectives in terms of quality of life is clear because most patients opt for treatment once presented with options such as chemotherapy or heart surgery. While do not resuscitate orders are common, these are generally signed when they are hypothetical rather than real. Group identity is used to define quality of life rather than individual identity; levels of productivity and quality of life are considered equal across the patient base. We will demonstrate how text analysis can improve upon the concept currently in use for quality of life.

2   Preprocessing Data

In claims data, prescriptions are separated from inpatient and outpatient treatments as well as office visits and home health care. Because all of this information is stored in different files in a one-to-many relationship with a patient's identification number, the most important aspect of using these databases is to convert them to a one-to-one relationship after filtering down to the condition under study. We take advantage of the data step and the use of summary statistics to do both. Each patient claim is identified by an ICD-9 code as to the primary reason for the medication or treatment. Osteoporosis, for example, is identified by the codes, 733.0x where x can vary from 0 to 9 (http://icd9cm.chrisendres.com/). Each of the datasets has a column for the primary code. We can use an if...then statement in a data step to isolate patients with a specific condition.

Once the different data sets have been filtered down to a specific condition, we need to convert them to a one-to-one relationship. We then choose one of the datasets to serve as the primary set and merge the datasets using a left or a right join, depending upon the order of the data sets. In addition, we have to be concerned about whether medication is discontinued, or if the patient switched to a different treatment medication.

Because the database has accurate dates for prescriptions, we can investigate in more detail the occurrence of medication switching using survival data mining. In order to do this, we need to transpose both date and medication. Doing a similar code to transpose the medication date, we then merge the two transposed datasets together so that both medication and date are in the same dataset.

We then need to search for the first prescription that involves switching, and the date when the switching occurs. If no switching occurs, we define the final date as a censoring value. The censoring variable can be modified to search for specific endpoint medications. For example, if we want to know whether the change is equal to the drug, Boniva, then we define Boniva=0 if medchange='Boniva' and =1 otherwise. Then we apply survival analysis, stratifying by the initial medication using the start of the year, 2006, as time=0. In doing this, we make the assumption that future medication choice depends on the present medication and not on the past medications.

Because SAS software (SAS Institute, Inc.; Cary, NC) is used so commonly in medical research and drug development, we provide the SAS code for the preprocessing in the appendix, using SAS version 9.2.

3   Comparison of Multiple Drugs for Best Value

We provide an example of a comparison of multiple medications for the treatment of osteoporosis. We want to see if there is a difference in the medical tests performed given the different medications to include this information in a comparison between drugs. In this example, we combine different datasets taken from the Medical Expenditure Panel Survey (http://www.meps.ahrq.gov/mepsweb/).

We want to see if patients taking different medications have different types of other treatments that can increase costs. We first looked at the costs for each type of care: medications, inpatient, outpatient, office visits, and home health care. We also looked at the issue of patient compliance in relation to the medications. It is possible that patients are more likely to comply with one medication over another, and compliance might reduce the overall costs in terms of treatment. Table 1 gives the costs of the medications used to treat osteoporosis along with the different payers.

Table 1. Total Cost for Osteoporosis Medications

Year	N Obs	Variable	Mean	Sum	N
2005	3733	selfpay	50.1955746	187380.08	3733
		medicare	2.9947924	11179.56	3733
		medicaid	10.4126493	38870.42	3733
		private	27.2414894	101692.48	3733
		va	0	0	3733
		total	94.2722127	351918.17	3733
2006	4179	selfpay	36.7708279	153665.29	4179
		medicare	27.6511079	115553.98	4179
		medicaid	2.7505288	11494.46	4179
		private	17.6373654	73706.55	4179
		va	0	0	4179
		total	88.2418689	368762.77	4179

Table 1 indicates that the average prescription went from $50 self-pay to $36 while Medicare again increased 10-fold and Medicaid paid 1/3 of the amount in 2006 that it paid in 2005 for these medications. Private insurance declined considerably from $101,692 in 2005 to $73,707 in 2006 for this cohort of patients. The results suggest that most of the patients prescribed these medications are in the Medicare eligible population. The patients were just shifted in terms of payment and payer for their continuing medication.

Table 2 gives the frequency count for the medication, Actonel, which is a once-a-week prescription. In a year's time, there should be 12 prescriptions, with each prescription equal to 4 doses. Possibly, there are 90-day prescriptions of 12 tablets, so we need to take this into consideration as well. We do this by computing the product of the frequency of the prescription by the average quantity per prescription by patient. Note that the most frequent number of prescriptions per patient is for just one. The patients who get just one prescription most probably had difficulty with the medication and discontinued its use.

Table 2. Frequency Count for Number of Actonel Prescriptions

FREQ	Frequency	Percent	Cumulative Frequency	Cumulative Percent
1	23	20.91	23	20.91
2	13	11.82	36	32.73
3	10	9.09	46	41.82
4	13	11.82	59	53.64
5	9	8.18	68	61.82
6	8	7.27	76	69.09
7	7	6.36	83	75.45
8	6	5.45	89	80.91
9	6	5.45	95	86.36
10	1	0.91	96	87.27
11	5	4.55	101	91.82
12	3	2.73	104	94.55
13	3	2.73	107	97.27
15	1	0.91	108	98.18
16	2	1.82	110	100.00

Figure 1 gives the spread of the number of doses for Boniva. Boniva is taken once per month. In a year's time, there should be 52/4, or 13 prescriptions per patient; however, only 6 patients have achieved that number.

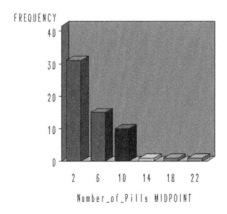

Fig. 1. Number of Doses for Boniva

The mode in Figure 1 is for 4 doses or fewer when it should be for 12 or 13. Again, it does not appear that patients are taking the full medication. It is possible that the patients are switching medications because of adverse effects, so we need to take switching into consideration as we define compliance.

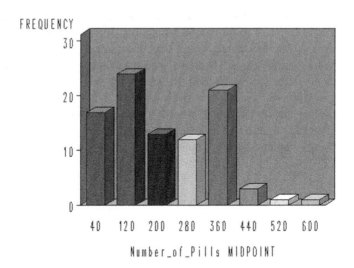

Fig. 2. Number of Doses for Evista

Evista is used daily, which suggests that a patient should have approximately 365 doses in a year's time. While there are many who have that number of doses, there are many more who do not, which suggests a lack of compliance with the medication requirements.

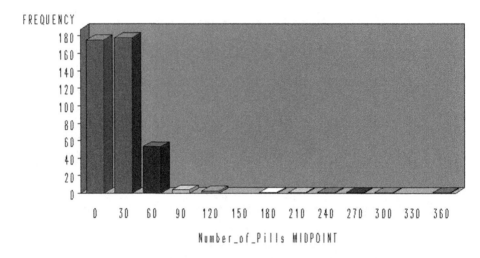

Fig. 3. Number of Doses for Fosamax

This medication (Fosamax), too, should have 52 doses in a year, although there is a daily dose (which appears to be taken by very few patients). There are some extreme outliers, but most patients are getting fewer than the 52 doses.

While this preliminary investigation indicates that most of the patients are not in compliance, concluding this result can be misleading. If a patient switches from Actonel to Fosamax during the middle of the year, that patient will appear to be out of compliance for both medications. Therefore, we must change the observational unit to reflect the total doses for each drug. First, we separate the patients with more than one medication from those with exactly one medication.

Table 3 shows the number of patients who switched medications. The number is fairly small. It is sufficiently large so that patients who switch need to be taken into consideration when defining compliance. Note that most of the switching is to Fosamax.

Table 3. Second Medication and Number Who Switched

RXNAME	Frequency	Percent	Cumulative Frequency	Cumulative Percent
Boniva	5	15.15	5	15.15
Evista	1	3.03	6	18.18
Fosamax	27	81.82	33	100.00

In order to work with medication combinations, we first need to standardize the value. Therefore, we compute a simple ratio for each medication taken, $c(med_i)$=number of doses prescribed/number of doses needed for full compliance. Then we add the sum of $c(med_i)$ for each medication. For example, suppose a patient takes Fosamax for 2/3 of a year and Boniva for the remaining 1/3 of a year. Then, compliance for Fosamax=36/52 and compliance for Boniva=3/12 for that patient. The sum of these values is equal to 36/52+3/12=0.69+0.25=0.94, or very close to one, the ideal identified as full compliance. Finally, we have to make certain that we distinguish between a once-a-day dose and a once-a-week dose. A patient who has 240 doses is on a once-a-day prescription.

We also want to look in the patient conditions listed with the prescriptions for these patients with medications for osteoporosis to ensure that they have been properly diagnosed. Therefore, we consider the ICD9 codes that are associated with each of the medications. For Actonel, there are 646 (out of a total of 996) primary codes given as 733, or Other disorders of bone and cartilage. The specific codes for osteoporosis are 733.01 (Senile osteoporosis or postmenopausal osteoporosis), V17.81 (Osteoporosis), 733.02 (Idiopathic osteoporosis), 733.03 (Disuse osteoporosis), 733.0 (Osteoporosis), and 733.00 (Osteoporosis, unspecified). However, there are other primary patient conditions listed for Actonel that include 714 (Rheumatoid arthritis and other inflammatory polyarthropathies), 715 (Osteoarthrosis and allied disorders), 716 (Other and unspecified arthropathies), 718 (Other derangement of joint), and 719 (Other and unspecified disorders of joint). Actonel is not approved for arthritis and is not considered effective for its treatment. It is possible that arthritis is primary and osteoporosis is secondary as a patient condition. It is also possible that Actonel is used off-label to treat arthritis. However, 733 is not listed as a secondary ICD9 code for Actonel. Either the Actonel is prescribed improperly, or the ICD9 code is inappropriately listed, or the use is off-label.

Evista similarly has 296 out of 690 primary ICD9 codes listed as 733, but unlike Actonel, it has 5 secondary codes also listed as 733. While there are also diagnoses listed for arthritis (715-716), there are 88 primary codes for V68 (Encounters for administrative purposes). This code suggests that the purpose of the encounter was to write a new prescription for a recurring medication.

For Fosamax, there are 1531 primary codes out of 2009 for osteoporosis. There are an additional 88 primary codes for arthritis, 46 primary codes for V68, and 61 for V82 (Special screening for other conditions). In contrast, none of the primary codes for estrogen are for osteoporosis or arthritis. The primary code listed is for 627 (Menopausal and postmenopausal disorders). It suggests that the estrogen prescriptions are not for osteoporosis.

We want to look at the relationship between the level of compliance to the need for treatment for bone fractures that result from the condition of osteoporosis. The number of such patients is quite small; 12 inpatients and 19 outpatients are identified as having treatment for bone breaks, while also having the condition of osteoporosis.

Note that for patient #8, the primary code is for infection; it is the secondary code that reveals the bone fracture related to the infection. This problem of infection is frequently related to orthopedic treatments.

Table 4. Osteoporosis Medications by Inpatient Fractures

Row number	RXName	Dose Strength	Quantity of Prescription	ICD9 Code	ICD9 Code	ICD9 Code
1	Actonel	35	12	821, Fracture of other and unspecified parts of femur	-1	-1
2	Actonel	35	12	821, Fracture of other and unspecified parts of femur	-1	-1
3	Fosamax	70	90	822, Fracture of patella	-1	-1
4	Evista	60	150	724, Other and unspecified disorders of back	733, Other disorders of bone and cartilage	807, Fracture of rib(s), sternum, larynx, and trachea
5	Actonel	35	24	827, Other, multiple, and ill-defined fractures of lower limb	-1	-1
6	Fosamax	70	4	808, Fracture of pelvis	922, Contusion of trunk	-1
7	Fosamax	70	28	820, Fracture of neck of femur	707, Chronic ulcer of skin	-1
8	Fosamax	70	12	041, Bacterial infection in conditions classified elsewhere and of unspecified site	805, Fracture of vertebral column without mention of spinal cord injury	787, Symptoms involving digestive system
9	Fosamax	70	8	824, Fracture of ankle	-1	-1
10	Fosamax	35	24	824, Fracture of ankle	-1	-1
11	Actonel	35	12	812, Fracture of humerus	-1	-1
12	Fosamax	70	4	820, Fracture of neck of femur	812, Fracture of humerus	814, Fracture of carpal bone(s)

The patients taking Actonel in this group appear to be complying with the number of doses for a once a month treatment. The patients treated with Fosamax do not seem to be complying with the medication. If this is the case (and as shown previously, it is also true for patients generally prescribed the medication), it would be worthwhile to determine just why patients are not complying with the medication and how compliance can be improved.

This table does suggest that there are patients at high risk for fractures who are not complying with their medications. We can see if this remains the case for outpatient visits for fractures (Table5).

Table 5. Osteoporosis Medications by Outpatient Fractures

Row number	RXName	Dose Strength	Quantity of Prescription	ICD9 Code	ICD9 Code	ICD9 Code
1	Fosamax	35	4	805, Fracture of vertebral column without mention of spinal cord injury	-1	-1
2	Fosamax	70	12	825, Fracture of one or more tarsal and metatarsal bones	-1	-1
3	Fosamax	70	156	824, Fracture of ankle	-1	-1
4	Fosamax	70	156	824, Fracture of ankle	-1	-1
5	Fosamax	70	156	824, Fracture of ankle	-1	-1
6	Fosamax	70	156	824, Fracture of ankle	-1	-1
7	Fosamax	70	156	824, Fracture of ankle	-1	-1
8	Fosamax	70	156	824, Fracture of ankle	-1	-1
9	Fosamax	70	156	824, Fracture of ankle	-1	-1
10	Fosamax	70	156	824, Fracture of ankle	-1	-1
11	Fosamax	70	156	824, Fracture of ankle	-1	-1
12	Fosamax	70	156	824, Fracture of ankle	-1	-1
13	Fosamax	70	156	824, Fracture of ankle	-1	-1
14	Fosamax	70	156	824, Fracture of ankle	-1	-1

Row number	RXName	Dose Strength	Quantity of Prescription	ICD9 Code		ICD9 Code	ICD9 Code
15	Actonel	30	32	823, Fracture of tibia and fibula		-1	-1
16	Actonel	30	32	823, Fracture of tibia and fibula		-1	-1
17	Actonel	30	32	823, Fracture of tibia and fibula		-1	-1
18	Fosamax	70	4	820, Fracture of neck of femur		812, Fracture of humerus	814, Fracture of carpal bone(s)
19	Fosamax	70	4	820, Fracture of neck of femur		812, Fracture of humerus	814, Fracture of carpal bone(s)

There is a red flag on the 156 doses of Fosamax to consider; this patient is taking the daily treatment. This list also suggests that patients receive multiple follow up visits for treatment and there are actually just 5 patients in the sample receiving outpatient treatment for fractures. Preprocessing needs to isolate episodes of treatment rather than just a list of treatments.

It would be of interest to determine whether patients who are taking the medications just as a preventative measure to avoid osteoporosis are the ones with limited compliance compared to patients who already have the disease, and who have complications related to the disease. It is said that "an ounce of prevention is worth a pound of cure". However, if the patients do not accept the prevention, it will do little good.

To examine some of these potential problems, we look to the physician visits and laboratory tests datasets restricted to the patients prescribed osteoporosis medications.

Table 6. Treatment Performed in Physician Visit by Medication (Percent of Patients)

Treatment Performed	IV Therapy	Lab Tests	X-Rays	MRI/CATSCAN	Medication Prescribed
Actonel	1.20	13.08	10.94	15.97	3.38
Boniva	0	13.54	3.09	3.09	4.64
Evista	0	22.49	4.54	13.84	6.51
Fosamax	0.22	18.88	6.72	11.58	4.22
	EKG	EEG	Other Test	Surgical Procedure	
Actonel	3.45	0.26	16.34	7.45	
Boniva	2.04	0	4.51	6.80	
Evista	3.30	0	24.21	21.51	
Fosamax	2.24	0.50	20.72	11.84	

There are differences in the percentage of patients with the type of treatment given the different medications. Patients taking Actonel are much more likely to have an X-Ray or an MRI; those taking Boniva are much less likely. It could be that patients with more serious conditions are given Actonel while Boniva is used more for prevention; or it could be that physicians prescribing Actonel are more knowledgeable about needed follow up to guard against side effects. It could also mean that patients taking Actonel are more likely to be tested for fractures. The EKG and EEG are heart-related, and are more likely with Actonel and Evista compared to Boniva and Fosamax. Surgical procedures, too, are more likely with Evista. Therefore, there are additional consequences that are related to the medication choice.

Of course, this is a non-terminal, treatable disease. Terminal illnesses will always be cheaper not to treat. If not treated, the patient dies and is removed from the healthcare system. It is this reason for a threshold value when performing comparative effectiveness analysis; the healthcare system will pay so much and no more. That is why cancer patients are problematic. They are terminal if not treated and it will cost less not to treat and reduce the time of survival. Therefore, these patients are at the mercy of the threshold value.

4 Effectiveness Analysis Using a Threshold Value

In this section, we investigate the problem of defining a patient's quality of life in relationship to treatments when the choice is not between drug A and drug B, but the effectiveness is measured against a financial threshold value, as has become common in cancer treatments as well as other chronic diseases for which few options are available for patients.

4.1 NICE

The National Health Service in Britain has been using comparative effectiveness analysis for quite some time. NICE stands for the National Institute for Health and Clinical Excellence. This organization has defined an upper limit on treatment costs, and if the cost exceeds this pre-set limit, then the treatment is denied. It does not matter if the drug is effective or not. That means that there are many beneficial drugs that are simply not available to patients in Britain where fully 25% of cancer patients are denied effective chemotherapy medications. (Devlin 2008; Mason and Drummond 2009) The number of chemotherapy drugs denied is increasing regardless of their effectiveness.

NICE is not comparing drug A to drug B for chemotherapy. Instead, the organization compares the cost of a drug to the value the organization places on your life. If it costs too much to keep you alive given your defined value, or to improve your life, then you are denied treatment. Similar types of rationing have also come to the United States. Oregon has become notorious in its Medicaid benefit, denying cancer drugs to patients, but making the same patients aware that assisted suicide is available. Oregon will not make available drugs that can prolong a patient's life; it will make available a drug to end it (which will then save additional medical costs). Currently, pharmaceutical companies have been subsidizing Oregon's Medicaid by providing these drugs to patients who have been denied by Medicaid. (Smith 2009) It has been suggested that euthanasia is cheaper than end of life care, and more cost-effective than treating many patients with terminal illnesses. (Sprague 2009)

Just recently, the Food and Drug Administration has considered retracting approval of a chemotherapy drug for breast cancer on the basis of cost effectiveness rather than effectiveness. In this case, the definition of effectiveness has changed. The drug was approved based upon an improvement in disease-free survival. The intent is to withdraw approval because effectiveness is now defined as overall survival. The public outcry resulted in a postponement of a decision to remove approval at least for 4 months. (Anonymous-WSJ 2010; Perrone 2010) However, as of December, 2010, the FDA has voted to disapprove the drug for breast cancer.

4.2   QALY

A comparative effective analysis starts with the perceived patient's utility given the disease burden. The QALY, or quality of life-adjusted years, is an estimate of the number of years of life gained given the proposed intervention. Each year of perfect health is assigned a value of 1.0. A patient in a wheelchair is given a correspondingly lower value as is a patient who is elderly; this value is not clearly defined and is rarely based upon patient input. (Prieto and Sacristan 2003)

Consider an example. Suppose a cancer drug for patients with liver cancer allows a patient to live an average of 18 months compared to not using the drug. However, as with most cancer drugs, there are potent side effects. Suppose that the analyst decides that the quality of life is only 40% of perfect health (giving a weight of 0.4). Then the drug gives 1.5*0.4=0.6 QALYs to the patient. Suppose that at the initial introduction of this drug, it costs $1000 per month, or about $18,000 for the anticipated additional life of the patient. Then the cost per QALY is equal to 18,000/0.6=$30,000 per year of life saved. According to the NICE organization, this drug then would be too costly regardless of the fact that there is no comparable drug that is effective in prolonging the patient's life. However, suppose the analyst uses a measure of 60% of perfect health. Then the drug gives 1.5*0.6=0.9 QALYs to the patient at a cost of $20,000, which brings the amount closer to the pre-set value defined by NICE. Therefore, this definition of a scale of perfect health is of enormous importance. In fact, NICE has often denied such a cancer drug because of its cost. (Anonymous-NICE 2004; anonymous-NICE 2008; Anonymous-bevacizumab 2009; Anonymous-MedicalNews 2009; Anonymous-NICEreview 2009; Anonymous-NICEreview 2009)

If a person is otherwise young and healthy and a drug costs $10,000 per year, then the QALY is $10,000. However, if a patient is older and has a chronic condition, then that patient's utility may be defined as exactly half that of a young and otherwise healthy person. In that case, the QALY is $20,000 for the same drug. If the patient is old and has two or more chronic conditions, then the patient's utility could be defined as 25% that of a young and healthy person. In that case, the QALY IS $40,000 per year of life saved. By defining $15,000 as the upper limit for treatment, it is easy to see how the definition of a person's utility can be used to deny care to the elderly.

However, the cost of treating the disease is not restricted to the cost of medications. Therefore, we must look at all aspects of treatment, including physician visits, hospital care, and home health care. We must also look at the impact of patient compliance on the overall cost of healthcare. If patients have specific diseases that can be treated, but who do not use the treatment, then outcomes will not be the same compared to patients who do comply. Also, patients who switch treatments may suffer from adverse events of the first treatment that are not present in the second treatment. Therefore, we must examine the totality of patient care.

4.3   Definition of Concepts

There are a number of concepts used in developing comparative effectiveness models. These concepts are particularly important when only one drug is compared to a threshold value. There are several ways that are currently in use to define a patient's quality of life. However, each method deals with a hypothetical situation rather than one that is real, bringing into question the validity of the entire process. The methods are listed below (McNamee, Glendinning et al. 2004; Puhan, Schunemann et al. 2007):

- Time Trade Off (TTO): Respondents are asked to choose between remaining in a state of ill health for a period of time, or being restored to perfect health but having a shorter life expectancy.

In other words, individuals are given the choice between taking a "happy pill" that will guarantee them perfect health for a period a time, after which they will drop dead versus spending a longer period of time in imperfect health. Would you be willing to take this happy pill if you had ten years of perfect health followed by death? Suppose you had 20 years? 30 years? At what point will you take this happy pill? If you refuse to take this pill, then you will have imperfect health of some type, say arthritis, diabetes, or asthma for, say 30 years. Is ten years of perfect health better than 30 years of imperfect health? It is not a real choice since such a "happy pill" does not exist, and probably never will.

- Standard gamble (SG): Respondents are asked to choose between remaining in a state of ill health for a period of time, or choosing a medical intervention, which has a chance of either restoring them to perfect health, or killing them.

This, too, is a hypothetical situation. A medical intervention is offered to a patient if the benefit outweighs the risk for a patient in ill health. There is no current intervention where the choice is perfect health or death. There is also no indication of the actual risk involved. Suppose there is a 1% chance of death versus a 25% chance of death. The patient decision, even as a hypothetical, may be different.

- Visual Analogue Scale (VAS): Respondents are asked to rate a state of ill health on a scale from 0 to 100, with 0 representing death and 100 representing perfect health.

This scale assumes that individual patients have a reasonable concept of perfect health. Since the term is very vague, it is not certain how patients are reflecting upon the terminology in order to make a reasonable assessment. Moreover, this scale does not allow patients to indicate their quality of life, which also should be taken into consideration.

4.4   Use of Text Analysis

Because the concepts of perfect health and quality of life are important to the definition of comparative effectiveness models, we also need to know how the concepts are interpreted by patients as they complete the basic surveys used to define quality of life. We used text analysis and open ended surveys to see how individuals view these basic concepts. Text analysis goes beyond simple frequency counts of words. It examines how words and concepts are linked within sentences.

Generally, a document is converted into a row in a matrix. This row has a column for any word contained within the dataset of documents. The matrix value is equal to the number of times that word occurs in the document. The matrix will consist mostly of zeros since the list of words is much longer than the list of documents. Therefore, the next step is to reduce the dimension of the matrix. This is done through the process of singular value decomposition. This feature is extremely valuable for calls into customer service, for example, for chart notes, and to examine advertisements from the competition.

There are variations to this general methodology depending upon what you want to discover. For example, if you want to determine what documents contain a specific word for flagging purposes, this can be done through filtering. However, if you want to look at connections within the text structure itself, you can find much greater meaning using the word structure itself. The basics of text analysis are as follows:

1. Transpose the data so that the observational unit is the identifier and all nominal values are defined in the observational unit.
2. Tokenize the nominal data so that each nominal value is defined as one token.
3. Concatenate the nominal tokens into a text string such that there is one text string per identifier. Each text string is a collection of tokens.
4. Use text mining to cluster the text strings so that each identifier belongs to one cluster.
5. Use the clusters defined by text mining in other statistical analyses.

The general process of text analysis is outlined below:

The SVD of an N x p matrix A having N documents and p terms is equal to $A=U\Sigma V$ where U and V are N x p and p x p orthogonal matrices respectively. U is the matrix of term vectors and V is the matrix of document vectors; Σ is a p x p diagonal matrix with diagonal entries $d_1 \geq d_2 \geq \ldots \geq d_p \geq 0$, called the singular values of Σ. The truncated decomposition of A is when SVD calculates only the first K columns of U, Σ and V. SVD is the best least squares fit to A. Each column (or document) in A can be projected onto the first K columns of U. Similarly, each row (or term) in A can be projected onto the first K columns of V. The columns projection (document projection) of A is a method to represent each document by K distinct concepts. So, for any collection of documents, SVD forms a K dimensional subspace that is a best fit to describe data.

Cluster analysis, also called data segmentation, has a variety of goals. All goals relate to grouping or segmenting a collection of objects into subsets or "clusters" such that those elements within each cluster are more closely related to one another than the objects assigned to different clusters. An object can be described by a set of measurements or by its relation to other objects.

In addition, another goal is to arrange the clusters into a natural hierarchy. The arranging involves successively grouping the clusters themselves so that at each level of the hierarchy, clusters within the same group are more similar to each other than those in different groups. Cluster analysis is used to form descriptive statistics to assess whether or not the data consist of a set of distinct subgroups; each subgroup representing objects with substantially different properties.

Central to all of the goals of cluster analysis is the notion of the degree of similarity or dissimilarity between the individual objects being clustered. A clustering method attempts to group the objects based on the definition of similarity supplied to it. Clustering algorithms fall into three distinct types: combinatorial algorithms, mixture modeling and mode seeking.

Text analysis has as its basis the Expectation Maximization Algorithm. The expectation maximization (*EM*) algorithm uses a different approach to clustering in two important ways:

1. Instead of assigning cases or observations to clusters to maximize the differences in means for continuous variables, the *EM* clustering algorithm computes probabilities of cluster memberships based on one or more probability distributions. The goal of the clustering algorithm is to maximize the overall probability or likelihood of the data, given the final clusters.

2. Unlike the classic implementation of k-means clustering, the general *EM* algorithm can be applied to both continuous and categorical variables.

The expectation-maximization algorithm is used to estimate the probability density of a given set of data. EM is a statistical model that makes use of the finite Gaussian mixtures model and is a popular tool for simplifying difficult maximum likelihood problems. The algorithm is similar to the K-means procedure in that a set of parameters is re-computed until a desired convergence value is achieved. The finite mixture model assumes all attributes to be independent random variables.

4.5   Text Analysis of Open Ended Questions

Approximately 100 pre-nursing students were surveyed and asked to define "health", "perfect health", and "quality of life". The results show that there is some ambiguity amongst the students, but the general consensus appears to be focused on physical functioning or lifestyle habits rather than social functioning and social networks. We first look at the definition of "health" using text analysis. Table 7 shows the terms that were used in text analysis to define the different clusters.

Table 7. Text Clusters Representing the Term, "Health"

Cluster Number	Description	Percentage
1	Physically, health, emotionally	0.08
2	Happy, ability, own, one life	0.28
3	Bad, good, +will, life	0.08
4	+enjoy, in alive, +thing, +not	0.164
5	Need, +do, +can, on, patient	0.2
6	+disease, independent, healthy, life	0.08
7	Physical, with, well, emotional, able	0.08

The concept that exists in clusters 1,2,5,6, and 7 primarily focus on physical functioning while clusters 3 and 4 focus on the ability to enjoy life. These clusters indicate that there are two prime considerations held by different groups of people when attempting to define "health". Moreover, these two groups would tend to approach surveys on quality of life in completely different ways. Similarly, Table 8 shows the text clusters defined for "quality of life".

Table 8. Text Clusters for "Quality of Life"

Cluster Number	Description	Percentage
1	+ condition, mental, medical, +problem, + illness	0.26
2	+absence, human body, human, within, +ability	0.14
3	Perfect health, perfect, +do, +not, health	0.24
4	When, diet, with, without, in	0.2
5	Stable, mentally, well, state, free	0.16

There is one group that defines "quality of life" by "perfect health" while a second cluster indicates that it is equivalent to the absence of illness. This cluster shows the ambiguity in the terms as the definition is circular. Two groups focus on a person's mental condition to define "quality of life". The final cluster defines quality as a lifestyle habit, primarily related to diet. Table 9 shows the clusters for the concept of "perfect health".

Table 9. Text Clusters for "Perfect Health"

Cluster Number	Description	Percentage
1	Disease, free, exercise, medical, eat	0.15
2	State, overall, individual, physical, well	0.30
3	Problem, health, not, living, health	0.1
4	Feeling, good diet, have, diet, lifestyle	0.04
5	Life, mentally, physically, do, basis	0.41

Three groups focus on the lack of disease and physical well being. Cluster number 5 includes mental well being while cluster number 4 looks at feeling good and lifestyle, including diet.

Another feature of text analysis is that links between terms can be visualized. Figure 4 examines the links to the term, health.

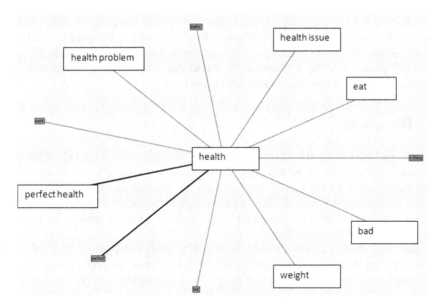

Fig. 4. Links to the Term, Health

Other than having some type of health problem, the links are to lifestyle concepts, eating and weight. There are no links to mental or social functioning. Similarly, Figure 5 examines the concept of life. Again, the emphasis appears to be upon physical functioning and the ability to do what one wants to do.

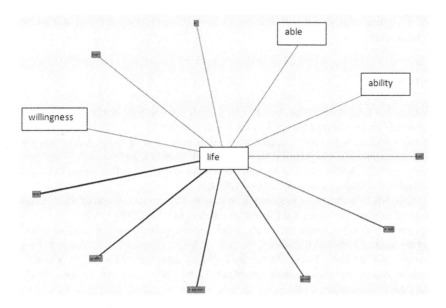

Fig. 5. Links to the Term, Life

Because of the critical nature of these concepts in the comparative effectiveness models, it is absolutely essential to discover how patients put meaning to the terms and to compare their understanding of the concepts to the understanding held by those who develop the models; if there are considerable differences in understanding, difficulties will arise when and if rationing occurs.

5   Discussion

The concepts used in comparative effectiveness analysis, such as quality of life and time trade off need to be examined closely; otherwise, the validity of the results are in doubt. Patient input should be as comprehensive as possible, and text analysis allows for them to demonstrate their different viewpoints with regard to the concepts. Patient understanding should also be compared to the understanding of those who perform comparative effectiveness analysis.

In addition, there is a considerable difference in using comparative models when comparing drug A to drug B to determine which drug provides both better cost and more benefit as opposed to comparing drug A to a threshold value. There should be some meaningful justification for the threshold. In addition, the full cost of treatment, including inpatient and outpatient treatments as well as physician visits and laboratory tests should be considered as the complete cost of treatment as opposed to just the cost of medication. The overall impact on the future development of medications and treatments should also be assessed.

If the quality of life is sufficiently lowered, it is almost always possible to exceed any fixed threshold value. The consequences of miscalculations can result in patient deaths because they are deprived of medications that are medically effective but not defined as cost effective. In other words, the perspective of the individual patient should be considered along with the perspective of society in terms of dollars spent upon healthcare.

References

Anonymous-bevacizumab, Bevacizumab and cetuximab for the treatment of metastatic colorectal cancer. National Health Service, London (2009)

Anonymous-MedicalNews, NICE acknowledge Alzheimer's model faulty, but do not plan to change recommendations, Medical News Today, UK (June 12, 2009)

Anonymous-NICE, NICE National Institute for Health and Clinical Excellence Guide to the methods of technology appraisal, London (2004)

Anonymous-NICE, Social value judgements: Principles for the development of NICE guidance. National Institute for Health and Clinical Excellence, London (2008)

Anonymous-NICEreview, Alzheimer's disease-donepezil, galantamine, rivastigmine (review) and memantine. NHS NICE (2009)

Anonymous-NICEreview, TA111 Alzheimer's disease-donepezil, galantamine, rivastigmine (review) and memantine: guidance (amended August 2009), National Institute for Health and Clinical Excellence (2009)

Anonymous-WSJ, The Avastin Mugging: The FDA rigs the verdict against a good cancer. The Wall Street Journal (August 18, 2010), http://www.WSJ.com

Devlin, K.: NHS patients denied drugs due to lack of common sense at NICE, say charities. Telegraph (November 27, 2008)

Mason, A.R., Drummond, M.F.: Public funding of new cancer drugs: Is NICE getting nastier? European Journal of Cancer 45, 1188–1192 (2009)

McNamee, P., Glendinning, S., et al.: Chained time trade-off and standard gamble methods. Applications in oesophageal cancer. European Journal of Health Economics 5(1), 81–86 (2004)

Perrone, M.: FDA delays decision on breast cancer drug Avastin, AP Associated Press, Washington, DC, September 17 (2010)

Prieto, L., Sacristan, J.A.: Problems and solutions in calculating quality-adjusted life years (QALYs). Health and Quality of Life Outcomes 1(80) (2003)

Puhan, M.A., Schunemann, H.J., et al.: The standard gamble showed better construct validity than the time trade-off. Journal of Clinical Epidemiology 60(10), 1029–1033 (2007)

Smith, W.J.: Save money by killing the sick: euthanasia as health care cost containment not such a parody as the author think (2009)

Sprague, C.: The economic argument for euthanasia (2009)

Appendix: SAS Code for Preprocessing

Many-to-One

```
TITLE;
TITLE1 "Summary Statistics";
TITLE2 "Results";
FOOTNOTE;
FOOTNOTE1    "Generated    by    the    SAS    System
(&_SASSERVERNAME,              &SYSSCPL)              on
%TRIM(%QSYSFUNC(DATE(),        NLDATE20.))              at
%TRIM(%SYSFUNC(TIME(), NLTIMAP20.))";
PROC MEANS DATA=WORK.SORTbyID
    FW=12
    PRINTALLTYPES
    CHARTYPE
    NWAY
    VARDEF=DF
          MEAN
          STD
          MIN
          MAX
          N      ;
    VAR  TOTTCH06  OBTTCH06  OPVTCH06  OPOTCH06  AMETCH06
AMATCH06   AMTTCH06   AMTOTC06   ERDTCH06   ZIFTCH06   IPFTCH06
DVTOT06    DVOTCH06    HHNTCH06    VISTCH06    OTHTCH06
RXTOT06;
    CLASS cost_Sum /    ORDER=UNFORMATTED ASCENDING;

RUN;
```

Merge Datasets

```
PROC SQL;
   CREATE                                    TABLE
SASUSER.QUERY_FOR_SUMMARYOFCONDITIONS_SA AS
         SELECT t1.patientID,
           t1.remaining variables from dataset,
           t2.variables from second dataset
   FROM   claims.summaryofconditions   AS   t1   RIGHT   JOIN
claims.h105 AS t2 ON (t1.patientID = t2.patientID);
   QUIT;
```

Transpose Data

```
proc transpose data=medications out=medicationbyid
    prefix=med_;
   id patientid;
run;
```

Defining Number of Prescriptions

```
data sasuser.survivaldata;
  set medicationbytranspose;
  array meds(379) med_1 - med_379;
  array dates(379) date_1 - date_379;
do j=1 to 379;
   if dates(j)=. then dates(j)='31dec2004'd;
   censor=1;
end;
do i=1 to 379;
  if i=1 then temp=meds(i);
  if meds(i) ne temp then do;
    med_num=i;
    date_num=dates(i);
   medchange=meds(i);
   censor=0;
    i=379;
  end;
 end;
run;
```

Create Censoring Variable

```
if date_num = . then date_num='12dec2006'd;
if (medchange eq ' ') then censor=1;
if (medchange eq 'Drug_1') then drug_1=0;
else drug_1=1;
if (medchange eq 'Drug_2') then Drug_2=0;
else drug_2=1;
finaldate=input(newlastdate,anydtdtm17.);
format finaldate datetime17.;
final=datepart(finaldate);
      format final date9.;
```

Survival Analysis

```
PROC LIFETEST DATA=sasuser.survival data ALPHA=0.05
;
    BY medchange;
    STRATA med_1;
    TIME Days * censor (1);

RUN;
```

Chapter 5
Portfolio Risk Management Modelling
by Bi-level Optimization

Todor Stoilov and Krasimira Stoilova

Bulgarian Academy of Sciences,
Institute of Information and Communication Technologies,
1113 Sofia, Acad. G. Bontchev str. BL.2, Bulgaria
todor@hsi.iccs.bas.bg

Abstract. The portfolio optimization theory targets the optimal resource allocation between sets of securities, available at the financial markets. Thus, the investment process is a task, which targets the maximization of the portfolio return and minimization of the portfolio risk. Because such an optimization problem becomes multi-criterion optimization one it lacks an unique solution. A balance between the portfolios risk and portfolio return has to be integrated in a common scalar criterion for the risk management. The book chapter considers a bi-level optimization paradigm for the investment process. The optimization process evaluates the optimal Sharp ratio of risk versus the return to identify the parameter of the investor's preferences to risk at the upper level. At the lower level of optimization the optimal portfolio is evaluated using the upper level defined investor's preferences. In that manner, the portfolio optimization results in an unique solution, which is determined according to the objective considerations and it is not based on subjective assumptions of the portfolio problem. As a result, the portfolio risk is minimized according to two arguments: the content of the portfolio with appropriate assets and by the parameter of investor's preferences to risk.

1   Introduction

The estimation and the forecast of the financial risk is currently one of the major tasks of the investment's process management. This problem is in the scope of statistics and probability modelling. The financial risk is always related with the portfolio management [20]. The uncertainty about future events makes the market behaviour unpredictable and prevents the assessment of the parameters of the financial markets under dynamical changes. The analysis of the market is performed under predefined assumptions, which are taken into consideration when allocating financial resources. Generally, such assumptions concern uncertainty in ideal mathematical behavior, constant and not changing environment influences. The formal models in investment apply mathematical analytical tools, which formalize both the behavior of the market players and future events associated with financial markets. The allocation of investment resources is formalized and the resulting mathematical methods strongly influence the working practice of financial institutions [4].

J. Lu, L.C. Jain, and G. Zhang: Handbook on Decision Making, ISRL 33, pp. 91–110.

According to the portfolio theory the decision maker makes his decisions taking into account the risk of the investment. The risk has a meaning of uncertainty. The term "risk" is used when the future is not determined and predictable. Currently, the portfolio optimization models are based on probability theory. However, the probabilistic approaches cannot fully formalize the real market behaviour. Another attempt for handling uncertainty of the financial market is the application of the fuzzy set theory [4, 21].

The most monumental contribution for the application of the modern mathematical models in finance and particularly in risk assessment gives the work of Markowitz [9]. The portfolio selection is the most impact-making development in modern mathematical finance management. The Markowitz theory of portfolio management deals with the individual investor. This theory makes combination of the probability theory and optimization. The investor's goal is to maximize the return and to minimize the risk of the investment decisions. The investor's return is formalized as the mean value of a random behaved function of the portfolio securities returns. The risk is formalized as a variance of these portfolio securities. These mathematical representations of return and risk allow defining a simple optimization problem which formalizes the portfolio management. The two important goals of the investor are to maximize the profit and to minimize the risk of the investment. The exact portfolio solution depends on the level of risk the investor can bear in comparison with the level of portfolio return. Thus, the relation between return and risk is always a major parameter, which has to be identified by the investor for practical utilization of the portfolio theory. In that manner, the decision making process of the investment is generally managed by the subjective assumptions of the investor for the risk/return relation of the portfolio. A decreasing of the subjective influence of the investment process can be achieved if the unknown investor's coefficient for undertaking risk is calculated according to the optimization problem. A formal model proposing a new bi-level optimization problem for the portfolio optimization is presented in the book chapter. The upper level evaluates the parameter of the investor's risk preference. Then, this parameter is used for the optimal resource allocation by minimizing risk and maximizing the portfolio return. Thus, in a common formal problem the portfolio management is performed with a lack of subjective influence in the process of resource allocation. The portfolio risk is minimized according to two types of arguments: the portfolio content and the parameter of the investor's risk preference.

2   Taxonomy of the Risk

The content of the term "risk" is hidden in the uncertainty of the future process, which influences the return or costs of the financial assets. The term "risk" is addressed to several categories of the financial world [3].

Market Risk. It is defined as a risk to financial portfolio, related to the dynamical changes of the market prices of equity, foreign exchange rates, interest rates, commodity prices. The financial firms generally take a market risk to receive profits. Particularly, they try to take a risk they intent to have and they actively manage the market risk.

Liquidity risk is defined as the particular risk from conducting transactions in markets with low liquidity as evidence low trading volume and large bid-ask spread. Under such conditions the attempt to sell assets may push prices lower and the assets have to be sold below their fundamental values or within a time frame, which is longer than expected.

Operational risk is defined as a risk of loss due to physical catastrophe, technical failure and human error in the operation of the firm.

Credit risk is defined as the risk that counterparty may become less likely to fulfill its obligations upon date.

Business risk is defined as the risk that changes in variables of a business plan. It will destroy that plan, including quantifiable risks for business cycle and demands, changes in the competitive behaviour and/or technology.

The term of the risk that is usually employed and easily formalized is the variance of the dynamically changed costs of the financial assets. Introduced by Markovitz, the assets characteristics are defined by their average return E_i and their risk, evaluated as the variance τ_i. The book chapter considers the market risk, which results in different values of the variances of the average returns. As an example, data for the rates of three currencies, USD, GBP and CHF, taken from the Bulgarian site http://econ.bg for a period of 15 days is given in Fig.1.

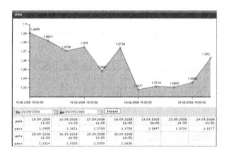
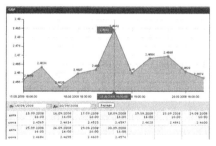

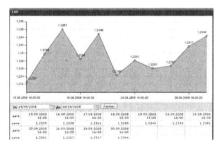

Fig. 1. Rates of USD, GBP CHF currency, taken from http://econ.bg

The data is employed to define the portfolio problem (see Table 1; BGN is the Bulgarian currency).

Table 1. Daily returns of three currencies

DATE	Rate of USD [BGN]	Rate of GBP [BGN]	Rate of CHF [BGN]	R_1 - daily return USD %	R_2 - daily return GBP %	R_3 - daily return CHF %
30.9.2008	1,3630	2,4574	1,2344	2,029	-0,142	0,219
29.9.2008	1,3359	2,4609	1,2317	0,406	-0,348	0,408
26.9.2008	1,3305	2,4695	1,2267	-0,068	0,045	0,049
25.9.2008	1,3314	2,4684	1,2261	0,279	0,341	-0,163
24.9.2008	1,3277	2,4600	1,2281	-3,363	-0,974	0,31
23.9.2008	1,3739	2,4842	1,2243	1,868	0,902	-0,858
19.9.2008	1,3487	2,4620	1,2349	-1,913	0,094	0,521
18.9.2008	1,3750	2,4597	1,2285	0,299	0,294	0,-615
17.9.2008	1,3709	2,4525	1,2361	-0,81	-0,688	0,512
16.9.2008	1,3821	2,4695	1,2298	-0,6	0,529	0,392
15.9.2008	1,3905	2,4565	1,2225			

These initial values are noted as R_i $(i=1,n; n=3)$, where i is the index of the asset. It is necessary to evaluate the average return E_i and the risk τ_i for each asset. The evaluation of the average return is found as the weighted sum $E_i = \sum_t P^t R_i^t$, where P^t is the probability that i has a return R_i^t at time t [17]. The values R_i^t for the USD currency are assumed to be probably equal to the probability $P^t = \dfrac{1}{N} = 0,1$; $N=10$ – number of days, used for the currency rate. Hence, the average return of the USD currency is calculated as

$$E_1 = 0,1(2,029 + 0,406 + ... - 0,6) = -0,1873 \quad .$$

In the same fashion,

$$E_2 = 0,1(-0,142 - 0,348 + ... + 0,529) = 0,0053$$
$$E_3 = 0,1(0,219 + 0,408 + ... + 0,392) = 0,15544 \quad ,$$

and the return vector of the three assets is

$$E^T = |- 0,1873 \quad 0,0053 \quad 0,1544 \; | \cdot$$

The risk of each asset is defined by the variance of the daily returns [3]:

$$\tau_i^2 = \sum_t P^t (E_i - R_t^i)^2 \quad ,$$

which results in $\tau_i^T = |1,5355 \quad 0,5357 \quad 0,4147 \; | \cdot$

These data are the input for the definition of the portfolio optimization problem. The average values E_i represent the mean value around which the daily returns R_i fluctuate. The risk τ_i is a quantitative assessment of the diapason in which R_i varies. Larger diapasons imply higher risk levels.

3   Portfolio Optimization Problem

The portfolio theory was developed as a decision support tool for the allocation of investments for the sells of financial assets (securities, bounds) from the stock exchange [1]. Such an allocation is called "investment" decision making. The investor treats each asset as a prospect for future income. Thus, the better combination of financial assets (securities) in the portfolio, the better return for the investor. The portfolio contains a set of securities. The portfolio optimization problem is defined as problem for optimal allocation of financial resources for trading financial assets. The problem of portfolio optimization targets the optimal resource allocation in the investment process [12]. The resource allocation is done by investing capital in financial assets (or goods), which will generate return for the investor after a period of time. The objective of the investment process is to maximize the return while keeping risk at minimum [11]. In 1952, Harry Markowitz suggested a simple and powerful approach to quantify risk. According to the portfolio theory [12] the analytical relations between the portfolio risk V_p, portfolio return E_p and the values of the investment per type of assets x_i are

$$E_p = \sum_{i=1}^{n} E_i x_i = E^T x$$

$$V_p = \sum_{j}^{n} \sum_{i=1}^{n} x_i x_j \, \mathrm{cov}(i, j) = x^T \, \mathrm{cov}(.)x \, ,$$

where
E_i - average value of the return of asset i ;
$E^T = (E_1,..., E_n)^T$ - vector with dimension 1 x n;
$\mathrm{cov}(i,j)$ – co-variation coefficient between the assets i and j .

The component $V_p = x^T \, \mathrm{cov}(.) \, x$ formalizes the quantitative assessment of the portfolio risk. The component $E_p = E^T x$ is the quantitative evaluation of the portfolio return. The portfolio problem solutions x_i , $i=1,n$ determine the relative amounts of the investment per security i .

The co-variation is calculated from previously available statistical data for the returns of assets i and j and it takes the form of a symmetrical matrix

$$\text{cov}(.) = \begin{vmatrix} \text{cov}(1,1) & \text{cov}(1,2) & \cdots & \text{cov}(1,n) \\ \text{cov}(2,1) & \text{cov}(2,2) & \cdots & \text{cov}(2,n) \\ \vdots & & & \\ \text{cov}(n,1) & \text{cov}(n,2) & \cdots & \text{cov}(n,n) \end{vmatrix}_{nxn}.$$

The components cov(i,j) are evaluated from the values $R_i^{(1)}, R_i^2, \cdots, R_i^{(N)}$ and $R_j^{(1)}, R_j^2, \cdots, R_j^{(N)}$, which concern the profit of assets i and j for discrete time moments (1), (2),..., (N). The co-variation coefficient between assets i and j is calculated as

$$\text{cov}(i,j) = \frac{1}{N}\left[\begin{matrix} (R_i^{(1)} - E_i)(R_j^{(1)} - E_j) + (R_i^{(2)} - E_i)(R_j^{(2)} - E_j) + \\ + \cdots (R_i^{(N)} - E_i)(R_j^{(N)} - E_j) \end{matrix} \right],$$

where

$$E_i = \frac{1}{N}\left[R_i^{(1)} + R_i^{(2)} + \cdots + R_i^{(N)} \right] \quad E_j = \frac{1}{N}\left[R_j^{(1)} + R_j^{(2)} + \cdots + R_j^{(N)} \right]$$

are the average profits of the assets i and j for the period $T = [1,2,...., N]$. Particularly, the value $\text{cov}(i,i) = \tau_i^2$ gives the variation of the return of asset i. The portfolio theory defines the so-called "standard" problem of optimization [12]:

$$\min_x [\frac{1}{2} x^T \text{cov}(.)x - \sigma E^T x]$$

$$x^T.\mathbf{1} = 1,$$

(1)

where cov(.) – a symmetric positively defined n x n square matrix,
E – a (n x 1) vector of the average profits of the assets for the period of time
$T = [1,2,...., N]$;

$$\mathbf{1} = \begin{vmatrix} 1 \\ \vdots \\ 1 \end{vmatrix}, \qquad \text{is a unity vector, } n \times 1;$$

σ – a parameter of the investor's preferences to undertake a risk in the investment process.

The constraint of the optimization problem presents the equation $x_1 + x_2 + \cdots + x_n = 1$, which formalizes the fact that the investment is not partly implemented and the full amount of the resources are devoted for the investments. If the right side of the constraint is less than 1, this means that the amount of the investment is not effectively used. The investment per different assets has to be performed for the total amount of the available investment resources, numerically presented as a relative value of 1. The solutions x_i , i=1, 2, ..., n give the relative values of the investment, which are allocated for the assets i, i=1, 2, ..., n.

The component of the target function $V_p = x^T \mathrm{cov}(.)x$ is the quantitative assessment of the portfolio risk. The component $E_p = E^T x$ is the quantitative value of the portfolio return. The target function of problem (1) aims to minimize the portfolio risk V_p and also maximize its return E_p. The parameter σ has a numerical value from the range $[0, +\infty]$. This coefficient quantitatively formalizes the investor's ability to undertake risk. If $\sigma = 0$, the investor is very cautious (even a coward) and his general task is to decrease the risk of the investment, $\min_x [x^T \mathrm{cov}(.)x]$. If $\sigma = +\infty$, the investor has forgotten the existence of the risk in the investments. His target is to obtain a maximal return from the investment. For that case, the relative weight of the return in the target function is most important, and then the optimization problem has an analytical form: $\min_x [-\sigma E^T x] \equiv \max_x [E^T x]$.

Thus, in the portfolio problem a new unknown parameter σ is introduced, which assesses the investor's preferences for undertaking risk in decision making. This parameter influences the portfolio problem, making it a parametric one. Respectively, for a new value of σ, the portfolio problem (1) has to be solved again. The trivial case when σ is not properly estimated the optimization problem has to be solved for a set of σ. The values σ introduce strong subjective influence to the solutions of the portfolio problem. Additionally, for practical reasons, the portfolio problem has to be solved multiple times with a set of values for the coefficient of the investor's preferences σ to undertake risk. Thus, for real time applications of investment, the estimation of σ and the solution of (1) become quite important.

The numerical assessment of σ is a subjective task for the financial analyzer. This coefficient strongly influences the definition and respectively the solutions of the portfolio problem. Respectively, σ also changes the final investment decision.

The portfolio theory uses the space risk-return $V_p = V_p(E_p)$ for the assessment of the portfolio characteristics found as combinations of admissible assets. The investors have to choose optimal portfolios from the upper set of admissible solutions named "efficiency frontier". This "efficiency frontier" is not evidently found. Points from this curve can be found by solving the portfolio optimization problem with different values of the parameter σ. The "efficient frontier" is evaluated point by point according to an iterative numerical procedure:

1. An initial value of σ for the investor's preferences is chosen. The zero value $\sigma = 0$, is a good starting point and this corresponds to the case of an investor, who is not keen on risky decisions;
2. the portfolio problem is solved with the chosen σ

$$\min_x [\frac{1}{2} x^T \mathrm{cov}(.)x - \sigma E^T x]$$
$$x^T \times 1 = 1$$

and the optimal solution $x(\sigma)$ is found;

3. evaluation of the portfolio risk and portfolio return:

$$V_p = x^T(\sigma)\text{cov}(.)x^T(\sigma) , \qquad E_p = E^T x(\sigma).$$

These values give a point into the space $V_p = V_p(E_p)$, which belongs to the efficient frontier;

4. new value of $\sigma_{new} = \sigma_{old} + \Delta$ is chosen, where Δ is determined by considerations for completeness in moving into the set $\sigma = [0, +\infty]$. Then, go to point 2.

Hence, for each solution of the portfolio optimization problem one point into the space $V_p = V_p(E_p)$, belonging to the curve of the efficiency frontier is found (see Fig.2).

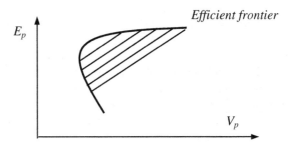

Fig. 2. Efficiency frontier of the portfolio optimization

For practical cases of individual investor, problem (1) is solved with a set of values of σ. Having a set of solutions $x(\sigma)$ the final value of $\sigma*$ for that investor is empirically estimated, which gives also the final optimal portfolio solution $x(\sigma*)$ as well. However, such an approach generates a contradiction between the manner of quantitative definition of problem (1) and the final decision for the investment. The portfolio theory insists that the value of $\sigma*$ has to be estimated before solving the problem. However, in practice $\sigma*$ is estimated after evaluating a set of portfolio problems (1) with different values for σ. Respectively, the subjective influence in definition of $\sigma*$ is quite obvious.

The formal model, which is developed in this book chapter targets at the decrease of the subjective influence in evaluating and assessing the parameter of investor's preference to risk σ. The idea of the model is to formalize the decision making process by two hierarchically interconnected optimization problems (Fig.3).

The optimization problem for evaluating σ is stated at the upper hierarchical level. This problem can be defined from considerations, which are not subjectively influenced. For example, this optimization problem, can target the evaluation of such a σ, which consequently will result in a "well" ratio between the portfolio risk and return.

On the lower hierarchical level the standard portfolio optimization problem is solved using σ, estimated from the upper optimization problem. Unfortunately, both optimization problems are interconnected by their arguments. The solution of the upper level problem influences as parameter the corresponding low level optimization problem and vice versa. Hence, a bi-level optimization problem is stated, which represent the decision making process in portfolio optimization.

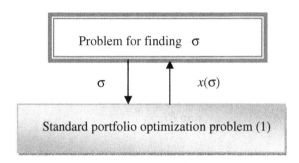

Fig. 3. Definition of bi-level portfolio optimization problem

4   Bi-level Hierarchical Optimization Problems

A general peculiarity of bi-level optimization problems is that by solving an appropriate optimization problem on the upper level, the evaluated solutions are used to define a set of parameters in the lower level optimization problems. The solutions of the last in turn define a set of parameters for the upper level problem. Thus, an interrelation between the solutions at the upper and lower level optimization problems influence the exact form of the optimization problems.

The general bi-level hierarchical optimization problem is made by the formulation of the Stackelberg game [16]. The Stackelberg problem can be interpreted as a game between two players, each of them making decisions [13, 14, 15]. The decisions of the leader (upper level problem) answer the questions: which is the best strategy for the leader, if he knows the goal function and the constraints of the follower (lower level problem) and how the leader has to choose his next decisions? When the leader evaluates his decisions, the follower chooses his own strategy for decision making for minimization of his target function. Respectively, the follower solves an optimization problem of mathematical programming form.

The formal presentation of the Stackelberg game in bi-level hierarchical forms is given as interconnected optimization sub-problems. The lower level optimization problem is in the form

$$\min_{y \in Y} f(x, y) \tag{2}$$

$$g(x, y) \leq 0 \tag{3}$$

where $x \in R^n$ is a coordination parameter, defined from the solutions of the upper level optimization problem, $y \in Y \subseteq R^m$ is the solution of the lower level optimization sub-problem, $f : R^n \times R^m \to R^1$ and $g : R^n \times R^m \to R^q$. This sub-problem is parameterized by the values of x. Let $P(x)$ denotes the optimal solution of problem (2) for given x:

$$P(x) = \left\{ y^* \in S(x) \mid f(x, y^*) = \min_{y \in Y} f(x, y^*), \quad g(x, y) \le 0 \right\},$$

where

$$S(x) = \{ y \in Y \mid g(x, y) \le 0 \} .$$

The optimal problem of the upper level for given lower level solution $y^* \in P(x)$ is

$$\min_{x \in X} F(x, y^*) \tag{4}$$

$$G(x, y^*) \le 0 \tag{5}$$

$$y^* \in P(x) \tag{6}$$

where $F : R^n \times R^m \to R^1$, $G : R^n \times R^m \to R^p$, $X \subseteq R^n$.

This problem is solved by the leader. The bi-level hierarchical problem, titled as Stackelberg game, is formulated as hierarchical system with two levels. The optimization sub-problem (2-3) is a slave one to the coordination problem (4)-(6). The particular constraint $P(x)$ determines the rational set of reactions of the slaver player. The feasible area of the coordination (4)-(6) is non-explicitly analytically defined

$$IR = \left\{ (x, y^*) \mid G(x, y^*) \le 0, \quad y^* \in P(x) \right\} .$$

The reaction of the slaver is evaluated from the set of rational reactions $P(x)$, while IR represents the feasible set for the decisions of the leader, among which he can search the optimal solution.

The book chapter considers a special form of the Stackelberg's problem:

$$\min_{x \in X, y^*} \left\{ F(x, y^*) / G(x, y^*) \le 0 \right\} \tag{7}$$

$$y^* \in P(x) = \left\{ y^* \in S(x) \mid f(x, y^*) = \min_{y \in Y} f(x, y^*), \quad g(x, y) \le 0 \right\}, \tag{8}$$

where the upper level is influenced by the reaction of the lower level by the minimal function $w(x)$, defined as

$$w(x) = \min_{y \in Y} f(x, y) \tag{9}$$

satisfying the definition set

$$g(x, y) \le 0 . \tag{10}$$

For this model the notation $w(x)$ refers to the minimal value of the goal function of the lower level $f(x,y)$, where the optimization is performed towards the argument y. The upper level problem can be formulated in a way, excluding y, substituting it in the target function and constraints explicitly with the minimal valued function $w(x)$:

$$\min_{x \in X} F(x, w(x)) \tag{11}$$

$$G(x, w(x)) \leq 0, \tag{12}$$

where $F : R^n \times R^1 \to R^1$ and $G : R^n \times R^1 \to R^p$. By the combination of relations (9)-(10) and (11)-(12) the bi-level hierarchical optimization problem is stated in the form

$$\min_{x \in X} F(x, w(x)) \tag{13}$$

$$G(x, w(x)) \leq 0 \tag{14}$$

$$w(x) = \min_{y \in Y} f(x, y) \tag{15}$$

$$g(x, y) \leq 0 \tag{16}$$

Both (7)-(8) and (13)-(16) are general nonlinear optimization problems. Due to methodological difficulties for the solution of hierarchically interconnected optimization problems, the classical application of the portfolio theory currently lacks a solution of the bi-level optimization problems. The portfolio problem is solved by quantitative assessment of $\sigma *$ in advance, without applying interconnected hierarchical optimization. The value of $\sigma *$ is estimated intuitively or empirically by an expert. Here, a methodology for the solution of bi-level portfolio problem is applied, derived as non-iterative coordination [17, 19]. The methodology for non-iterative coordination in hierarchical systems defines analytical approximations of the inexplicit function $w(x)$, used by the upper and lower optimization problems. Thus, analytical relations between the investor's preferences for the risk σ and the solutions x_i are derived [18]. Such relations support fast solution of the bi-level problem and respectively support real time decision making. The upper level problem is defined with a target function, which minimizes the Sharp ratio: portfolio risk versus portfolio return. The argument of this optimization problem is the investor's preferences for the risk σ. Applying the non-iterative methodology [17, 19] analytical relations between the portfolio problem's parameters E_p, V_p, the portfolio solutions x_i and the parameter of the investor's preference σ are derived. These relations speed up the decision making process and the investment decisions can be made in real time.

5   Solution of Portfolio Bi-level Problem

The solutions of the initial problem (1) x_i have to be described as analytical functions of the σ parameter. For that case the initial problem (1) is rewritten in the form

$$\min_x[\frac{1}{2}x^TQx+R^Tx]$$

$$Ax = C \, ,$$

(17)

where the correspondence between problems (1) and (17) is:

$$Q=\text{cov}(.), \qquad R=-\sigma E, \qquad A=1, \qquad C=1.$$

If the value of the coefficient σ is asserted, problem (1) has a solution, denoted like $x(\sigma)$. For the case when σ changes, the solution of the portfolio problem x is an inexplicit analytical function of σ:

$$x=x(\sigma).$$

The portfolio risk

$$V_p(\sigma) = x^T(\sigma)\text{cov}(.)x(\sigma)$$

and the portfolio return

$$E_p(\sigma) = E^T x(\sigma)$$

are also implicit functions of σ.

Problem (2) can be solved using the method of the non-iterative coordination, which gives possibility to derive approximations of the implicit analytical relations of the portfolio parameters $V_p(\sigma)$, $E_p(\sigma)$, $x(\sigma)$ towards the argument σ. Using relation (15) from [19], the analytical solution of problem (2) is

$$x^{opt} = -Q^{-1}[R - A^T(AQ^{-1}A^T)^{-1}(AQ^{-1}R + C)]\,.$$

(18)

Using this relation, the analytical descriptions of the portfolio risk and return become

$$V_p = x^{Topt}Qx^{opt} =$$
$$= \{-[(C^T + R^TQ^{-1}A^T)(-AQ^{-1}A^T)^{-1}A + R^T]Q^{-1}\}Q$$
$$\{-Q^{-1}[R - A^T(AQ^{-1}A^T)^{-1}(AQ^{-1}R+C)]\}$$

After several transformations it follows

$$V_p = R^TQ^{-1}[R - A^T(AQ^{-1}A^T)^{-1}A]Q^{-1}R + C^T(AQ^{-1}A^T)^{-1}C.$$

The analytical relation of the portfolio return is obtained as the linear relation towards x_{opt} or

$$E_p = E^T x = R^T x^{opt} = R^T \left\{ -Q^{-1} \left[R - A^T (AQ^{-1}A^T)^{-1}(AQ^{-1}R+C) \right] \right\} =$$
$$= -R^T Q^{-1}[R - A^T (AQ^{-1}A^T)^{-1} AQ^{-1}R] + R^T Q^{-1} A^T (AQ^{-1}A^T)^{-1} C$$

Finally

$$V_p = R^T Q^{-1} \left[R - A^T (AQ^{-1}A^T)^{-1} A \right] Q^{-1} R + C^T (AQ^{-1}A^T)^{-1} C \tag{19}$$

$$E_p = -R^T Q^{-1} [R - A^T (AQ^{-1}A^T)^{-1} AQ^{-1}R] + R^T Q^{-1} A^T (AQ^{-1}A^T)^{-1} C \tag{20}$$

Relations (19) and (20) can be expressed in terms of the initial portfolio problem (1). Thus, explicit analytical relations for the portfolio risk V_p, portfolio return E_p and the optimal solution of the portfolio problem x_{opt} are derived towards the coefficient of the investor's risk preference σ. For the current problem (1), taking into account the correspondence between problems (1) and (17), it follows

$$x^{opt}(\sigma) = Q^{-1} \left\{ \left[E - A^T (AQ^{-1}A^T)^{-1} AQ^{-1}E \right] \sigma + A^T (AQ^{-1}A^T)^{-1}C \right\} \tag{21}$$

$$V_p(\sigma) = E^T Q^{-1} [R - A^T (AQ^{-1}A^T)^{-1} A] Q^{-1} E \sigma^2 + C^T (AQ^{-1}A^T)^{-1} C \tag{22}$$

$$E_p(\sigma) = E^T x^{opt}(\sigma) = E^T Q^{-1} \left\{ R - A^T (AQ^{-1}A^T)^{-1} A \right] Q^{-1} E \sigma + A^T (AQ^{-1}A^T)^{-1} C \right\}. \tag{23}$$

To simplify the notations, the following coefficients are introduced:

$$\alpha = E^T Q^{-1} \left[R - A^T (AQ^{-1}A^T)^{-1} A \right] Q^{-1} E \tag{24}$$

$$\beta = C^T (AQ^{-1}A^T)^{-1} C$$

$$\gamma = E^T Q^{-1} A^T (AQ^{-1}A^T)^{-1} C \quad,$$

where the parameters α, β and γ are scalars. Relations (22) and (23) become

$$V_p(\sigma) = \alpha\sigma^2 + \beta, \qquad E_p(\sigma) = \alpha\sigma + \gamma. \tag{25}$$

The new derived relations (21)-(24) describe in analytically explicit form the functional relations between the portfolio parameters for risk, return and optimal solution towards the coefficient of the investor's preferences to risk σ. Hence, the solution of the portfolio problem (1) is calculated using relations (21)-(23) without the implementation of optimization algorithms for the solution of the low level optimization problem. This considerably speeds up the problem solution of (1). Hence, the portfolio optimization problem can be solved in real time, with no iterative calculations, which benefits the decision making in the fast dynamic environment of the stock exchange.

On the upper optimization level it is necessary to evaluate the parameter of investor's preferences σ, under which the better (minimal) value of Sharp ratio (the relation Risk/Return) is optimized. The problem for the evaluation of σ in a formal way is stated as

$$\min_{\sigma \geq 0} \left\{ \frac{Risk(\sigma)}{Portfolio\_return(\sigma)} = \frac{V_p(\sigma)}{E_p(\sigma)} \right\} .$$

According to relation (25) the analytical form of the problem is

$$\min_{\sigma} \left\{ \frac{V_p(\sigma)}{E_p(\sigma)} = \frac{\alpha\sigma^2 + \beta}{\alpha\sigma + \gamma} = \eta(\sigma) \right\} . \tag{26}$$

This problem evaluates the parameter of the investor's preferences σ according to objective considerations. Thus, the portfolio optimization problem is stated as bi-level optimization procedure (Fig. 4). The advantage for the evaluation of σ comes from the fact that the estimation of σ is done by overcoming the subjective influences of the investor, and it is found from a real optimization problem.

The solution σ_{opt} of such a problem is found according to the relations

$$\sigma_{opt} \equiv \arg\left\{ \min\left[0, \quad \frac{d\eta(\sigma)}{d\sigma} = 0 \right] \right\}$$

$$\sigma_{opt} = \min\left[0, \quad \frac{d}{d\sigma}(\frac{\alpha\sigma^2 + \beta}{\alpha\sigma + \gamma}) = 0 \right]$$

$$\frac{d\eta(\sigma)}{d\sigma} = \frac{2\alpha\sigma(\alpha\sigma + \gamma) - \alpha(\alpha\sigma^2 + \beta)}{(\alpha\sigma + \gamma)^2} = 0 .$$

The following condition must hold:

$$\alpha\sigma + \gamma = E_p > 0, \qquad \alpha \neq 0 . \tag{27}$$

Then

$$\sigma_1^{opt} = \frac{-\gamma + \sqrt{\gamma^2 + \alpha\beta}}{\alpha} = \frac{-1 + \sqrt{1 + \dfrac{\alpha}{\gamma}\dfrac{\beta}{\gamma}}}{\dfrac{\alpha}{\gamma}} .$$

For the particular case when the value of C is a digital number ($C=1$ for relative assessment of the investment), then

$$\sigma^{opt} = \max\left[0, \quad \frac{\gamma}{2}(-1 + \sqrt{\frac{E^T Q^{-1} E}{E^T Q^{-1} A^T (AQ^{-1} A^T)^{-1} AQ^{-1} E}}) \right] \tag{28}$$

This relation gives analytical way of calculation of the optimal parameter for risk preferences of the investor. For that reason the solution of the upper level optimization problem is reduced to analytical relation (28), applied for the calculation of σ_{opt}.

6   Assessment of the Bi-level Calculations

An illustration of the solution of a set of bi-level optimization problems is given below. A set of optimization problems is defined with a maximal amount of 13 securities, traded at the Bulgarian stock exchange, $n=13$. The portfolio optimization problems has been defined and solved with variable number of securities n ($n=2, 3, …,13$). Respectively, the corresponding matrices for the portfolios problems were chosen from the largest matrices $Q|_{13x13}$ and $E|_{13x1}$, which were defined from the Bulgarian stock exchange data as

$Q_{13}=$ [0.786167   -0.162559   -0.071959   -0.164858   0.047331
          -0.603850   0.166243   -1.047479   -0.539301   -0.603588
          -0.891402   0.008453   -0.133797;

          -0.162559   2.986604   0.793027   0.007331   0.026117
          0.098366   0.045948   -0.111006   0.655404   -0.071133
          0.035753   -0.293635   -0.410321;

          -0.071959   0.793027   0.555667   0.039004   -0.008898
          0.174713   0.013785   -0.060925   0.161520   0.066703
          0.170407   -0.029705   -0.249643;

          -0.164858   0.007331   0.039004   0.848372   -0.727659
          0.746201   0.241088   -1.953395   -1.625136   -0.313008
          -1.802728   -0.786455   0.865357;

          0.047331   0.026117   -0.008898   -0.727659   1.819902
          -1.231908   0.170827   -1.672948   -0.841580   -1.216701
          -1.527534   0.263455   -0.134094;

          -0.603850   0.098366   0.174713   0.746201   -1.231908
          5.813142   0.356443   0.135327   -1.384723   -0.966858
          0.370034   1.723733   -0.238691;

          0.166243   0.045948   0.013785   0.241088   0.170827
          0.356443   1.200181   0.075195   0.246796   0.080846
          1.063820   0.504612   -0.220170;

          -1.047479   -0.111006   -0.060925   -1.953395   -1.672948
          0.135327   0.075195   12.288469   0.077632   -0.607134
          3.229930   -1.407467   -0.200097;

-0.539301	0.655404	0.161520	-1.625136	-0.841580
-1.384723	0.246796	0.077632	5.300921	1.053221
2.486395	1.651404	-3.039378;		
-0.603588	-0.071133	0.066703	-0.313008	-1.216701
-0.966858	0.080846	-0.607134	1.053221	2.517762
0.653272	-0.035333	-0.194460;		
-0.891402	0.035753	0.170407	-1.802728	-1.527534
0.370034	1.063820	3.229930	2.486395	0.653272
29.446252	1.644501	0.462070;		
0.008453	-0.293635	-0.029705	-0.786455	0.263455
1.723733	0.504612	-1.407467	1.651404	-0.035333
1.644501	11.792594	1.208723;		
-0.133797	-0.410321	-0.249643	0.865357	-0.134094
-0.238691	-0.220170	-0.200097	-3.039378	-0.194460
0.462070	1.208723	16.782235];		

$$E_{13}^{T} = [\ 1.147143 \quad 1.805000 \quad 1.084717 \quad 1.239130 \quad 1.713784$$
$$2.571667 \quad 1.099146 \quad 2.230377 \quad 1.554639 \quad 1.217075$$
$$2.512333 \quad 2.510000 \quad 1.844951].$$

The problems with lower dimension $n<13$ are defined by sequential removal of the leading row and column from Q. Respectively, the lower order matrices E were generated by removal of the leading component of vector E.

The target of the experiments was to evaluate 30 points of the efficient frontier for each optimization problem. Then, having the efficient frontier, the optimization procedure continues with finding the portfolio, which has minimal Sharp ratio (risk versus return). For that case the parameter of the investor's preferences for risk σ_{opt} is calculated, using (28).

The sequence of the solution of the portfolio problem is the following:

- Analytical definition of the portfolio problem (1) with $n=13$;
- Evaluation of the scalar values of the intermediate parameters $\alpha(n)$, $\beta(n)$, $\gamma(n)$ from (24);
- Starting the calculations of the efficient frontier with initial value $\sigma^*=0$;
- Evaluation of the portfolio parameters $V_p = V_p(\sigma^*, \alpha(n), \beta(n), \gamma(n), E_p = E_p(\sigma^*, \alpha(n), \beta(n), \gamma(n))$, according to (25). Thus, one point from the efficient frontier in the space risk/return $V_p^*(E_p^*)$ is found;
- New value of the coefficient σ is chosen, $\sigma^{**}=\sigma^*+1/30$.

These steps are performed for 30 points of the graphics $V_p = V_p(E_p)$.

Following problem (26), the optimal value of the parameter of the investor's prefe-rences σ_{opt} is identified, evaluated as a solution of an upper level optimization problem:

$$\min_{\sigma}\left\{\frac{Risk(\sigma)}{Re\,turn(\sigma)}\right\} = \min_{\sigma}\left\{\frac{V_p(\sigma)}{E_p(\sigma)} = \frac{x^T(\sigma)Qx(\sigma)}{E^Tx(\sigma)}\right\}, \tag{29}$$

where $x(\sigma)$ is an implicit function, defined by the solution of the portfolio optimiza-tion problem (1) for different values of σ. Problem (29) introduces an objective crite-rion for the choice and estimation of the coefficient of the investor's preferences. This problem makes advantages for the estimation of σ in comparison with its subjective choice from the financial analyzer, which is performed according to the classical model of the portfolio optimization.

The optimal value of σ_{opt} is numerically calculated using (28).

Figure 4 presents the graphics $\dfrac{V_p(\sigma)}{E_p(\sigma)}$ for different optimization problems with varying dimensions $n=2,3,..7$. These graphics explicitly demonstrate the minimum (towards σ) of the ratio of portfolio risk versus return. The corresponding value σ_{opt} is found according to objective considerations, coming from the upper level optimization problem for minimization of Sharp ratio:

$$\sigma^{opt} = \arg\left\{\min_{\sigma}\left\{\frac{V_p(\sigma)}{E_p(\sigma)} = \frac{x^T(\sigma)Qx(\sigma)}{E^Tx(\sigma)}\right\}\right\}. \tag{30}$$

Problem (30) uses objective target function, which is the Sharp ratio. Thus, the argu-ment σ is calculated as a solution of a well defined and consistent optimization prob-lem. In comparison with the classical portfolio theory the value of σ is not assessed by subjective consideration of the financial analyzer, which is an advantage of the bi-level portfolio problem. The solution of problem (30) can be expressed also analytically, according to relation (28).

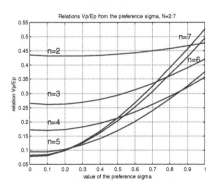

Fig. 4. Relation of the ratio risk/return from σ for different problem dimensions ($n=2, 3, ..., 7$)

As illustration the corresponding values of risk $V_p(x(\sigma_{opt}))$, portfolio return $E_p(x(\sigma_{opt}))$ and its optimal value ratio $\dfrac{V_p(\sigma)}{E_p(\sigma)}$ are given in figures 5-7. For the case of portfolio problems with dimensions $n=[2; 3; 4; 5; 6; 7]$, the optimal values of σ_{opt}, risk V_p and return E_p are the following

$$\sigma_{opt} = [\,0.2161;\,0.1307;\,0.0851;\,0.0464;\,0.0406;\,0.0384\,];$$

$$V_p(\sigma^{opt}) = [\,0.5715;\;0.2903;\,0.197;\;0.1199;\,0.1118;\;0.1083\,];$$

$$E_p(\sigma^{opt}) = [1.3223;\,1.1103;\,1.1568;\,1.292;\,1.3767;\,1.4091];$$

$$V_p/(E_p) = [0.4322;\,0.2614;\,0.1703;\,0.0928;\,0.0812;\,0.0769].$$

The graphical interpretation of these results is given in figures 5 - 8, where the notation sigma-opt is used for the value σ_{opt}.

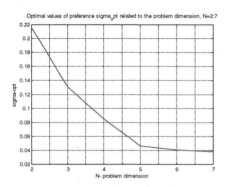

Fig. 5. Relation of σ_{opt} and the problem dimension $n=2,3,\ldots,7$

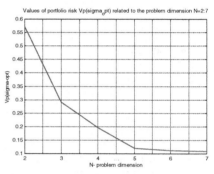

Fig. 6. Relation of $V_p(\sigma^{opt})$ and the problem dimension $n=2,3,\ldots,7$

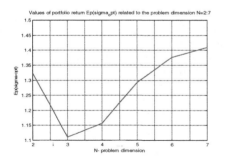

Fig. 7. Relation of $E_p(\sigma^{opt})$ and the problem dimension $n=2,3,\ldots,7$

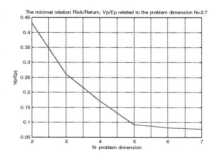

Fig. 8. Relation of $V_p/(E_p)$ and the problem dimension $n=2,3,\ldots,7$

These results prove the consistency of the definition of the portfolio optimization problem as a bi-level optimization one. On the upper level the optimal value of the parameter of the investor's preferences is calculated, according to objective optimization criteria. In the current case it an optimization problem for the minimization of the ratio risk/return (Sharpe Ratio) has been chosen. The optimal value of the parameter σ was derived analytically as a solution of the upper level optimization problem. This overcomes the weakness of the classical definition of the portfolio optimization problem, which assumes subjective estimation of σ.

7   Conclusion

The book chapter developed a new formal model of the portfolio problem, which was presented as bi-level optimization one. The risk of the investment was minimized twice by optimal content of portfolio securities and optimal assessment of the parameter of risk preference. The classical description of the portfolio problem is like single level optimization with predefined parameter for risk preference σ. This parameter has to be estimated by the financial analyzer and the portfolio theory insists σ to be given before solving the portfolio problem. The estimation of σ is a source of subjective influence for the problem definition and the evaluated optimal solution. Currently, the portfolio problem is solved for a set of values of σ by means to estimate the influence of σ to the problem solutions. In this research the process of decision making was presented as a two level optimization system. The upper level defined the optimal value of the parameter of risk preferences of the investor σ by minimizing the Sharp ratio (portfolio risk versus portfolio return). The lower optimization level used σ and solved the portfolio optimization problem. The bi-level formalism defined in an unique way the most appropriate value of σ by optimizing the Sharp ratio. In that manner, the bi-level formalism achieved two benefits: suppressed the subjective assessment of the investor's risk preferences and calculated and applied the optimal value of σ by minimizing the Sharp ratio. These two outcomes considerably improved the bi-level definition of the portfolio problem in comparison with the classical single level optimization problem.

Additionally, this work developed and applied a special method for solving the optimization problem, titled non-iterative coordination. It allowed to define explicitly and analytically the upper level optimization problem for solving σ and to derive explicit analytical relations between the portfolio problem solutions and σ, $x(\sigma)$. These relations speed up the optimal problem solution and the definition of the efficient frontier of portfolios. Thus, the decision making process can be performed in real time which can respond to the fast dynamic changes of the security market and reduce the risk of the investment.

Acknowledgement. This work is partly supported by project INPORT DVU01/0031 funded by "Science Researches" of Ministry of Education and Science in Bulgaria.

References

1. Bodie, Z., Kane, A., Marcus, A.: Investments. Naturela, Sofia (2000)
2. Campbell, J., Chacko, G., Rodriguez, J., Viceira, L.: Strategic Asset Allocation in a Continuous-Time VAR Model, pp. 1–21. Harvard University, Cambridge (2002)
3. Christoffersen, P.F.: Elements of financial risk management. Elsevier (2003)
4. Fang, Y., Lai, K.K., Wang, S.: Fuzzy portfolio optimization. Springer, Heidelberg (2008)
5. Ivanova, Z., Stoilova, K., Stoilov, T.: Portfolio optimization-Internet Information Service. Academician Publisher M. Drinov, Sofia (2005) (in Bulgarian)
6. Kohlmann, M., Tang, S.: Minimization of risk and linear quadratic optimal control theory. SIAM J. Control Optim. 42, 1118–1142 (2003)
7. Korn, R.: Continuous-Time Portfolio Optimization Under Terminal Wealth Constraints. ZOP-Mathematical Methods of Operations Research 42, 69–92 (1995)
8. Magiera, P., Karbowski, A.: Dynamic portfolio optimization with expected value-variance criteria. Bucharest, 308–313 (2001)
9. Markowitz, H.M.: Portfolio selection. J. of Finance 7, 77–91 (1952)
10. Mateev, M.: Analysis and assessment of investment risk. University publisher Economy, Sofia (2000) (in Bulgarian)
11. Sharpe, W., Alexander, G., Bailey, J.: Investments. Prentice Hall, England Cliffs (1999)
12. Sharpe, W.: Portfolio theory & Capital markets. Mc Grow Hill (2000)
13. Shimizu, K., Ishizuka, Y., Bard, J.: Nondifferentiable and Two-Level Mathematical Programming. Kluwer Academic Publishers (1997)
14. Simaan, M.: Stackelberg Optimization of Two-Level Systems. IEEE Trans. Systems, Man and Cybernetics, SMC 7, 554–556 (1997)
15. Simaan, M., Cruz, J.B.: On the Stackelberg Strategy in Nonzero-sum Games. J. of Optimiz. Theory & Applic. 11, 535–555 (1973)
16. Stackelberg, H.: The Theory of the Market Economy. Oxford University Press (1952)
17. Stoilov, T., Stoilova, K.: Noniterative Coordination in Multilevel Systems. Kluwer Academic Publisher, Dordrecht (1999)
18. Stoilova, K., Stoilov, T.: Noniterative Coordination Application in Solving Portfolio Optimisation Problems. In: Proceedings of the International Conference Automatics and Informatics, Sofia, pp. 159–162 (2003)
19. Stoilova, K.: Predictive Noniterative Coordination in Hierarchical Two-level Systems. Comptes Rendus De l'Académie Bulgare Des Sciences 58, 523–530 (2005)
20. Thomas, L.C.: A survey of credit and behavioural scoring: forecasting finan-cial risk of lending to consumers. Int. J. of Forecasting 16, 149–172 (2000)
21. Zadeh: Fuzzy sets. Information and Control 8, 338–353 (1965)

Chapter 6
Possibilistic Decision-Making Models for Portfolio Selection Problems

Peijun Guo

Faculty of Business Administration, Yokohama National University
79-4 Tokiwadai, Hodogaya-Ku, Yokohama, 240-8501, Japan
guo@ynu.ac.jp

Abstract. The basic assumption for using probabilistic decision-making models for portfolio selection problems, such as Markowitz's model, is that the situation of stock markets in future can be correctly reflected by security data in the past, that is, the mean and covariance of securities in future is similar to the past one. It is hard to ensure this kind of assumption for the real ever-changing stock markets. Possibilistic decision-making models for portfolio selection problems are based on possibility distributions, which are used to characterize experts' knowledge. A possibility distribution is identified using the returns of securities associated with possibility grades provided by experts. Based on the obtained possibility distribution, we construct a possibilistic portfolio selection decision-making model as a quadratic programming problem. Because experts' knowledge is very valuable, it is reasonable that possibilistic decision-making models are useful in real investment environment.

1   Introduction

It is well known that modern portfolio analysis is initiated by Markowitz [13]. In Markowitz's mean-variance model, returns of securities are assumed to be random variables, and the investors are assumed to seek a portfolio of securities which maximizes the return and minimizes the risk of their investment, where the return is quantified by the mean, and the risk is characterized by the variance of a portfolio of securities. Many researches have been done as the extensions of Markowitz's mean-variance models [2, 8, 12, 14, 19] within probability framework. The basic assumption of probability distribution based portfolio selection models is that the situation of stock market in future can be correctly reflected by securities data in the past. However, there are many uncertain factors and imprecise information so that it is hard to believe that such assumption can hold in the real ever-changing stock markets. Under such circumstances, it is benefit to take advantage of the experts' knowledge about stock market, which can be represented by the judgment with the historical data of securities to evaluate the future security returns. Tanaka and Guo initially characterized the expert knowledge on stock return prediction by the possibility distribution which is identified from the past time data plus possibility grades judged by an expert,

J. Lu, L.C. Jain, and G. Zhang: Handbook on Decision Making, ISRL 33, pp. 111–123.
springerlink.com   © Springer-Verlag Berlin Heidelberg 2012

and a possibilistic decision-making model for portfolio selecting is presented based on the possibility theory [16-17]. Tanaka et al [18] proposed two kinds of portfolio selection models based on fuzzy probabilities and exponential possibility distributions. Guo et al [4-5] considered a decision-making model for the portfolio selection problem with multiple decision makers. Fuzzy mathematical programming based portfolio selection models have been studied in the papers [1, 3, 6-7, 9-11, 15].

In this chapter, the methods for estimating the possibility distributions of security returns is introduced, where the dual possibility distributions, called upper and lower possibility distributions, are identified from the security data in the past and the possibility degrees given by experts. These two possibility distributions reflect two extreme opinions on the prediction of stock markets. The upper possibility distribution can be regarded as an optimistic viewpoint and the lower distribution as a pessimistic one in the sense that the upper possibility distribution always gives a higher possibility grade than the lower one. Based on the identified possibility distributions, possibilistic portfolio selection models are given.

This chapter is organized as follows. In Section 2, Markowitz's portfolio selection model is introduced. In Section 3, the concepts of upper and lower possibility distributions are addressed. In Section 4, the methods for identifying dual possibility distributions are introduced. In Section 5, possibilistic decision-making models for portfolio selection problems are given. In section 6, a numerical example is used to show the proposed methods. Finally, some concluding remarks are included.

2   Markowitz's Portfolio Selection Model

Assume that there are n securities denoted by S_j ($j=1,...,n$). The return of the security S_j is denoted as x_j and the proportion of total investment fund devoted to this security is denoted as r_j. Thus, the following equation holds.

$$\sum_{j=1}^{n} r_j = 1 \tag{1}$$

Since the returns of the securities x_j ($j=1,...,n$) vary from time to time, those are assumed to be random variables which can be represented by the pair of the average vector and the covariance matrix. For instance, it is assumed that the observation data on returns of securities over m periods are given. At the discrete time i, the vector of n returns is denoted as $\mathbf{x}_i = [x_{i1}, \cdots, x_{in}]^t$. Thus, the total data over m periods are denoted as the following matrix.

$$\begin{pmatrix} x_{11} & x_{12} & \cdots & x_{1n} \\ x_{21} & x_{22} & \cdots & x_{2n} \\ \vdots & \vdots & \ddots & \vdots \\ x_{m1} & x_{m2} & \cdots & x_{mn} \end{pmatrix}, \tag{2}$$

where x_{ij} denoting the return of the jth security at the time i is defined as {(Closing price of the jth security at the time i)-(Its closing price at time i-1)+(Its dividends at the time i)}/(Its closing price at time i-1). The average vector of returns over m periods denoted as $\mathbf{x}^0 = [x_1^0, \cdots, x_n^0]^t$ is defined as

$$\mathbf{x}^0 = \begin{bmatrix} \sum_{i=1}^{m} x_{i1} / m \\ \vdots \\ \sum_{i=1}^{m} x_{in} / m \end{bmatrix}. \tag{3}$$

The variance-covariance matrix $\mathbf{Q} = [q_{ij}]$ is defined as

$$q_{ij} = \sum_{k=1}^{m} (x_{ki} - x_i^0)(x_{kj} - x_j^0)/m \quad (i=1,\ldots,n, j=1,\ldots,n). \tag{4}$$

Therefore, random variables x_j ($j=1,\ldots,n$) can be represented by the average vector $\mathbf{x}^0$ and the covariance matrix $\mathbf{Q}$, denoted as ($\mathbf{x}^0$, $\mathbf{Q}$). Now, the return associated with a portfolio $\mathbf{r} = [r_1, \cdots, r_n]^t$ is given by

$$z = \mathbf{r}^t \mathbf{x}. \tag{5}$$

The average and variance of z are given as:

$$E(z) = E(\mathbf{r}^t \mathbf{x}) = \mathbf{r}^t E \mathbf{x} = \mathbf{r}^t \mathbf{x}^0, \tag{6}$$

$$V(z) = V(\mathbf{r}^t \mathbf{x}) = \mathbf{r}^t \mathbf{Q} \mathbf{r}. \tag{7}$$

Since the variance of a portfolio return is regarded as the risk of investment, the best investment is one with the minimum variance (7) subject to a given average return c. This is the famous Markowitz's model [13]. It can be formalized as the following quadratic programming (QP) problem.

$$\min_{\mathbf{r}} \ \mathbf{r}^t \mathbf{Q} \mathbf{r} \tag{8}$$

$$\text{s.t.} \quad \mathbf{r}^t \mathbf{x}^0 = c,$$

$$\sum_{i=1}^{n} r_i = 1,$$

$$r_i \geq 0.$$

By changing the value of c in the QP problem (8), we can obtain the corresponding minimum variance of the portfolio return with the expected return given.

3   Upper and Lower Possibility Distributions

Generally speaking, the vagueness and ambiguity of human understanding, the ignorance of cognition and the diversity of evaluation are always contained in human knowledge. A possibility distribution is a kind of representation of knowledge and information where the center reflects the most possible case and the spread reflects the others with relatively low possibilities. The area of the possibility distribution can be regarded as a sort of measure of ambiguity. In some case, it is difficult to directly give some possibility distribution to represent expert knowledge. However, we can always obtain the possibility grades of discrete data from an expert reflecting his judgment on some specified events. In portfolio selection problems, experts can choose some typical patterns from the past security data and give them associated possibility grades to reflect their judgment about the situation of stock markets in the future. The higher the possibility grades of security data, the more similar to the future.

The knowledge from one expert can be represented by a data set $\{(\mathbf{x}_i, h_i) \mid i=1,...,m\}$ where $\mathbf{x}_i = [x_{i1}, \cdots, x_{in}]^t$ is an n-dimensional vector characterizing the returns of n securities of the ith sample. h_i is an associated possibility grade given by an expert to reflect his judgment to which degree this return vector will occur in the future. m is the number of samples. The data set $(\mathbf{x}_i, h_i)$ $(i=1,...,m)$ can be approximated by a dual data sets $(\mathbf{x}_i, h_{li})$ and $(\mathbf{x}_i, h_{ui})$ $(i=1,...,m)$ with the condition $h_{li} \le h_i \le h_{2i}$. Assume that the values h_{li} and h_{2i} are from a class of the functions $G(\mathbf{x}, \boldsymbol{\theta})$ with the parameter vector $\boldsymbol{\theta}$. Let $G(\mathbf{x}, \boldsymbol{\theta}_l)$ and $G(\mathbf{x}, \boldsymbol{\theta}_u)$ correspond to h_{li} and h_{2i} $(i=1,...,m)$, respectively and simply denote as $\pi_l(\mathbf{x})$ and $\pi_u(\mathbf{x})$. Given the data set $(\mathbf{x}_i, h_i)$ $(i=1,...,m)$, the objective of estimation is to obtain two optimal parameter vectors $\boldsymbol{\theta}_u^*$ and $\boldsymbol{\theta}_l^*$ from the parameter space to approximate $(\mathbf{x}_i, h_i)$ from upper and lower directions according to some given measure. Moreover, the dual optimal parameter vectors ($\boldsymbol{\theta}_u^*$, $\boldsymbol{\theta}_l^*$) make the relation $G(\mathbf{x}, \boldsymbol{\theta}_l^*) \le G(\mathbf{x}, \boldsymbol{\theta}_u^*)$ hold for any arbitrary n-dimensional vector $\mathbf{x}$.

Suppose that the function $G(\mathbf{x}, \boldsymbol{\theta})$ is an exponential function $\exp\{-(\mathbf{x}-\mathbf{a})^t \mathbf{D}_A^{-1}(\mathbf{x}-\mathbf{a})\}$, simply denoted as $(\mathbf{a}, \mathbf{D}_A)_e$. Then the following formulas hold.

$$\pi_l(\mathbf{x}_i) = \exp\{-(\mathbf{x}_i - \mathbf{a})^t \mathbf{D}_l^{-1}(\mathbf{x}_i - \mathbf{a})\} = (\mathbf{a}, \mathbf{D}_l)_e \quad (i=1,...,m), \tag{9}$$

$$\pi_u(\mathbf{x}_i) = \exp\{-(\mathbf{x}_i - \mathbf{a})^t \mathbf{D}_u^{-1}(\mathbf{x}_i - \mathbf{a})\} = (\mathbf{a}, \mathbf{D}_u)_e \quad (i=1,...,m), \tag{10}$$

$$\pi_l(\mathbf{x}_i) \le h_i \le \pi_u(\mathbf{x}_i) \quad \text{and} \quad \pi_l(\mathbf{x}) \le \pi_u(\mathbf{x}), \tag{11}$$

where $\mathbf{a} = [a_1, a_2, \cdots, a_n]^t$ is a center vector, $\mathbf{D}_u$ and $\mathbf{D}_l$ are positive definite matrices, denoted as $\mathbf{D}_u > 0$ and $\mathbf{D}_l > 0$, respectively. It can be seen that in the above exponential function, the vector $\mathbf{a}$ and matrices $\mathbf{D}_u$ and $\mathbf{D}_l$ are parameters to be solved. Different parameters $\mathbf{a}$, $\mathbf{D}_u$ and $\mathbf{D}_l$ lead to different values $\pi_l(\mathbf{x}_i)$ and $\pi_u(\mathbf{x}_i)$ which approximate the given possibility degree h_i of $\mathbf{x}_i$ to the different extent.

Definition 1. Given the formulas (9), (10) and (11), the inconsistency index of these two approximations (9) and (10), denoted as κ, is defined as follows:

$$\kappa = -\ln\sqrt[m]{\prod_{i=1}^{m} \frac{\pi_l(\mathbf{x}_i)}{\pi_u(\mathbf{x}_i)}} = (\sum_{i=1}^{m}(\ln\pi_u(\mathbf{x}_i) - \ln\pi_l(\mathbf{x}_i)))/m$$

$$= \sum_{i=1}^{m}(\mathbf{x}_i - \mathbf{a})^t \mathbf{D}_l^{-1}(\mathbf{x}_i - \mathbf{a}) - \sum_{i=1}^{m}(\mathbf{x}_i - \mathbf{a})^t \mathbf{D}_u^{-1}(\mathbf{x}_i - \mathbf{a})$$

(12)

It is known from Definition 1 that the smaller the parameter κ is, the closer to h_i the values $\pi_l(\mathbf{x}_i)$ and $\pi_u(\mathbf{x}_i)$ are from lower and upper directions, respectively.

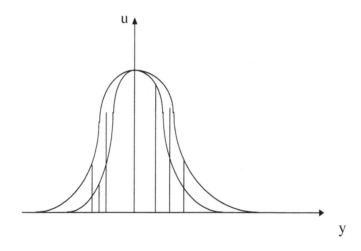

Fig. 1. The concept of upper and lower possibility distributions (The upper curve is the upper possibility distribution and the lower curve is the lower possibility distribution)

Definition 2. Denote the optimal solutions of $\mathbf{a}$, $\mathbf{D}_u$ and $\mathbf{D}_l$ as $\mathbf{a}_*$, $\mathbf{D}_{*u}$ and $\mathbf{D}_{*l}$, respectively, which minimize κ with the constraint (11). The following functions

$$\pi_{*l}(\mathbf{x}) = \exp\{-(\mathbf{x} - \mathbf{a}_*)^t \mathbf{D}_{*l}^{-1}(\mathbf{x} - \mathbf{a}_*)\},$$

(13)

$$\pi_{*u}(\mathbf{x}) = \exp\{-(\mathbf{x} - \mathbf{a}_*)^t \mathbf{D}_{*u}^{-1}(\mathbf{x} - \mathbf{a}_*)\},$$

(14)

are called lower and upper exponential possibility distributions of the vector $\mathbf{x}$, respectively. For simplicity afterwards we write $\pi_u(\mathbf{x})$ and $\pi_l(\mathbf{x})$ instead of

$\pi_{*u}(\mathbf{x})$ and $\pi_{*l}(\mathbf{x})$. The concept of upper and lower possibility distributions is illustrated in Fig. 1. It can be seen from Fig. 1 that the given possibility degrees are completely included into the boundary of upper and lower possibility distributions. The upper possibility distribution can be regarded as an optimistic viewpoint and the lower distribution as a pessimistic one in the sense that the upper possibility distribution always gives a higher possibility grade than the lower one. The difference between the dual possibility distributions reflects the inconsistency of expert knowledge.

4   Identification of Upper and Lower Possibility Distributions

The model to identify the upper and lower possibility distributions can be formulated to minimize the inconsistency index as follows [5]:

$$\min_{\mathbf{a},\mathbf{D}_u,\mathbf{D}_l} \quad \sum_{i=1}^{m}(\mathbf{x}_i-\mathbf{a})^t\mathbf{D}_l^{-1}(\mathbf{x}_i-\mathbf{a}) - \sum_{i=1}^{m}(\mathbf{x}_i-\mathbf{a})^t\mathbf{D}_u^{-1}(\mathbf{x}_i-\mathbf{a}) \tag{15}$$

$$\text{s. t.} \quad (\mathbf{x}_i-\mathbf{a})^t\mathbf{D}_l^{-1}(\mathbf{x}_i-\mathbf{a}) \geq -\ln h_i \quad (i=1,\ldots,m),$$

$$(\mathbf{x}_i-\mathbf{a})^t\mathbf{D}_u^{-1}(\mathbf{x}_i-\mathbf{a}) \leq -\ln h_i \quad (i=1,\ldots,m),$$

$$\mathbf{D}_u-\mathbf{D}_l \geq 0,$$

$$\mathbf{D}_l > 0.$$

In the following, let us consider how to obtain the center vector $\mathbf{a}$ and the positive matrices $\mathbf{D}_l$ and $\mathbf{D}_u$. It is straightforward that upper and lower exponential distributions should have the same center vector. Otherwise, the relation $\pi_u(\mathbf{x}) \geq \pi_l(\mathbf{x})$ can not always hold. Because the vector $\mathbf{x}$ with the highest possibility grade should be closest to the center vector $\mathbf{a}$ among all $\mathbf{x}_i$ $(i=1,\ldots,m)$, the center vector $\mathbf{a}$ can be approximately estimated as

$$\mathbf{a}=\mathbf{x}_{i^*}, \tag{16}$$

where $\mathbf{x}_{i^*}$ denotes the vector whose grade is $h_{i^*} = \max_{k=1,\ldots,m} h_k$. The associated possibility grade of $\mathbf{x}_{i^*}$ is revised to be 1 because it becomes the center vector. Taking the transformation $\mathbf{y}=\mathbf{x}-\mathbf{a}$, the problem (15) is changed into the following problem.

$$\min_{\mathbf{D}_u,\mathbf{D}_l} \quad \sum_{i=1}^{m}\mathbf{y}_i^t\mathbf{D}_l^{-1}\mathbf{y}_i - \sum_{i=1}^{m}\mathbf{y}_i^t\mathbf{D}_u^{-1}\mathbf{y}_i \tag{17}$$

$$\text{s. t.} \quad \mathbf{y}_i^t\mathbf{D}_l^{-1}\mathbf{y}_i \geq -\ln h_i \quad (i=1,\ldots,m),$$

$$\mathbf{y}_i^t\mathbf{D}_u^{-1}\mathbf{y}_i \leq -\ln h_i \quad (i=1,\ldots,m),$$

$$\mathbf{D}_u-\mathbf{D}_l \geq 0,$$

$$\mathbf{D}_l > 0.$$

The formula (17) is a nonlinear optimization problem due to the last two constraints. To cope with this difficulty, we use principle component analysis (PCA) to rotate the given data $(\mathbf{y}_i, h_i)$ to obtain a positive definite matrix easily. The data $\mathbf{y}_i$ ($i=1,\ldots,m$) can be transformed by linear transformation $\mathbf{T}$ whose columns are eigenvectors of the matrix $\Sigma = [\sigma_{ij}]$, where σ_{ij} is defined as

$$\sigma_{ij} = \{\sum_{k=1}^{m}(x_{ki} - a_i)(x_{kj} - a_j)h_k\}/\sum_{k=1}^{m} h_k \tag{18}$$

Using the linear transformation $\mathbf{T}$, the data $\mathbf{y}_i$ is transformed into $\{\mathbf{z}_i = \mathbf{T}'\mathbf{y}_i\}$. Then the formulas (9) and (10) can be rewritten as follows:

$$\pi_l(\mathbf{z}_i) = \exp\{-\mathbf{z}_i{}'\mathbf{T}'\mathbf{D}_u^{-1}\mathbf{T}\mathbf{z}_i\} \quad (i=1,\ldots,m), \tag{19}$$

$$\pi_u(\mathbf{z}_i) = \exp\{-\mathbf{z}_i{}'\mathbf{T}'\mathbf{D}_l^{-1}\mathbf{T}\mathbf{z}_i\} \quad (i=1,\ldots,m). \tag{20}$$

Since $\mathbf{T}$ is obtained by PCA, $\mathbf{T}'\mathbf{D}_u^{-1}\mathbf{T}$ and $\mathbf{T}'\mathbf{D}_l^{-1}\mathbf{T}$ can be assumed to be diagonal matrices as follows:

$$\mathbf{C}_u = \mathbf{T}'\mathbf{D}_u^{-1}\mathbf{T} = \begin{pmatrix} c_{u1} & & 0 \\ & \cdot & \\ & & \cdot \\ 0 & & c_{un} \end{pmatrix}, \tag{21}$$

$$\mathbf{C}_l = \mathbf{T}'\mathbf{D}_l^{-1}\mathbf{T} = \begin{pmatrix} c_{l1} & & 0 \\ & \cdot & \\ & & \cdot \\ 0 & & c_{ln} \end{pmatrix}. \tag{22}$$

The model (17) can be rewritten as the following LP problem:

$$\min_{C_l, C_u} \sum_{i=1}^{m} \mathbf{z}_i{}'\mathbf{C}_l \mathbf{z}_i - \sum_{i=1}^{m} \mathbf{z}_i{}'\mathbf{C}_u \mathbf{z}_i \tag{23}$$

$$\text{s. t.} \quad \mathbf{z}_i{}'\mathbf{C}_u \mathbf{z}_i \le -\ln h_i \ (i=1,\ldots,m),$$

$$\mathbf{z}_i{}'\mathbf{C}_l \mathbf{z}_i \ge -\ln h_i \ (i=1,\ldots,m),$$

$$c_{lj} \ge c_{uj},$$

$$c_{uj} \ge \varepsilon \ (j=1,\ldots,n),$$

where the condition $c_{lj} \ge c_{uj} \ge \varepsilon > 0$ makes the matrix $\mathbf{D}_u - \mathbf{D}_l$ semi-positive definite and matrices $\mathbf{D}_u$ and $\mathbf{D}_l$ positive. Denote the optimal solutions of (23) as $\mathbf{C}_u^*$ and $\mathbf{C}_l^*$. Thus, we have

$$\mathbf{D}_u^* = \mathbf{TC}_u^{*-1}\mathbf{T}^t,$$

$$\mathbf{D}_l^* = \mathbf{TC}_l^{*-1}\mathbf{T}^t. \tag{24}$$

For simplicity afterwards we write $\mathbf{D}_u$ and $\mathbf{D}_l$ instead of $\mathbf{D}_u^*$ and $\mathbf{D}_l^*$.

5   Possibilistic Decision-Making Models for Portfolio Selecting

Assume that $\mathbf{x}$ is governed by an n-dimensional possibility distribution $(\mathbf{a}, \mathbf{D}_A)_e$, denoted as $\mathbf{X} \sim (\mathbf{a}, \mathbf{D}_A)_e$, the possibility distribution of the possibility number Y with $Y = \mathbf{r}'\mathbf{X}$, denoted as $\pi_B(\mathbf{y})$, is defined by the extension principle as follows:

$$\pi_B(y) = \max_{\{\mathbf{x}| y = \mathbf{r}'\mathbf{x}\}} \exp\{-(\mathbf{x}-\mathbf{a})'\,\mathbf{D}_A^{-1}(\mathbf{x}-\mathbf{a})\}, \tag{25}$$

where $\mathbf{r}$ is an n-dimensional vector. Solving the optimization problem (25), the possibility distribution of Y can be obtained as (See Appendix)

$$\pi_B(y) = \exp\{-(y-\mathbf{r}'\mathbf{a})^2(\mathbf{r}'\mathbf{D}_A\mathbf{r})^{-1}\} = (\mathbf{r}'\mathbf{a}, \mathbf{r}'\mathbf{D}_A\mathbf{r})_e, \tag{26}$$

where $\mathbf{r}'\mathbf{a}$ is the center value and $\mathbf{r}'\mathbf{D}_A\mathbf{r}$ is the spread value of the possibility number Y. $Y \sim (\mathbf{r}'\mathbf{a}, \mathbf{r}'\mathbf{D}_A\mathbf{r})_e$ is called the one-dimensional realization of $\mathbf{X} \sim (\mathbf{a}, \mathbf{D}_A)_e$.

The portfolio return can be written as

$$z = \mathbf{r}'\mathbf{x} = \sum_{j=1,\dots,n} r_j x_j, \tag{27}$$

where r_j denotes the proportion of the total investment funds devoted to the security S_j and x_j is its return.

Because the return vector $\mathbf{x}$ is governed by the dual possibility distribution $\mathbf{X}_f \sim <(\mathbf{a}_c, \mathbf{D}_{cu})_e, (\mathbf{a}_c, \mathbf{D}_{cl})_e>$, using the formula (26) the dual distributions of a possibility portfolio return Z, denoted as $Z \sim <\pi_{Z_u}(z), \pi_{Z_l}(z)>$, are obtained as follows:

$$\pi_{Z_u}(z) = \exp\{-(z-\mathbf{r}'\mathbf{a}_c)^2(\mathbf{r}'\mathbf{D}_{cu}\mathbf{r})^{-1}\} = (\mathbf{r}'\mathbf{a}_c, \mathbf{r}'\mathbf{D}_{cu}\mathbf{r})_e, \tag{28}$$

$$\pi_{Z_l}(z) = \exp\{-(z-\mathbf{r}'\mathbf{a}_c)^2(\mathbf{r}'\mathbf{D}_{cl}\mathbf{r})^{-1}\} = (\mathbf{r}'\mathbf{a}_c, \mathbf{r}'\mathbf{D}_{cl}\mathbf{r})_e, \tag{29}$$

where $\mathbf{r}'\mathbf{a}_c$ is the center value, $\mathbf{r}'\mathbf{D}_{cu}\mathbf{r}$ and $\mathbf{r}'\mathbf{D}_{cl}\mathbf{r}$ are the spreads of a possibility portfolio return Z based on upper and lower possibility distributions.

Considering the dual possibility distributions, the following two quadratic programming problems to minimize the spread of possibility portfolio return are given where the spread of possibility portfolio return is regarded as the measure of risk [4].

$$\min_{\mathbf{r}} \ \mathbf{r}'\mathbf{D}_{cu}\mathbf{r} \tag{30}$$

$$\text{s. t. } \mathbf{r}'\mathbf{a}_c = c,$$

$$\sum_{i=1}^{n} r_i = 1,$$

$$r_i \geq 0,$$

$$\min_{\mathbf{r}} \ \mathbf{r}'\mathbf{D}_{cl}\mathbf{r} \tag{31}$$

$$\text{s. t. } \mathbf{r}'\mathbf{a}_c = c,$$

$$\sum_{i=1}^{n} r_i = 1,$$

$$r_i \geq 0,$$

where c is the expected center value of a possibility portfolio return which should comply with the constraint $\min_{i=1,\ldots,n} a_i \leq c \leq \max_{i=1,\ldots,n} a_i$ to guarantee the existence of the solution in (30) and (31). Because $\mathbf{D}_{cu}$ and $\mathbf{D}_{cl}$ are positive definite matrices, (30) and (31) are convex programming problems.

Theorem [5]. The spread of the possibility portfolio return based on the lower possibility distribution is not larger than the one based on the upper possibility distribution.

The nondominated solutions with considering two objective functions, i.e., the spread and the center of a possibility portfolio in the possibilistic portfolio selection models (30) and (31) can form two efficient frontiers.

6   Numerical Example

In order to show the above-proposed approaches, a numerical example for the stock market analysis is given. Table 1 lists the security data with associated possibility grades for stock return prediction given by an expert. Based on the approach introduced in Section 4, the dual possibility distributions, denoted as $\mathbf{X} \sim < (\mathbf{a}, \mathbf{D}_u)_e, (\mathbf{a}, \mathbf{D}_l)_e >$ was obtained as follows:

$$\mathbf{a} = [0.513, 0.098, 0.285, 0.714]^t,$$

$$\mathbf{D}_u = \begin{bmatrix} 82.803 & -259.691 & 59.369 & -63.082 \\ -259.691 & 825.545 & -188.827 & 202.048 \\ 59.369 & -188.827 & 44.471 & -46.040 \\ -63.082 & 202.048 & -46.040 & 50.606 \end{bmatrix},$$

$$\mathbf{D}_l = \begin{bmatrix} 0.695 & 0.111 & -0.255 & 0.149 \\ 0.111 & 0.133 & 0.154 & -0.085 \\ -0.255 & 0.154 & 1.144 & 0.073 \\ 0.149 & -0.085 & 0.073 & 0.648 \end{bmatrix}.$$

Using models (30) and (31), the portfolios based on upper and lower possibility distributions with $c=0.5$ was shown in Fig. 2.

Table 1. Security data with an expert's knowledge

Years	Possibility	Sec.1	Sec.2	Sec.3	Sec.4
1977(1)	0.192	-0.305	-0.173	-0.318	-0.477
1978(2)	0.901	0.513	0.098	0.285	0.714
1979(3)	0.54	0.055	0.2	-0.047	0.165
1980(4)	0.517	-0.126	0.03	0.104	-0.043
1981(5)	0.312	-0.28	-0.183	-0.171	-0.277
1982(6)	0.623	-0.003	0.067	-0.039	0.476
1983(7)	0.676	0.428	0.3	0.149	0.225
1984(8)	0.698	0.192	0.103	0.26	0.29
1985(9)	0.716	0.446	0.216	0.419	0.216
1986(10)	0.371	-0.088	-0.046	-0.078	-0.272
1987(11)	0.556	-0.127	-0.071	0.169	0.144
1988(12)	0.54	-0.015	0.056	-0.035	0.107
1989(13)	0.709	0.305	0.038	0.133	0.321
1990(14)	0.535	-0.096	0.089	0.732	0.305
1991(15)	0.581	0.016	0.09	0.021	0.195
1992(16)	0.669	0.128	0.083	0.131	0.39
1993(17)	0.482	-0.01	0.035	0.006	-0.072
1994(18)	0.661	0.154	0.176	0.908	0.715

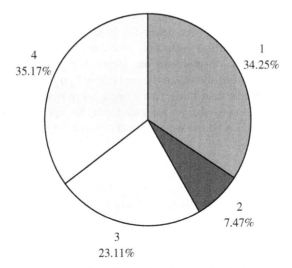

Portfolio based on the upper distribution

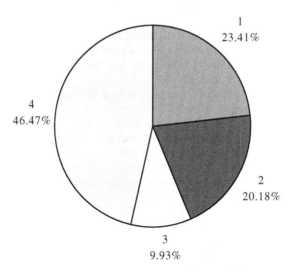

Portfolio based on the lower distribution

Fig. 2. Portfolios based on upper and lower possibility distributions with $c=0.5$

7   Conclusions

Probability distribution based portfolio selection models basically assume that the situation of stock markets in future can be characterized by the past security data. It is difficult to ensure such assumption in real investment problems. Possibilistic decision-making models for portfolio selection problems take advantage of the past security data and experts' judgment, which can gain an insight into a change of stock markets. This chapter addresses an approach for obtaining dual possibility distributions, i.e., the upper and the lower possibility distributions to reflect the different viewpoints of experts in portfolio selection problems. Possibilistic decision-making models for portfolio selection problems are formalized as quadratic programming problems. It is clear that portfolio returns based on lower possibility distributions have smaller possibility spreads than those based on upper possibility distributions. Because portfolio experts' knowledge is characterized by the upper and lower possibility distributions, the obtained portfolio would reflect portfolio experts' judgment.

References

1. Bilbao-Terol, A., Pérez-Gladish, B., Arenas-Parra, M., Rodríguez-Uría, M.V.: Fuzzy compromise programming for portfolio selection. Applied Mathematics and Computation 173, 251–264 (2006)
2. Dantzig, G.B., Infanger, G.: Multi-stage stochastic linear programs for portfolio optimization. Annals of Operations Research 45, 59–76 (1993)
3. Fang, Y., Lai, K.K., Wang, S.Y.: Portfolio rebalancing model with transaction costs based on fuzzy decision theory. European Journal of Operational Research 175(2), 879–893 (2006)
4. Guo, P., Tanaka, H.: Decision analysis based on fused double exponential possibility distributions. European Journal of Operational Research 148, 467–479 (2003)
5. Guo, P., Zeng, D., Shishido, H.: Group decision with inconsistent knowledge. IEEE Transactions on SMC, Part A 32, 670–679 (2002)
6. Guo, P., Tanaka, H.: Possibilistic data analysis and its application to portfolio selection problems. Fuzzy Economic Review 3/2, 3–23 (1998)
7. Inuiguchi, M., Ramík, J.: Possibilistic linear programming: a brief review of fuzzy mathematical programming and a comparison with stochastic programming in portfolio selection problem. Fuzzy Sets and Systems 111, 3–28 (2000)
8. Konno, H., Wijayanayake, A.: Portfolio optimization under D.C. transaction costs and minimal transaction unit constraints. Journal of Global Optimization 22, 137–154 (2002)
9. Lacagnina, V., Pecorella, A.: A stochastic soft constraints fuzzy model for a portfolio selection problem. Fuzzy Sets and Systems 157, 1317–1327 (2006)
10. León, T., Liern, V., Vercher, E.: Viability of infeasible portfolio selection problems: A fuzzy approach. European Journal of Operational Research 139, 178–189 (2002a)
11. León, T., Liern, V., Vercher, E.: Two fuzzy approaches for solving multi-objective decision problems. Computational Economics 19, 273–286 (2002b)
12. Li, D., Ng, W.L.: Optimal dynamic portfolio selection: multiperiod mean-variance formulation. Mathematical Finance 10(3), 387–406 (2000)
13. Markowitz, H.M.: Portfolio Selection: Efficient Diversification of Investments. John Wiley, New York (1959)
14. Mossin, J.: Optimal multiperiod portfolio policies. Journal of Business 41(2), 215–229 (1968)
15. Ortí, J.J., Sáez, J., Terceño, A.: On the treatment of uncertainty in portfolio selection. Fuzzy Economic Review 7, 59–80 (2002)
16. Tanaka, H., Guo, P.: Possibilistic Data Analysis for Operations Research. Physica-Verlag, Heidelberg (1999)

17. Tanaka, H., Guo, P.: Portfolio selection based on upper and lower exponential possibility distributions. European Journal of Operational Research 114, 115–126 (1999)
18. Tanaka, H., Guo, P., Turksen, I.B.: Portfolio selection based on fuzzy probabilities and possibility distributions. Fuzzy Sets and Systems 111, 387–397 (2000)
19. Yoshimoto, A.: The mean-variance approach to portfolio optimization subject to transaction costs. Journal of the Operational Research Society of Japan 39, 99–117 (1996)

Appendix

Give a possibilistic linear system

$$\mathbf{Y} = \mathbf{TX},$$ (1)

where T is a matrix and $rank[\mathbf{T}] = m$ and the m-dimensional possibilistic vector $\mathbf{X}$ is governed by the following exponential possibility distribution

$$\Pi_A(\mathbf{x}) = \exp\{-(\mathbf{x}-\mathbf{a})^t \mathbf{D}_A^{-1}(\mathbf{x}-\mathbf{a})\} = (\mathbf{a}, \mathbf{D}_A)_e.$$ (2)

The possibility distribution of the possibilistic vector Y denoted as $\Pi_Y(\mathbf{y})$, is

$$\Pi_Y(\mathbf{y}) = (\mathbf{Ta}, \mathbf{TD}_A \mathbf{T}^t)_e.$$ (3)

Proof. The possibility distribution of the possibilistic vector Y can be obtained by the extension principle as

$$\Pi_Y(\mathbf{y}) = \max_{\{\mathbf{x}|\mathbf{y}=\mathbf{Tx}\}} \exp\{-(\mathbf{x}-\mathbf{a})^t \mathbf{D}_A^{-1}(\mathbf{x}-\mathbf{a})\}.$$ (4)

This optimization problem (4) can be reduced to the minimization problem of Lagrangian function as follows:

$$L(\mathbf{x},\mathbf{k}) = (\mathbf{x}-\mathbf{a})^t \mathbf{D}_A^{-1}(\mathbf{x}-\mathbf{a}) + \mathbf{k}^t (\mathbf{y}-\mathbf{Tx}),$$ (5)

The necessary and sufficient conditions for optimality, that is, $\partial L / \partial \mathbf{x} = \mathbf{0}$, $\partial L / \partial \mathbf{k} = \mathbf{0}$, yield the optimal $\mathbf{x}^*$ as

$$\mathbf{x}^* = \mathbf{a} + \mathbf{D}_A \mathbf{T}^t (\mathbf{TD}_A \mathbf{T}^t)^{-1}(\mathbf{y}-\mathbf{a})$$ (6)

Substituting (6) to (4), we have

$$\Pi_Y(\mathbf{y}) = \exp\{-(\mathbf{y}-\mathbf{Ta})^t (\mathbf{TD}_A \mathbf{T}^t)^{-1}(\mathbf{y}-\mathbf{Ta})\},$$ (7)

where there exists $(\mathbf{TD}_A \mathbf{T}^t)^{-1}$ because $rank[\mathbf{T}] = m$ is assumed. The parametric representation of $\Pi_Y(\mathbf{y})$ is

$$\Pi_Y(\mathbf{y}) = (\mathbf{Ta}, \mathbf{TD}_A \mathbf{T}^t)_e.$$ (8)

Chapter 7
Searching Musical Representative Phrases Using Decision Making Based on Fuzzy Similarities

Emerson Castañeda, Luis Garmendia, and Matilde Santos

Facultad de Informática, Universidad Complutense de Madrid, 28040-Madrid, Spain
emecas@ieee.org, lgarmend@fdi.ucm.es, msantos@dacya.ucm.es

Abstract. A new method to find representative phrases from a musical score is given in this paper. It is based on the computation and use of a fuzzy proximity relation on a set of phrases. This relation is computed as a conjunction of values given by a W-indistinguishability on a set of variation of notes in the phrases, where W is the Lukasiewicz t-norm. Different fuzzy logics are used and compared in order to show their influence on the final decision. The proposed method to find the most representative phrase has been proved successful on different musical scores.

1 Introduction

The concept of musical "motif" stands for a short musical phrase on which a composer develops the whole musical score. The "motif" is a melodic element that is important throughout the work and that can be varied to generate more musical phrases. This work uses a practical approximation to the criteria of Overill [8] for searching musical motifs based on the analysis of the different phrases. The motif of a score is found using a "fuzzy pattern machine model" that uses indistinguishability operators and proximity fuzzy relations to compare the phrases.

In [8] the author discusses the importance in music analysis of establishing the occurrences of a musical motif and its variants. He presents it as a tedious and time-consuming process; therefore, it is a task that can be carried out by a computer using several models that must include the design of which variants are to be included in the search. The number of variants that are considered have been found to have a profound effect on the computer time required. He presents two models that are based on recurrence relations and closed analytic expression of fuzzy pattern matching.

Each one of the Overill [8] models assumes the existence of an atomic exact matching operation that can be represented in a formula to be evaluated and tabulated as a function of some independent parameters. These results allow a prior estimation of the relative run times of different music searches. Both proposed models are also equally capable of handling inversion, retrogradation or inverted retrogradation of a motif [1]. Nevertheless, both models are only concerned with pitch, without taking into account other musical issues such as the duration.

Finally, Overill concludes that from the music analysis, the traditional approaches such as the two models analyzed in his paper, have many drawbacks and limitations,

J. Lu, L.C. Jain, and G. Zhang: Handbook on Decision Making, ISRL 33, pp. 125–149.

mainly due to the complexity and the computer time they require, which makes them less useful for practical applications.

Some other approaches present a method to find representative phrases from a musical score [3][4]. They are based on the computation of a fuzzy proximity relation on the set of phrases. Two musical phrases can be considered 'similar' when the variations between the first and the second notes are 'equivalent', AND the variations between the second and the third notes are 'equivalent', AND ..., so on and so forth. That is, two phrases are similar if the conjunction of the distances between couples of consecutive notes are similar. This conjunction can be modeled using different mathematical operators, specifically triangular norms. Therefore, two phrases can be similar even if the starting tone is different, because the comparison system works by evaluating relative distances between notes allowing transportations on the scale.

Once the fuzzy relation "proximity" has been computed on a set of phrases, a method that automatically selects some phrases is defined by computing a fuzzy set representing the characteristic of being 'similar to the other phrases' on the set of phrases. The chosen representative phrase is the one with highest membership degree in such fuzzy set. As an example, this method is applied to find the representative phrase of the musical score shown in Figure 1.

Fig. 1. A few phrases of Invention # 1 of J. Bach

This chapter is organized as follows. Section 2 includes some preliminaries about fuzzy logic and decision making in uncertain environments with imprecise information. Also the concepts of proximity and similarity which are basis for the proposed searching methodology are defined. Section 3 presents the concept of specificity measurement that is used to evaluate the reliability of results produced by the decision method. The definitions that are used to compute the distance between consecutive notes of a phrase are described in section 4. It is also proven that the negation of a distance is a W-indistinguishability operator. This is applied to the variations of consecutive notes for each couple of phrases in order to compute the proximity on the set of phrases. It is then possible to obtain how similar the phrases sound. The experimental procedure, step by step experiments and results are explained in sections 5 to 8. An example to show how the algorithm for searching musical motifs performs starts in section 5. The details on how to build a proximity relationship on the set of phrases using 3 different t-norms and the OWA operator with 3 levels of tolerance are shown in section 6, in order to compare different aggregations' operators.

Section 7 details the method to choose a representative phrase, and finally, section 8 includes the last step of the application related to the calculation of the specificity measurement. The chapter finalizes with the conclusions in section 9.

2   Fuzzy Logic in Decision Making with Uncertainty

Fuzzy logic is useful when dealing with vague, uncertain, and complex environments. The imprecise information that characterizes the elements of a universe can be interpreted as a linguistic variable and modeled with fuzzy sets.

Given a universe of discourse E, a fuzzy set [13] is a mapping μ: E $\rightarrow$ [0, 1] gives a membership degree to every element of E in the interval [0, 1].

A semantic label is assigned to this fuzzy set and its membership degree is used to measure a characteristic of the elements of the universe E.

It is also well known that an algebra on fuzzy sets allows to define an extension of the logic operators AND, OR, and NOT, using triangular norms (t-norms), triangular conorms (t-conorms) and negation operators respectively [7]. The t-norm can be defined as follows [9][10],

For all x, y, z in [0,1], a binary operation T: [0, 1] $\times$ [0, 1] $\rightarrow$ [0, 1] is a t-norm if it satisfies the following axioms:

$$T(1, x) = x, T(0, x) = 0 \text{ for all x in } [0, 1] \tag{1}$$

$$T(x, y) = T(y, x) \text{ –symmetry-} \tag{2}$$

$$T(x, T(y, z)) = T(T(x, y), z) \text{ –associativity-} \tag{3}$$

$$\text{If } x \leq x' \text{ and } y \leq y' \text{ then } T(x, y) \leq T(x', y') \text{ –monotonicity-} \tag{4}$$

The t-conorm operator can be defined in a similar way, but having S(1, x) = 1, S(x, 0) = x, and similar axioms (2), (3), and (4).

The most common continuous logic operators are shown in Table 1.

Table 1. Most used t-norms and t-conorms in fuzzy logic

Logic	t-norm	t-conorm
Zadeh	min(x,y)	max(x,y)
Product	x*y	x + y - xy
Łukasiewicz	max(0, x+y-1)	min(1, x+y)

The usual negation operator is defined as a mapping $N: [0, 1] \to [0, 1]$ with $N(x) = 1-x$ for any $x \in [0, 1]$. It is possible to define the fuzzy set "NOT A" from a fuzzy set "A" and a negation operator N as follows:

$$\mu_{\text{NOT A}}(x) = N(\mu_A(x)) \text{ for every x in E.} \tag{5}$$

When it is necessary to gather several concepts, informations or fuzzy sets in a single fuzzy set, aggregation operators are useful. For example, operators between t-norms and t-conorms, such as averages, are aggregation operators [2].

On the other hand, fuzzy relations $R: E \times E \to [0, 1]$ have many applications to make fuzzy inference with uncertain, imprecise, or incomplete knowledge. For example, it can be used to model similarity or implication relations in order to generate rules and therefore to infer some conclusions [13]. A fuzzy rule can be expressed as,

IF "x is A" THEN "y is B"

where A and B are fuzzy sets (may be defined as a conjunction, aggregations or disjunction of other fuzzy sets), and x and y are measurable variables of the elements of the universe E. The main characteristic of the approximate reasoning is that from the knowledge of "x is A' ", for example 'x is almost A', and the rule stated above, and by applying the compositional rule of inference it is possible to learn that "y is B' " [14], for example, inferring that "x is almost B".

2.1   Proximity and Similarity

Let $E = \{e1, ..., en\}$ be a finite set and R a fuzzy relation on E. The relation degree for every pair of elements ei and ej in E is denoted eij. So $eij = R(ei, ej)$. Some common fuzzy relations properties are the following:

- A fuzzy relation R is reflexive if $e_{ii} = 1$ for all $1 \leq i \leq n$.
- A fuzzy relation R is α -reflexive if $e_{ii} \geq \alpha$, for all $1 \leq i \leq n$.
- The relation R is symmetric if $e_{ij} = e_{ji}$ for all $1 \leq i, j \leq n$.

A reflexive and symmetric fuzzy relation is called a fuzzy **proximity** relation.

Let T be a t-norm [9]. A fuzzy relation $R: E \times E \to [0, 1]$ is **T-transitive** if and only if $T(R(a, b), R(b, c)) \leq R(a, c)$ for every a, b, c in E. So R is T transitive if $T(e_{ik}, e_{kj}) \leq e_{ij}$ for every $1 \leq i, j, k \leq n$.

A **T-indistinguishability** relation is a reflexive, symmetric and T-transitive fuzzy relation.

Finally, a fuzzy **similarity** is a reflexive, symmetric and min-transitive fuzzy relation.

Similarities are therefore particular cases of T-indistinguishabilities, where the t-norm is the minimum [11].

It is possible to establish the following proposition: Let d be a normalized distance in E. Let $N: [0, 1] \to [0, 1]$ be the usual negation operator. Then $S: E \times E \to [0, 1]$ defined as $S(x, y) = N(d(x, y))$ is a **W-indistinguishability**, where W is the Łukasiewicz t-norm.

Proof

If d is a distance,

$$d(x, x) = 0 \text{ for all } x \text{ in } E \tag{6}$$

$$d(x, y) = d(y, x) \text{ for all } x, y \text{ in } E. \tag{7}$$

$$d(x, y) \leq d(x, z) + d(z, y) \text{ (triangular inequality)} \tag{8}$$

Then S is a W-indistinguishability because:

$$S(x, x) = N(d(x, x)) = N(0) = 1 \text{ for all } x \text{ in } E \text{ (S is reflexive)} \tag{9}$$

$$S(x, y) = N(d(x, y)) = N(d(y, x)) = S(y, x) \text{ for all } x, y \text{ in } E \text{ (symmetry)} \tag{10}$$

$$\text{By the triangular inequality of } d, d(x, y) \leq W^*(d(x, z), d(z, y)) \tag{11}$$

where W^* is the Łukasiewicz t-conorm, that is, the dual t-conorm of W [9]. Then applying the operator N,

$$N(d(x, y)) \geq N(W^*(d(x, z), d(z, y))) = W(N(d(x, z)), N(d(z, y))) \tag{12}$$

and then

$$S(x, y) \geq W(S(x, z), S(z, y)) \text{ for all } x, y, z \text{ in } E. \tag{13}$$

Then S is W-transitive and therefore is a W-indistinguishability.

3 Measure of Specificity on Fuzzy Sets

The concept of specificity provides a measurement of the degree of having just one element in a fuzzy set or a possibility distribution. It is strongly related to the inverse of the cardinality of a set.

Specificity values were introduced by Yager showing their usefulness as a measure of "tranquility" when making a decision. Yager stated the specificity-correctness tradeoff principle. The output information of an expert system or any other knowledge based system should be both specific and correct if it is to be useful. Yager suggested the use of specificity in default reasoning, in possibility qualified statements and in data mining processes, giving several possible manifestations of this measure.

Let X be a set with elements $\{x_i\}$ and let $[0, 1]^X$ be the class of fuzzy sets of X. A measure of specificity [12] is a mapping Sp: $[0, 1]^X \rightarrow [0, 1]$ such that:

$$Sp(\mu) = 1 \text{ if and only if } \mu \text{ is a singleton } (\mu = \{x_1\}) \tag{14}$$

$$Sp(\varnothing) = 0 \tag{15}$$

$$\text{If } \mu \text{ and } \eta \text{ are normal fuzzy sets in X and } \mu \subset \eta, \text{ then } Sp(\mu) \geq Sp(\eta) \tag{16}$$

Given a measurement of specificity Sp on a fuzzy set and given a T-indistinguishability S, the expression $Sp(\mu/S)$ is a **measure of specificity under T-indistinguishabilities** [6] when it verifies the following four axioms:

$$Sp(\{x\} / S) = 1 \tag{17}$$

$$Sp(\varnothing / S) = 0 \tag{18}$$

$$Sp(\mu / Id) = Sp(\mu) \tag{19}$$

$$Sp(\mu / S) \geq Sp(\mu) \tag{20}$$

Yager introduced the linear measurement of specificity on a finite space X as:

$$Sp(\mu) = a_1 - \sum_{j=2}^{n} w_j a_j \tag{21}$$

where a_j is the j_{th} greatest membership degree of μ and $\{w_j\}$ is a set of weights that verifies,

$$w_j \in [0, 1] \sum_{j=2}^{n} w_j = 1 \tag{22}$$

$w_j \geq w_i$ for all $1 < j < i$.

This is the operator which is going to be applied to musical composition to know how useful the decision of choosing a phrase is. If there is only one representative phrase, the specificity of the fuzzy set "similar to other phrases" is maximal. By adding some more information on the proximity on phrases, some groups of similar phrases can be clustered in just one phrase.

4   Intelligent Algorithm for Searching Musical Motifs

A first step to find musical motifs is to separate a musical score into phrases. Then, the phrases are compared to each other in order to evaluate the proximity degree of every couple of phrases. From the proximity and the concept "similar to other phrases' defined in a fuzzy set, it is possible to identify the motifs from the set of candidate phrases [5].

An approximation of the pre-searching method described by [8] is used as a starting point for searching the musical motifs. The algorithm in pseudo code can be written as:

a) A score is separated into phrases.
b) The proximity degree of every couple of phrases is computed.

c) A fuzzy set *'candidate to be a motif'* is computed on the set of phrases by aggregating the proximity degree of each phrase with other phrases.
d) The most representative phrase is the one with highest membership degree on the fuzzy set 'candidate to be a motif'.

The selection of the set of phrases can be done in different ways. The whole process depends on the chosen way of separation of phrases.

The pre-searching method takes into account the following criteria:

    1)   The variation of tones into a phrase.
    2)   The distance of the intervals between notes into a phrase.

The following definitions are given to establish a notation and a description of the given methodology.

4.1   Phrases and Variation Points

A phrase of a score is a sorted set of notes. For each couple of consecutive notes there is a measurable variation. It is possible to define a point for each couple of consecutive notes represented by (x, y).

A variation point for a note is the pair $p_i = $ [tone, variation], where the tone is represented using the positive integer in the standard MIDI for the respective note, and the variation is the difference between two tones (number of semitones of difference with the next note).

4.2   Distance between Ordered Notes

Let $P_n = \{ p_1, p_2, \ldots, p_n\}$ be a set of notes, that is, a musical phrase.

A function that computes the n-1 distances between consecutive notes is defined as follows:

$$D(P_n) = [\ d(p_1, p_2\ ),\ d(\ p_2, p_3\ ),\ \ldots, d(\ p_{n-1}, p_n\ )\ ] \tag{23}$$

where d is a distance, for example, the Euclidean distance.

4.3   A W-Indistinguishability S of Consecutive Notes

Let Sr be a W-indistinguishability operator $Sr : R \times R \rightarrow [0,1]$ defined by

$$Sr\ (a\ ,\ b) = (\ r_{max} - d(\ |\ a - b\ |\ )\ /\ r_{max}\ \text{where the range}\ r_{max}\ \text{is}\ |\ a_{max} - b_{max}\ | \tag{24}$$

A function of real numbers $Sv: R_n \times R_n \rightarrow R_n$ is used to compute the W-indsitinguishabilities between the (n-1) variation points of two phrases X and Y. It is defined as follows,

$$Sv(\{x_1, x_2, \ldots, x_{n1}\}, \{y_1, y_2, \ldots y_{n1}\}) = \{Sr(x_1, y_1), Sr(x_2, y_2),\ \ldots,\ Sr(x_n,\ y_n)\ \} \tag{25}$$

To define a proximity degree S: $P_n \times P_{n'} \rightarrow [0,1]$ of two musical phrases P_n and P_n, a conjunction operator T on the W-indistinguishability degrees of the distances between variation points given by D is computed as follows.

$$S (P_{ni}, P_{nj}) = T(Sv (D(P_{ni}), D(P_{nj}))) \tag{26}$$

Note that T is an n-ary t-norm operator (a conjunction operator) defined from a binary t-norm through the associative property:

$$T(x_1, x_2, ..., x_n) = T(x_1, T(x_2, T(..., T(..., x_n)))) \tag{27}$$

4.4   Choosing Operators for Different Meanings of "Representative Phrases"

Once a proximity relationship on the set of phrases is obtained, it is necessary to translate it into musical concepts. The way in which a fuzzy set describing the representative concept on the set of phrases based on the proximity relation can be done using different operators. Depending on how is it understood the concept of representative phrase:

- If a phrase is representative when it is similar to ALL other phrases, it is possible to define the representativeness of a phrase by using a conjunction of the proximity values with the rest of phrases through a t-norm. So a phrase is representative when it is similar to a phrase 1 AND it is similar to phrase 2 AND…it is similar to phrase n.
- If the phrase is representative when it is similar to ANY other phrase, then it is possible to use a disjunction, for example, the MAX t-conorm. So a phrase is representative when it is similar to a phrase 1 OR it is similar to phrase 2 OR…it is similar to phrase n (all other phrases but itself).
- If the phrase is representative when it is similar to SOME other phrase, it is possible to use aggregation operators, which are in between conjunction and disjunctions. For example, an Ordered Weighted Averaging function (OWA), or an average.

R. R. Yager´s Ordered weighted averaging functions (OWAs) are also a class of averaging aggregation functions. The difference in the weighted arithmetic mean is that the weights are not associated with the inputs but with the magnitude. In some applications all inputs are equivalent, and the importance of one specific input is determined by its absolute value.

5   Experiments and Results

In this section, a step by step example is developed to show how the proposed algorithm performs. The eight first notes of Figure 1 are considered to evaluate if there is a possible motif. The score is divided into phrases of 8 notes.

The first phrase P^n_1 includes the first 8 notes in the score; the initial silence is omitted. The rest of the phrases P^n_i are formed taking 8 consecutive notes starting from the 9th note in the superior line (first voice); then it is shifted a note for every phrase for the two voices separately until there are 10 more phrases. All the resulted phrases after the normalization of the durations of their notes are shown in Figure 2.

Note that in same way that the models described in [8], this method is concerned only with pitch, without taking into account the real duration of the notes in order to simplify the pre-searching. On the other hand, based on the combination of tools and techniques, this method is capable of handling inversion, retrogradation and inverted retrogradation of motifs.

Fig. 2. Normalized phrases of invention #1

The 8 notes represented by their scale and duration of the 11 phrases are:

$$P^n_1 = [\ (C5 \ , \ 2) \ , \ (D5, \ 2) \ , \ (E5 \ , \ 1) \ , \ (F5 \ , \ -3) \ , \ (D5, \ 2) \ , \ (E5 \ , \ -4) \ , \ (C5 \ , \ 7) \ , \ (G5 \ , \ 0) \ ]$$
$$P^n_2 = [\ (C6 \ , \ -1) \ , \ (B5 \ , \ 1) \ , \ (C6 \ , \ 2) \ , \ (D6, \ -7) \ , \ (G5, \ 2) \ , \ (A5, \ 2) \ , \ (B5 \ , \ 1) \ , \ (C6 \ , \ -3) \ ]$$
$$P^n_3 = [\ (B5 \ , \ 1) \ , \ (C6 \ , \ 2) \ , \ (D6, \ -7) \ , \ (G5, \ 2) \ , \ (A5, \ 2) \ , \ (B5 \ , \ 1) \ , \ (C6 \ , \ -3) \ , \ (A5 \ , \ 0) \ ]$$
$$P^n_4 = [\ (C6 \ , \ 2) \ , \ (D6, \ -7) \ , \ (G5, \ 2) \ , \ (A5, \ 2) \ , \ (B5 \ , \ 1) \ , \ (C6, \ -3) \ , \ (A5 \ , \ 2) \ , \ (B5 \ , \ 0) \ ]$$
$$P^n_5 = [\ (D6 \ , \ -7) \ , \ (G5, \ 2) \ , \ (A5 \ , \ 2) \ , \ (B5 \ , \ 1) \ , \ (C6, \ -3) \ , \ (A5 \ , \ 2) \ , \ (B5 \ , \ -4) \ , \ (G5 \ , \ 0) \ ]$$
$$P^n_6 = [\ (G5 \ , \ 2) \ , \ (A5, \ 2) \ , \ (B5 \ , \ 1) \ , \ (C6, \ -3) \ , \ (A5, \ 2) \ , \ (B5 \ , \ -4) \ , \ (G5 \ , \ 7) \ , \ (D6 \ , \ 0) \ ]$$
$$P^n_7 = [\ (C4 \ , \ 2) \ , \ (D4, \ 2) \ , \ (E4 \ , \ 1) \ , \ (F4 \ , \ -3) \ , \ (D4, \ 2) \ , \ (E4, \ -4) \ , \ (C4 \ , \ 7) \ , \ (G4 \ , \ 0) \ ]$$
$$P^n_8 = [\ (D4 \ , \ 2) \ , \ (E4 \ , \ 1) \ , \ (F4 \ , \ -3) \ , \ (D4, \ 2) \ , \ (E4, \ -4) \ , \ (C4 \ , \ 7) \ , \ (G4 \ , \ -12 \ , \ (G3 \ , \ 0) \ ]$$
$$P^n_9 = [\ (E4 \ , \ 1) \ , \ (F4 \ , \ -3), \ (D4, \ 2) \ , \ (E4, \ -4) \ , \ (C4, \ 7) \ , \ (G4, \ -12, \ (G3, \ -) \ , \ (- \ , \ 0) \ ]$$
$$P^n_{10} = [\ (F4 \ , \ -3), \ (D4, \ 2) \ , \ (E4, \ -4) \ , \ (C4, \ 7) \ , \ (G4, \ -12), \ (G3, -) \ , \ (- \ , \ -) \ , \ (- \ , \ 0) \ ]$$
$$P^n_{11} = [\ (D4 \ , \ 2) \ , \ (E4, \ -4), \ (C4, \ 7) \ , \ (G4, \ -12), \ (G3, -) \ , \ (- \ , \ -) \ , \ (- \ , \ -) \ , \ (G4 \ , \ 0) \ ]$$

In this case, only eleven of the initial phrases for the two voices are analyzed in order to simplify the example and to show some details. This procedure can be applied to the whole musical score (more than 400 phrases), using different lengths to find the best motif.

The variation points of the eleven normalized phrases of Invention # 1 of J. Bach (Figure 2) are represented in the following matrix using the traditional musical notation for the different classes of pitch, with the first seven letters of the Latin alphabet and a number after the letter which represents the octave. The selected phrases include notes between 4th and 6th octaves.

The n-1 distances $D(P^n)$ between the variation points of every phrase are calculated and shown in the following Table 2:

Table 2. Distance $D(P^n)$ between notes of every phrase

$D(P^n{}_0)$ = [	2,00	2,24	4,12	5,83	6,32	11,70	15,65]
$D(P^n{}_1)$ = [	2,24	1,41	9,22	11,40	2,00	2,24	1,41]
$D(P^n{}_2)$ = [	1,41	9,22	11,40	2,00	2,24	4,12	5,83]
$D(P^n{}_3)$ = [	9,22	11,40	2,00	2,24	4,12	5,83	2,24]
$D(P^n{}_4)$ = [	11,40	2,00	2,24	4,12	5,83	6,32	11,70]
$D(P^n{}_5)$ = [	2,00	2,24	4,12	5,83	6,32	11,70	15,65]
$D(P^n{}_6)$ = [	2,00	2,24	4,12	5,83	6,32	11,70	15,65]
$D(P^n{}_7)$ = [	2,24	4,12	5,83	6,32	11,70	20,25	22,47]
$D(P^n{}_8)$ = [	4,12	5,83	6,32	11,70	20,25	16,97	9,00]
$D(P^n{}_9)$ = [	5,83	6,32	11,70	20,25	16,97	0,00	10,00]
$D(P^n{}_{10})$ = [	6,32	11,70	20,25	16,97	0,00	0,00	7,00]

The next step is to normalize Table 2 using the maximum distance, which in this case is 22.47 (see row 8, last column). The normalized table is shown in Table 3.

Table 3. Normalized distances D(Pn) between notes of every phrase

Phrase	Normalized Values						
1	0,09	0,1	0,18	0,26	0,28	0,52	0,7
2	0,1	0,06	0,41	0,51	0,09	0,1	0,06
3	0,06	0,41	0,51	0,09	0,1	0,18	0,26
4	0,41	0,51	0,09	0,1	0,18	0,26	0,1
5	0,51	0,09	0,1	0,18	0,26	0,28	0,52
6	0,09	0,1	0,18	0,26	0,28	0,52	0,7
7	0,09	0,1	0,18	0,26	0,28	0,52	0,7
8	0,1	0,18	0,26	0,28	0,52	0,9	1
9	0,18	0,26	0,28	0,52	0,9	0,76	0,4
10	0,26	0,28	0,52	0,9	0,76	0	0,45
11	0,28	0,52	0,9	0,76	0	0	0,31

The next step is to compute the W-indistinguishability of every phrase with respect to the other 10 phrases. This is obtained by computing the distance of every variation point of the phrases (values of Table 3) and the other 10 rows, and applying the usual negation operator $N(x) = 1 - x$ of distances, to obtain W-indistinguishabilities between variation points of different phrases.

For example, Table 4 shows the proximity of phrase 6 with the rest of the phrases using different continuous conjunction operators (t-norms of Table 1). Every value in the rows of Table 4 is calculated using the negation operator $N(x) = 1 - x$ to the subtraction of 2 values from Table 3. Those are obtained by comparing phrase by phrase and variation point by variation point for each one of the eleven phrases. For example, the value in column 1 and row 5 of Table 4 (0.58) is computed using the negation $N(x) = 1 - x$, where $x = (0.09-0.51)$, normalized values of the first variations points of phrases 6 and 5 respectively (Table 3).

The last 3 columns of Table 4 are the values for every t-norm of Table 1 (product, minimum, Łukasiewicz) applied to each one of the rows. For example, the value obtained by the t-norm product for the phrase 5 (0.299) is the product of $(0.58 * 0.99 * 0.92 * 0.92 * 0.98 * 0.76 * 0.82)$, that is, the multiplication of all the values of the corresponding row.

Table 4. W-indistinguishabilities of the variation points of Phrase 6 with the other phrases

	Phrase 6							product	min	w
1	1	1	1	1	1	1	1	1	1	1
2	0,99	0,96	0,77	0,75	0,81	0,58	0,37	0,095	0,366	0,000
3	0,97	0,69	0,68	0,83	0,82	0,66	0,56	0,115	0,563	0,000
4	0,68	0,59	0,91	0,84	0,9	0,74	0,4	0,082	0,403	0,000
5	0,58	0,99	0,92	0,92	0,98	0,76	0,82	0,299	0,582	0,000
7	1	1	1	0,92	1	1	1	1	1	1
8	0,99	0,92	0,92	0,92	0,76	0,62	0,7	0,254	0,620	0,000
9	0,91	0,84	0,9	0,92	0,38	0,77	0,7	0,130	0,380	0,000
10	0,83	0,82	0,66	0,92	0,53	0,48	0,75	0,078	0,479	0,000
11	0,81	0,58	0,28	0,92	0,72	0,48	0,61	0,026	0,282	0,000

6   Building a Proximity Relationship on the Set of Phrases

Two phrases can be considered 'similar' when the variation between the first and the second notes are 'similar', AND the variation between the second and the third notes are 'similar', AND ..., so on and so forth. Such concept of 'similar' is replaced by 'W-indistinguible' in this paper's proposal.

The calculation of the conjunction of W-indistinguishabilities of the variation points for each phrase regarding the others defines a proximity relationship on the set of phrases. The final values of the proximity on the set of phrases are shown in Tables 5, 6 and 7 where the conjunction (AND) is implemented by the three different t-norms of Table 1. Tables 8, 9 and 10 also show the proximity values using the OWA operator with different percentages. Table 8 shows the OWA operator at 85%, taking out the least significant variation point, Table 9 shows the OWA operator at 71% taking out the two least significant variation points, and finally, Table 10 shows the OWA operator at 57%, taking out the three least significant variation points.

These three cases are equivalent to using a vector of weighs, $W_i = 1/6$, for the 6 highest membership degrees values of each one of the phrases. The second case would correspond to using a vector of weighs $W_i = 1/5$ for the 5 highest membership degree values of every phrase, and the last case uses a vector $W_i = 1/4$ for the 4 highest membership degree values of every phrase.

Table 5. Proximity of phrases using the t-norm minimum

	1	2	3	4	5	6	7	8	9	10	11
1	1	0,366	0,563	0,403	0,582	1	1	0,620	0,380	0,358	0,282
2	0,366	1	0,582	0,556	0,542	0,366	0,366	0,063	0,188	0,334	0,509
3	0,563	0,582	1	0,582	0,556	0,563	0,563	0,259	0,198	0,188	0,334
4	0,403	0,556	0,582	1	0,579	0,403	0,403	0,100	0,282	0,198	0,188
5	0,582	0,542	0,556	0,579	1	0,582	0,582	0,380	0,358	0,282	0,198
6	1	0,366	0,563	0,403	0,582	1	1	0,620	0,380	0,358	0,282
7	1	0,366	0,563	0,403	0,582	1	1	0,620	0,380	0,358	0,282
8	0,620	0,063	0,259	0,099	0,380	0,620	0,620	1	0,400	0,099	0,099
9	0,380	0,188	0,198	0,282	0,358	0,380	0,380	0,400	1	0,245	0,099
10	0,358	0,334	0,188	0,198	0,282	0,358	0,358	0,099	0,245	1	0,245
11	0,282	0,509	0,334	0,188	0,198	0,282	0,282	0,099	0,099	0,245	1

The proximity values using the t-norm minimum are in the range of 0.063 to 1, with an average value of 0.46; t-norm minimum gives the highest values in the set of proximities.

Table 6. Proximity of phrases using the t-norm product

	1	2	3	4	5	6	7	8	9	10	11
1	1	0,095	0,115	0,082	0,299	1	1	0,269	0,104	0,030	0,014
2	0,095	1	0,240	0,113	0,099	0,095	0,095	0,004	0,027	0,066	0,105
3	0,115	0,240	1	0,241	0,113	0,115	0,115	0,019	0,024	0,030	0,098
4	0,082	0,113	0,241	1	0,249	0,082	0,082	0,007	0,027	0,015	0,027
5	0,299	0,099	0,113	0,249	1	0,299	0,299	0,059	0,050	0,033	0,016
6	1	0,095	0,115	0,082	0,299	1	1	0,269	0,104	0,030	0,014
7	1	0,095	0,115	0,082	0,299	1	1	0,269	0,104	0,030	0,014
8	0,269	0,004	0,019	0,007	0,059	0,269	0,269	1	0,133	0,007	0,002
9	0,104	0,027	0,024	0,027	0,050	0,104	0,104	0,133	1	0,085	0,004
10	0,030	0,066	0,030	0,015	0,033	0,030	0,030	0,007	0,085	1	0,084
11	0,014	0,105	0,098	0,027	0,016	0,014	0,014	0,002	0,004	0,084	1

The proximity values using the t-norm product are in the interval 0.002 to 1 with an average value of 0.23; t-norm product gives medium values of proximities.

Table 7. Proximity of phrases using the t-norm Lukasiewicz

	1	2	3	4	5	6	7	8	9	10	11
1	1	0,000	0,000	0,000	0,000	1	1	0,000	0,000	0,000	0,000
2	0,000	1	0,000	0,000	0,000	0,000	0,000	0,000	0,000	0,000	0,000
3	0,000	0,000	1	0,000	0,000	0,000	0,000	0,000	0,000	0,000	0,000
4	0,000	0,000	0,000	1	0,000	0,000	0,000	0,000	0,000	0,000	0,000
5	0,000	0,000	0,000	0,000	1	0,000	0,000	0,000	0,000	0,000	0,000
6	1	0,000	0,000	0,000	0,000	1	1	0,000	0,000	0,000	0,000
7	1	0,000	0,000	0,000	0,000	1	1	0,000	0,000	0,000	0,000
8	0,000	0,000	0,000	0,000	0,000	0,000	0,000	1	0,000	0,000	0,000
9	0,000	0,000	0,000	0,000	0,000	0,000	0,000	0,000	1	0,000	0,000
10	0,000	0,000	0,000	0,000	0,000	0,000	0,000	0,000	0,000	1	0,000
11	0,000	0,000	0,000	0,000	0,000	0,000	0,000	0,000	0,000	0,000	1

The proximity values calculated by the t-norm of Lukasiewicz are only 0s and 1s. It gives the lowest values of proximities.

Table 8. Proximity of phrases using OWA 85%

	1	2	3	4	5	6	7	8	9	10	11
1	1	0,579	0,663	0,592	0,761	1	1	0,697	0,704	0,479	0,479
2	0,579	1	0,653	0,592	0,592	0,579	0,579	0,198	0,344	0,606	0,542
3	0,663	0,653	1	0,653	0,592	0,663	0,663	0,282	0,428	0,344	0,606
4	0,592	0,592	0,653	1	0,582	0,592	0,592	0,358	0,504	0,428	0,344
5	0,761	0,592	0,592	0,582	1	0,761	0,761	0,521	0,526	0,504	0,428
6	1	0,579	0,663	0,592	0,761	1	1	0,697	0,704	0,479	0,479
7	1	0,579	0,663	0,592	0,761	1	1	0,697	0,704	0,479	0,479
8	0,697	0,198	0,282	0,358	0,521	0,697	0,697	1	0,620	0,380	0,311
9	0,704	0,344	0,428	0,504	0,526	0,704	0,704	0,620	1	0,620	0,245
10	0,479	0,606	0,344	0,428	0,504	0,479	0,479	0,380	0,620	1	0,620
11	0,479	0,542	0,606	0,344	0,428	0,479	0,479	0,311	0,245	0,620	1

The proximity values obtained by applying the OWA 85% are in the range of 0.198 to 1, with an average of 0.608. It eliminates the first level of the lowest values.

Table 9. Proximity of phrases using OWA 71%

	1	2	3	4	5	6	7	8	9	10	11
1	1	0,752	0,676	0,679	0,824	1	1	0,761	0,739	0,526	0,504
2	0,752	1	0,803	0,679	0,676	0,752	0,752	0,568	0,662	0,618	0,751
3	0,676	0,803	1	0,840	0,679	0,676	0,676	0,579	0,568	0,803	0,781
4	0,679	0,679	0,840	1	0,903	0,679	0,679	0,663	0,579	0,568	0,741
5	0,824	0,676	0,679	0,903	1	0,824	0,824	0,592	0,663	0,579	0,568
6	1	0,752	0,676	0,679	0,824	1	1	0,761	0,739	0,526	0,504
7	1	0,752	0,676	0,679	0,824	1	1	0,761	0,739	0,526	0,504
8	0,761	0,568	0,579	0,663	0,592	0,761	0,761	1	0,761	0,445	0,358
9	0,739	0,662	0,568	0,579	0,663	0,739	0,739	0,761	1	0,761	0,380
10	0,526	0,618	0,803	0,568	0,579	0,526	0,526	0,445	0,761	1	0,761
11	0,504	0,751	0,781	0,741	0,568	0,504	0,504	0,358	0,380	0,761	1

The proximity values using the OWA operator at 71% gives values between 0.36 and 1 with an average value of 0.71, OWA 71%. This operator eliminates the first and second levels of the lowest values.

Table 10. Proximity of phrases using OWA 57%

	1	2	3	4	5	6	7	8	9	10	11
1	1	0,773	0,689	0,739	0,916	1	1	0,916	0,766	0,663	0,579
2	0,773	1	0,903	0,689	0,689	0,773	0,773	0,774	0,803	0,781	0,752
3	0,689	0,903	1	0,903	0,739	0,689	0,689	0,752	0,774	0,814	0,817
4	0,739	0,689	0,903	1	0,916	0,739	0,739	0,676	0,699	0,655	0,788
5	0,916	0,689	0,739	0,916	1	0,916	0,916	0,739	0,676	0,719	0,719
6	1	0,773	0,689	0,739	0,916	1	1	0,916	0,766	0,663	0,579
7	1	0,773	0,689	0,739	0,916	1	1	0,916	0,766	0,663	0,579
8	0,916	0,774	0,752	0,676	0,739	0,916	0,916	1	0,854	0,739	0,479

Table 10. (*continued*)

9	0,766	0,803	0,774	0,699	0,676	0,766	0,766	0,854	1	0,854	0,739
10	0,663	0,781	0,814	0,655	0,719	0,663	0,663	0,739	0,854	1	0,854
11	0,579	0,752	0,817	0,788	0,719	0,579	0,579	0,479	0,739	0,854	1

The proximity values obtained by OWA 57% are in the range of 0.48 to 1 with an average value of 0.79, OWA 57%. This operator eliminates the first, second, and third levels of the lowest values.

7   A Method to Choose a Representative Phrase

Once all the previous measurements have been calculated, the proposed method tries to find the representative phrases from the information of Tables 5, 6, 7, 8, 9 and 10. By the aggregation of every row in the proximity matrix and by using the arithmetic mean, a fuzzy set: '*proximity with the rest of phrases*' is defined on the set of phrases. Then, the phrase or phrases with certain membership degree (that exceed a threshold), are chosen as the most representative phrases. The normalized mean values are presented in Tables 11 and 12 and Figures 3 and 4.

Table 11. Fuzzy set "proximity with other phrases" on the set of phrases using t-norms

Phrase	Avg Product		Avg Min		Avg W	
1	0,301	1,000	0,555	1,000	0,200	1,000
2	0,094	0,312	0,387	0,697	0,000	0,000
3	0,111	0,369	0,439	0,790	0,000	0,000
4	0,093	0,308	0,369	0,665	0,000	0,000
5	0,152	0,504	0,464	0,835	0,000	0,000
6	0,301	1,000	0,555	1,000	0,200	1,000
7	0,301	1,000	0,555	1,000	0,200	1,000
8	0,104	0,345	0,326	0,587	0,000	0,000
9	0,066	0,220	0,291	0,524	0,000	0,000
10	0,041	0,137	0,267	0,480	0,000	0,000
11	0,038	0,125	0,252	0,453	0,000	0,000

Table 12. Fuzzy set "proximity with other phrases" on the set of phrases using OWAs

Phrase	Avg OWA85%		Avg OWA71%		Avg OWA57%	
1	0,695	1,000	0,746	1,000	0,804	1,000
2	0,526	0,757	0,701	0,940	0,771	0,959
3	0,555	0,798	0,708	0,949	0,777	0,966
4	0,524	0,753	0,701	0,939	0,754	0,938
5	0,603	0,867	0,713	0,956	0,794	0,988
6	0,695	1,000	0,746	1,000	0,804	1,000
7	0,695	1,000	0,746	1,000	0,804	1,000
8	0,476	0,685	0,625	0,837	0,776	0,965
9	0,540	0,777	0,659	0,883	0,770	0,957
10	0,494	0,711	0,611	0,819	0,740	0,921
11	0,453	0,652	0,585	0,785	0,688	0,856

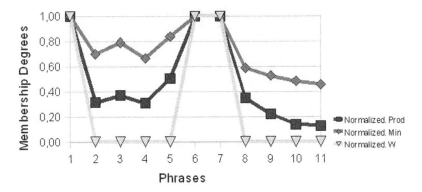

Fig. 3. Fuzzy set "proximity of every phrase i with the rest of phrases", using t-norms connectives

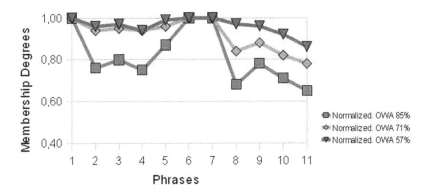

Fig. 4. Fuzzy set "proximity of every phrase i with the rest of phrases", using OWAs aggregations

It is possible to conclude that the representative phrases in this case are 1, 6, and 7, with mean values over 30%, 55% and 20% respectively for each t-norm. These phrases are shown in Figure 5.

It is also possible to identify the second set of representative phrases, 3 and 5, by looking at Tables 11 and 12. They are also representative with values over 40% and 10% for the t-norms Min and Product respectively. A musical representation of phrases 3 and 5 is shown in Figure 6.

The case using OWAs is similar to the t-norms, but in this case, the membership degrees for each phrase have undergone a big increment that is inversely proportional to the percentage of the OWA. This is because the elements with lower values are taken off from the calculations. In the case of OWA 85% the representative phrases are still 1, 6, and 7; phrases 3 and 5 are an additional subset of representative phrases with values over 79%. In the cases of OWA 71% and OWA 57% all of the values are over 78% as they tend to 1, so the important differences have been lost especially for the range of phrases between 1 and 7.

Fig. 5. A musical representation of phrases 1, 6 and 7

Fig. 6. A musical representation of phrases 3 and 5

Phrase 7 is descending an octave and phrase 6 is ascending in 7 semitones (see how they confirm proximity relations in Tables 5, 6, 7). On the other hand, phrases 3 and 5 have a high level of proximity, and in practice, it is easy to see that these phrases contain an important part of representative phrases 1, 6 and 7.

8   Computing the Specificity Measure of the Fuzzy Set "Similar to Other Phrases" and the Inference Independent Sets Using the Proximity on Phrases

After choosing the representative phrases, following the presented procedure, the next step consists in computing the specificity measure of every one of the fuzzy sets obtained in section 7 (Table 11 and Table 12). This is a mechanism to evaluate the decision's reliability from the perspective of the data that have been used in the selection of phrases. When the specificity is one, there is just one representative phrase to choose.

Figure 7 shows all the fuzzy sets obtained in section 7. Every fuzzy set is obtained by the aggregation of the proximity values of each one of the t-norm tables (Tables 5 to 7) and each one of the OWA tables (Tables 8 to 10).

Table 13 shows the 11 values for every one of the 6 fuzzy sets and their calculated values of specificity (last column), using the formula of lineal specificity (21).

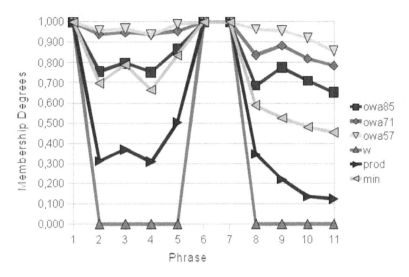

Fig. 7. Fuzzy set of values of proximity of phrases, OWA and t-norm cases

Table 13. Normalized membership degree "proximity with other phrases" on the set of phrases, OWA and t-norm cases

	1	2	3	4	5	6	7	8	9	10	11Sp
owa85	1,00	0,75	0,79	0,75	0,86	1,00	1,00	0,68	0,77	0,71	0,65	0,2
owa71	1,00	0,94	0,94	0,93	0,95	1,00	1,00	0,83	0,88	0,81	0,78	0,08
owa57	1,00	0,95	0,96	0,93	0,98	1,00	1,00	0,96	0,95	0,92	0,85	0,04
w	1,00	0,00	0,00	0,00	0,00	1,00	1,00	0,00	0,00	0,00	0,00	0,8
prod	1,00	0,31	0,36	0,30	0,50	1,00	1,00	0,34	0,22	0,13	0,12	0,56
min	1,00	0,69	0,79	0,66	0,83	1,00	1,00	0,58	0,52	0,48	0,45	0,29

The inference independent sets [6] aims to gather in one class all the similar phrases, so the decision is easier, as we now choose between a few cases representing some similar phrases. Those sets of phrases are calculated for every one of the initial fuzzy sets that are shown in Table 13. For each case, one of the proximity matrixes of Table 5 to 10 is used. The result is a table of 6 new fuzzy sets obtained from the original fuzzy sets that in some cases have changes in their values.

Table 14 is the result of computing the proximity values from Table 8 with the fuzzy sets in Table 13. This is how the inference independent sets for the OWA 85% case are obtained. A graphical view of the results is shown in Figure 8. These results should be compared with Figure 7. In this specific case, all the values begin with at least 25%. This is a special case where the fuzzy sets obtained by product t-norm and Lukasiewicz t-norm have the same membership degrees.

Table 14. New membership degree "proximity with other phrases" on the set of phrases for OWA 85%

	1	2	3	4	5	6	7	8	9	10	11	Sp
owa85	1,00	0,75	0,79	0,75	0,86	1,00	1,00	0,69	0,77	0,71	0,65	0,19
owa71	1,00	0,94	0,94	0,93	0,95	1,00	1,00	0,83	0,88	0,81	0,78	0,08
owa57	1,00	0,95	0,96	0,93	0,98	1,00	1,00	0,96	0,95	0,92	0,85	0,04
w	1,00	0,57	0,66	0,59	0,76	1,00	1,00	0,69	0,70	0,47	0,47	0,30
prod	1,00	0,57	0,66	0,59	0,76	1,00	1,00	0,69	0,70	0,47	0,47	0,30
min	1,00	0,69	0,79	0,66	0,83	1,00	1,00	0,69	0,70	0,48	0,47	0,26

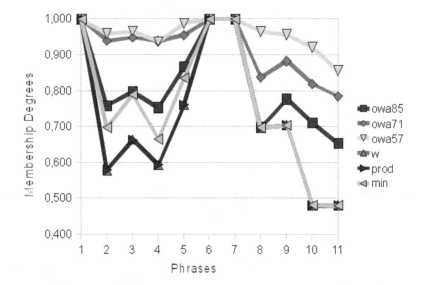

Fig. 8. New fuzzy sets of values of proximity under OWA 85% proximity

Table 15 and Figure 9 show the results of computing the proximity values from Table 9 with the fuzzy sets in Table 13 for the OWA 71% operator. In this case, all values are over 50%. Again, the fuzzy sets of product t-norm and Lukasiewicz t-norm have the same membership degrees.

Table 15. New membership degree "proximity with other phrases" on the set of phrases under OWA 71% proximity

	1	2	3	4	5	6	7	8	9	10	11	Sp
owa85	1,00	0,75	0,79	0,77	0,86	1,00	1,00	0,76	0,77	0,71	0,65	0,19
owa71	1,00	0,94	0,94	0,93	0,95	1,00	1,00	0,83	0,88	0,81	0,78	0,08
owa57	1,00	0,95	0,96	0,93	0,98	1,00	1,00	0,96	0,95	0,92	0,85	0,04
w	1,00	0,75	0,67	0,67	0,82	1,00	1,00	0,76	0,73	0,52	0,50	0,25
prod	1,00	0,75	0,67	0,67	0,82	1,00	1,00	0,76	0,73	0,52	0,50	0,25
min	1,00	0,75	0,79	0,73	0,83	1,00	1,00	0,76	0,73	0,59	0,57	0,22

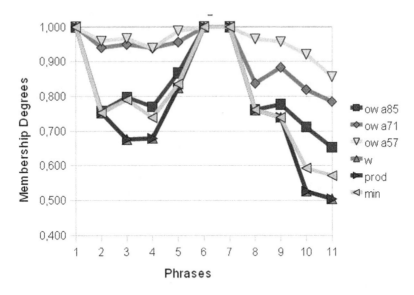

Fig. 9. New fuzzy sets of values of proximity under OWA 71% proximity

Table 16 and Figure 10 show the result of computing the proximity values from Table 10 with the fuzzy sets in Table 13 for the OWA 57% case. All the values are over 57%; again they are special cases where the fuzzy sets of product t-norm and Lukasiewicz t-norm have the same membership degrees.

Table 16. New membership degree "proximity with other phrases" on the set of phrases under OWA 57% proximity

	1	2	3	4	5	6	7	8	9	10	11	Sp
owa85	1,00	0,77	0,79	0,78	0,91	1,00	1,00	0,91	0,77	0,71	0,65	0,16
owa71	1,00	0,94	0,94	0,93	0,95	1,00	1,00	0,91	0,88	0,81	0,78	0,08
owa57	1,00	0,95	0,96	0,93	0,98	1,00	1,00	0,96	0,95	0,92	0,85	0,04
W	1,00	0,77	0,68	0,73	0,91	1,00	1,00	0,91	0,76	0,66	0,57	0,19
prod	1,00	0,77	0,68	0,73	0,91	1,00	1,00	0,91	0,76	0,66	0,57	0,19
min	1,00	0,77	0,79	0,75	0,91	1,00	1,00	0,91	0,76	0,66	0,60	0,18

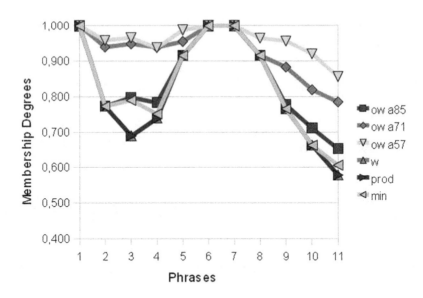

Fig. 10. New fuzzy sets of values of proximity under OWA 57% proximity

Table 17 and Figure 11 show the result of computing the proximity values from Table 7 with the fuzzy sets in Table 13 for the Lukasiewicz t-norm case. The values are still the same, distributed between 0 and 1.

Table 17. New membership degree "proximity with other phrases" on the set of phrases under Lukasiewicz t-norm % proximity

	1	*2*	*3*	*4*	*5*	*6*	*7*	*8*	*9*	*10*	*11*	*Sp*
owa85	1,00	0,75	0,79	0,75	0,86	1,00	1,00	0,68	0,77	0,71	0,65	0,20
owa71	1,00	0,94	0,94	0,93	0,95	1,00	1,00	0,83	0,88	0,81	0,78	0,08
owa57	1,00	0,95	0,96	0,93	0,98	1,00	1,00	0,96	0,95	0,92	0,85	0,04
W	1,00	0,00	0,00	0,00	0,00	1,00	1,00	0,00	0,00	0,00	0,00	0,80
prod	1,00	0,31	0,36	0,30	0,50	1,00	1,00	0,34	0,22	0,13	0,12	0,56
min	1,00	0,69	0,79	0,66	0,83	1,00	1,00	0,58	0,52	0,48	0,45	0,29

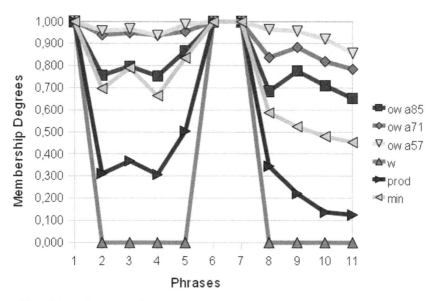

Fig. 11. New fuzzy sets of values of proximity under Lukasiewicz t-norm proximity

Table 18 shows the results of computing the proximity values from Table 6 with the fuzzy sets in Table 13 for the product t-norm proximity. The fuzzy sets are shown in Figure 12. The values are distributed between 0 and 1. Note that in this case the only fuzzy set that changes is the Lukasiewicz t-norm.

Table 18. New membership degree "proximity with other phrases" on the set of phrases under product t-norm % proximity

	1	2	3	4	5	6	7	8	9	10	11	Sp
owa85	1,00	0,75	0,79	0,75	0,86	1,00	1,00	0,68	0,77	0,71	0,65	0,20
owa71	1,00	0,94	0,94	0,93	0,95	1,00	1,00	0,83	0,88	0,81	0,78	0,08
owa57	1,00	0,95	0,96	0,93	0,98	1,00	1,00	0,96	0,95	0,92	0,85	0,04
W	1,00	0,09	0,11	0,08	0,29	1,00	1,00	0,26	0,10	0,03	0,01	0,69
prod	1,00	0,31	0,36	0,30	0,50	1,00	1,00	0,34	0,22	0,13	0,12	0,56
Min	1,00	0,69	0,79	0,66	0,83	1,00	1,00	0,58	0,52	0,48	0,45	0,29

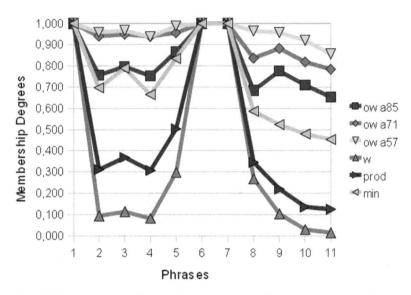

Fig. 12. New fuzzy sets of values of proximity under Product t-norm proximity

Table 19 and Figure 13 summarize the proximity values obtained from Table 5 with the fuzzy sets in Table 13 for the minimum t-norm proximity. The values in this case are distributed between 29% and 100%. Again, the fuzzy sets of product t-norm and Lukasiewicz t-norm have same membership degrees.

Table 19. New membership degree "proximity with other phrases" on the set of phrases under min t-norm % proximity

	1	*2*	*3*	*4*	*5*	*6*	*7*	*8*	*9*	*10*	*11*	*Sp*
owa85	1,00	0,75	0,79	0,75	0,86	1,00	1,00	0,68	0,77	0,71	0,65	0,20
owa71	1,00	0,94	0,94	0,93	0,95	1,00	1,00	0,83	0,88	0,81	0,78	0,08
owa57	1,00	0,95	0,96	0,93	0,98	1,00	1,00	0,96	0,95	0,92	0,85	0,04
W	1,00	0,36	0,56	0,40	0,58	1,00	1,00	0,62	0,38	0,35	0,28	0,44
prod	1,00	0,36	0,56	0,40	0,58	1,00	1,00	0,62	0,38	0,35	0,28	0,44
Min	1,00	0,69	0,79	0,66	0,83	1,00	1,00	0,62	0,52	0,48	0,45	0,29

Additional information is the number of times that every of the membership degrees has increased during each one of the inference independent sets calculation. It allows us to identify the most susceptible elements to be inferred from the proximity

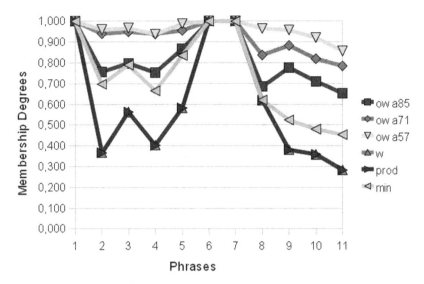

Fig. 13. New fuzzy sets of values of proximity under Minimum t-norm proximity

relations and other elements in the fuzzy set. Those elements are considered susceptible to be inferred because their increments are part of an inference independent set.

The phrase with the highest number of variations is phrase 8 with a total of 17 variations, followed by the set of phrases 2, 9 and 11 with 12 variations. The set composed by phrases 5 and 10 has 11 variations, and phrase 3 with just 9 variations. All of these are listed in Table 20.

Figure 14 is a graphical representation of the number of variations of every phrase for every case of inference independent set calculation process and an accumulate total of variations per phrase.

Table 20. Number of variations in the calculation of inference independent sets

Num of Variations	1	2	3	4	5	6	7	8	9	10	11
owa85	0	2	2	2	2	0	0	4	3	2	3
owa71	0	3	2	4	2	0	0	4	3	3	3
owa57	0	4	2	4	4	0	0	5	3	3	3
w	0	0	0	0	0	0	0	0	0	0	0
pro	0	1	1	1	1	0	0	1	1	1	1
min	0	2	2	2	2	0	0	3	2	2	2
Total	0	12	9	13	11	0	0	17	12	11	12

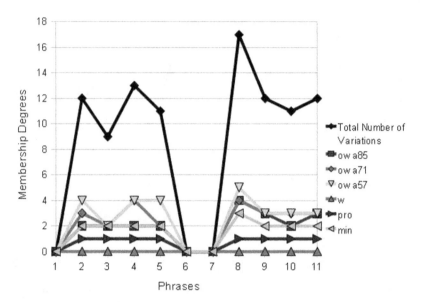

Fig. 14. Variations in membership degrees during inference independent sets calculation

9   Conclusions and Remarks

A method to search musical representative phrases using a W-indistinguishability operator and fuzzy proximity relations on a set of phrases is proposed and illustrated.

An algorithm for searching musical motifs is followed step by step. A musical score is separated in phrases, six cases of proximity relations are calculated, six fuzzy set of phrases candidates to be a motif are computed by aggregating the proximities with different operators. Finally, the results are evaluated by the calculation of the independent inference sets to choose among classes of similar phrases, instead of choosing single phrases, by the determination of the specificity measurements under the knowledge of the proximities. Three of the proximities are computed using t-norms and the other three were computing using different OWA operators.

On the other hand, Yager's specificity measure of fuzzy sets is considered in order to evaluate the reliability in the decisions of selecting the representative phrases. Different fuzzy logic operators were applied to compute each one of the proximities, and the determination of the representative phrases process was successfully carried out.

References

1. Andreatta, M.: On group-theoretical methods applied to music: some compositional and implementational aspects. In: Mazzola, G., Noll, T., Lluis-Puebla, E. (eds.) Perspectives in Mathematical and Computational Music Theory, pp. 169–193. Electronic Publishing Osnabrück, Osnabrück (2004)

2. Beliakov, G., Pradera, A., Calvo, T.: Aggregation functions: a guide for practitioners. Studies in fuzziness and soft computing. Springer, Heidelberg (2007)
3. Castañeda, E., Garmendia, L., Santos, M.: Searching musical representative phrases using W-indistinguishabilities and proximities. In: Proc. EUROFUSE 2009 Workshop Preference Modeling and Decision Analysis, pp. 213–218 (2009b)
4. Castañeda, E., Garmendia, L., Santos, M.: Intelligent System for Computer Aided Musical Composition. In: Intelligent Decision Making Systems, pp. 13–18. World Scientific (2009a)
5. Castañeda, E., Garmendia, L.: Searching musical representative phrases using similarities. In: International Conference on the Logic of Soft Computing and 5th Workshop of the ERCIM Working Group of Soft Computing. LCS-ERCIM2006 satellite International Congress of Mathematicians, ICM (2006)
6. Garmendia, L., Yager, R.R., Trillas, E., Salvador, A.: Measures of Specificity of Fuzzy Sets Under T-Indistinguishabilities. IEEE Transactions on Fuzzy Systems 14(4), 568–572 (2006)
7. Klement, E., Mesiar, R., Pap, E.: Triangular norms. Kluwer, Dordrecht (2000)
8. Overill, R.E.: On the combinatorial complexity of fuzzy pattern matching in music analysis. Computers and the Humanities 27(2), 105–110 (1993)
9. Schweizer, B., Sklar, A.: Probabilistic metric spaces. North-Holland, Amsterdam (1983)
10. Trillas, E., Valverde, L.: On mode and implication in approximate reasoning. In: Gupta, M.M., et al. (eds.) Approximate Reasoning in Expert Systems. Elsevier, North-Holland, Amsterdam (1985)
11. Valverde, L.: On the structure of F-indistinguishability operators. Fuzzy Sets and Systems 17, 313–328 (1985)
12. Yager, R.R.: Ordinal measures of specificity. International Journal of General Systems 17, 57–72 (1990)
13. Zadeh, L.A.: Fuzzy sets. Information and Control 8, 338–353 (1965)
14. Zadeh, L.A.: Similarity relations and fuzzy orderings. Inform. Sci. 3, 177–200 (1971)
15. Zadeh, L.A.: The concept of linguistic variable and its application to approximate reasoning, parts I, II, III. Inform. Sci. 8, 199–249; 8, 301–357; 9, 43–48 (1975)

Chapter 8
A Risk-Based Multi-criteria Decision Support System for Sustainable Development in the Textile Supply Chain

Besoa Rabenasolo[1,2] and Xianyi Zeng[1,2]

[1] Ecole Nationale Supérieure des Arts et Industries Textiles,
2 allée Louise et Victor Champier, 59056 Roubaix, France
[2] University of Lille Nord de France, 59000 Lille, France
{besoa.rabenasolo,xianyi.zeng}@ensait.fr

Abstract. The modern textile/clothing industry is facing to a great number of important challenges related to sustainability. These challenges include environmental disasters and hazards to human health caused by toxic materials, resource exhaustion (water, energy, raw materials), and social impacts caused by delocalization, counterfeiting and other elements. In this context, in order to guarantee the development of the textile/clothing industry in a sustainable and optimal way when developing or producing new textile products, industrial companies need to optimize their production organization by minimizing risks occurring not only at the level of materials and processes but also in the whole international textile supply chain. In this chapter, we propose a risk-based multi-criteria decision support system for evaluating risks of different textile materials and the corresponding suppliers using the criteria of sustainable development (environment protection, recycling capacity, energy saving, human health and safety, and social impacts). Some evaluation criteria have been normalized by recognized international organizations. A method of data aggregation with multiple fuzzy criteria is applied in order to select the most appropriate textile material and its supplier. A new method combining expert knowledge on enterprise strategy and sensitivity criteria will be used for determining the linguistic weights of these fuzzy criteria. In this decision support system, the aggregated evaluation index takes into account the criteria of sustainable development and the specific application context as well as company strategies on the management of its supply chain.

1 Introduction

Sustainable development is a common concern for all industrial sectors nowadays. Governments and industrial companies are required to make decisions using relevant and explicit criteria in order to minimize risks at all levels and select appropriate strategies and projects for sustainable development [1]. Finding a systematic and suitable approach for planning sustainability in a company or a supply chain has become a basic requirement for policy makers, product designers, and production and supply chain managers [2].

J. Lu, L.C. Jain, and G. Zhang: Handbook on Decision Making, ISRL 33, pp. 151–170.
springerlink.com © Springer-Verlag Berlin Heidelberg 2012

Sustainable development is a complex multidimensional concept dealing with environment, economics, human health and social impact. According to the well-known report *Our Common Future* [3], it is defined as "development that meets the needs of the present without compromising the ability of future generations to meet their own needs". It requires the unification of economics and ecology as well as social and cultural development [2]. In the process of product development, overseeing risk management is a key issue for sustainable development. The potential risks are related to the product, internal company operations or management, and supply chain management [4]. At the product level, for example, the product could contain hazardous materials which cause harm to the environment when disposed of landfill at the end of the product life or during its production process. It may be difficult to reuse or recycle the product when it becomes waste. At the company management level, accidents such as fires and explosions could occur with unsustainable operations. Maintaining low-cost labor could lead to serious social impacts and decrease of product quality. At the supply chain management level, some economic loss could be related to a bad choice of partners, a bad strategy of delocalization, or to counterfeiting. In practice, the development of a decision-making system based on the evaluation of multi-risks existing at all levels is important for realizing sustainable planning in an industrial company and its related supply chain. It can provide efficient support to decision-makers in industrial companies by suitably combining all sustainable criteria.

Risk evaluation and management in an industrial supply chain have been widely studied by researchers. An overview on supply chain risk management (SCRM) is given in [5], where SCRM is defined as the management of supply chain risks through coordination or collaboration among the supply partners so as to ensure profitability and continuity. The supply chain risks mainly include uncertain economic cycles, uncertain consumer demands, and unpredicted natural and man-made disasters. In [6], a general risk management procedure has been integrated into the design, planning, and performance evaluation process of supply chain networks through Petri net simulation. In [7], a multiperiod supply chain network model with risk management has been proposed for integrating social impacts and responsibilities of manufacturers, retailers and consumers. This model explicitly defines the cost functions, emission functions, risk functions, and demand functions of the supply chain network, and describes the behavior of the various decision makers in it. Based on this model, a solution for multicriteria decision making has been proposed in [8]. Environmental risk assessment and related decision making strategies that minimize risks have been discussed in [9]. It proposes a basic analytical framework that couples multicriteria decision analysis with adaptive management. In practice, systematic multicriteria decision analysis tools are usually considered as an appropriate method in risk management because it requires balancing scientific findings with multi-faceted input from many stakeholders with different values and objectives.

Although risk analysis and risk management applying decision making tools has been extensively studied by researchers, the principal focus of its development to date has been on the technical challenges of characterising and modelling the environmental behaviour and biological action of chemicals, whereas issues concerning its broader socio-political context have been generally neglected. Problem definition, risk analysis and decision making have, therefore, tended to be dominated by experts and by expert opinion. Fresh insights from the social sciences advocate a pluralistic, inclusive approach, with experts participating alongside other stakeholders in a consensual decision making process [10].

Sustainable development with risk evaluation and management is also an important factor in the success of the textile industry. Now it is commonly recognized that the European textile industry is no longer a manpower industry but a capital-intensive industry that must constantly develop its competitiveness through research to develop high-tech, niche products that meet consumers' ever-changing requirements [11]. One important action in textile industry is the development of eco-textiles, i.e. new materials with eco-friendly characteristics such as recycling of fabrics, use of non-toxic chemicals, reduction of waste at source, improvement of energy efficiency, and optimization of effluent treatment. We need to consider these characteristics in an integral way instead of studying them separately. In the design and production of new eco-materials, the risks at all levels of raw materials, manufacturing processes, and business process in the textile supply chain should be systematically taken into account. In the current situation, most existing work on sustainable development in textile industry is related to the development of new eco-materials, the company's final decision is still largely driven by cost minimization and other economic aspects, and human factors are rarely concerned. As human factors play an important role in economics and social and cultural contexts, they also constitute major components in sustainable development and risk management.

In this chapter, we propose a multicriteria decision support system for evaluating the risks of a number of textile materials and their corresponding suppliers using the criteria of sustainable development. The standardized environment-related criteria and human-factor-related criteria are aggregated using linguistic weights. These weights are generated by combining fuzzy sensitivity criteria and expert knowledge on enterprise strategies related to supply chain development, marketing, application contexts of finished products and other issues. Compared with the existing evaluation work of sustainable development, the proposed decision support system has the following advantages: 1) it is strongly related to the specific application context of textile materials; 2) it takes into account not only the well-known environment-related criteria (quantity of hazardous materials, recycling capacity, energy saving, …) but also social and human factors in the textile supply chain (human health and safety, low-cost labor impact, delocalization impact, …); 3) it has the capacity to process linguistic variables, which are frequently encountered in this kind of evaluation work.

This chapter is organized as follows. In Section 2, we formalize the proposed decision support system. Especially, we introduce the structure of risk evaluation for the textile supply chain and explicitly give the list of all evaluation indicators. The procedure of aggregation of all evaluation indicators is described in Section 3. In Section 4, we present the method using fuzzy sensitivity analysis for determining the weights of the evaluation criteria and adjust these weights by introducing the expert knowledge on enterprise strategy. In Section 5, we show the effectiveness of the proposed system using an illustrative example for selecting the most suitable textile material and its supplier, meeting the requirements of all the indicators. The conclusion is given in Section 6.

2   Formalization of the Proposed Decision Support System

The proposed decision support system aims at selecting the most suitable textile materials and its supplier by optimizing the predefined criteria of sustainable development. It includes the following modules: 1) a database of representative materials and all the risk criteria related to environment, recycling, energy saving as well as human and social factors. For each criterion, the description of its evaluation procedure is also included in this database. 2) A user interface for the input of material parameters and evaluation results. 3) A procedure for generating and adjusting the weights of the evaluation criteria. 4) A procedure for aggregating evaluation data of different criteria.

In this system, the structure of the evaluation procedure, the evaluation criteria and the corresponding evaluation data are formalized as follows.

2.1   Structure of the Evaluation Procedure

Sustainable planning or decision making for sustainable development is generally considered as a hierarchical evaluation structure, in which the criteria are distributed at different levels whose details increase from top to bottom [2]. In our evaluation procedure, the first level of the hierarchy always describes the overall purpose of the problem, and is usually perceived in a very abstract form. As we move down the hierarchy, we try to define the immediate evaluation nodes in more meaningful and tangible terms. The bottommost level contains concrete evaluation indicators which can be either measured directly using physical instruments or evaluated by experts. The environmental indicators are evaluated with well-known lifecycle assessment methods. The overall objective of the first level varies with values of evaluation indicators. The general scheme of this hierarchical evaluation structure is described in Fig. 1.

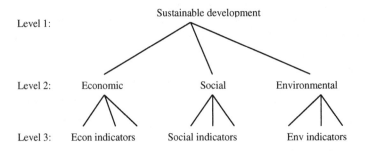

Fig. 1. General scheme of the hierarchical evaluation structure

In this paper, for simplicity, we formalize the hierarchical structure for evaluating textile materials into three levels. The readers can refer to [14, 15, 16] in order to obtain the details of the subsequent levels of the environmental, social and enterprise evaluation frameworks. All the concrete evaluation indicators are organized into three categories: economic, social and environmental. However, in a specific application, materials or products are evaluated using only one part of indicators. The general principle and methods we propose in this chapter can be easily extended to any specific application in which users can choose suitable evaluation indicators according to their purpose.

The related notations are defined below.

At Level 3, the economic indicators are denoted as I_1^{eco}, I_2^{eco}, ..., I_p^{eco} (the number of the economic indicators is p). The social indicators are denoted as I_1^{soc}, I_2^{soc}, ..., I_q^{soc} (the number of the social indicators is q). The environmental indicators are denoted as I_1^{env}, I_2^{env}, ..., I_r^{env} (the number of the environmental indicators is r). The weights relating Level 2 to Level 3 are W_i^{eco} ($i=1, ..., p$), W_j^{soc} ($j=1, ..., q$), W_k^{env} ($k=1, ..., r$). The evaluation indicators and their weights can be numerical or linguistic values.

At Level 2, the evaluation criteria for the economic, social and environmental categories can be calculated from the indicators at Level 3 and the weights relating Level 2 to Level 3. They are denoted as R^{eco}, R^{soc} and R^{env} respectively. The weights relating Level 2 to Level 1 are W^{eco}, W^{soc} and W^{env}.

The overall evaluation criterion denoted as R can be calculated from the criteria at Level 2 and the weights relating Level 2 to Level 1.

2.2 List of Evaluation Indicators

The most commonly used evaluation indicators for sustainable development in the textile supply chain are in the environmental category. In [12], ten normalized environmental indicators have been measured and calculated using the lifecycle assessment method CML 2001 [14]. They are

- Eutrophication Potential (EP – I_1^{env})
- Global Warming Potential (GWP – I_2^{env})
- Ozone Layer Depletion Potential (ODP – I_3^{env})
- Acidification Potential (AP – I_4^{env})
- Fresh Aquatic Eco-Toxicity Potential (FAETP – I_5^{env})
- Human Toxicity Potential (HTP – I_6^{env})
- Marine Aquatic Eco-Toxicity Potential (MAETP – I_7^{env})
- Photochemical Ozone Creation Potential (POCP – I_8^{env})

- Terrestrial Eco-Toxicity Potential (TETP – I_9^{env})
- Water Consumption (WC – I_{10}^{env})

For each textile material, these environment-related indicators are numerical or cardinal ($r = 10$). The objective of the environmental category is the minimization of all these indicators.

Among these indicators, the most important indicators selected by industrial companies include: I_2^{env} (Global Warming Potential), which is related to energy usage and measured in equivalent kg CO_2, I_6^{env} (Human Toxicity Potential), which causes risks to health and is measured in kg DCB eq., and I_{10}^{env} (Water Consumption), considered as the water-resource depletion indicator and measured in kg.

In this paper, the environmental indicators are combined with social and economic impacts related to risk criteria of industrial product design [4]. The list of additional evaluation indicators is given below. They include supply chain costs and risks related to the production processes. For the economic category ($p = 8$), the chosen indicators include

- Raw Material Cost (RMC – I_1^{eco})
- Machine Consumption Cost (MCC – I_2^{eco})
- Transportation Cost (TC – I_3^{eco})
- Labor Cost (LC – I_4^{eco})
- Energy Consumption Potential (ECP – I_5^{eco})
- Quality Risk Potential (QRP – I_6^{eco})
- Benefit Loss Potential (BLP – I_7^{eco})
- Time Consumption (TC – I_8^{eco})

The sum of the first five indicators constitutes the cost of finished products. The objective of the economic category is to obtain the maximum benefit and the best product quality within a very short time and with a minimal cost of the finished products.

Notice that some indicators of the economic category can also have impacts on the environmental indicators. For example, energy consumption potential (ECP) and transportation cost have impacts on global warming, quality risk potential on the generation of waste. In practice, the interaction between these two categories is complex. One economic action aiming at minimizing one environmental indicator can cause negative effects. For example, transportation cost can be decreased by setting up a short transportation distance, thus reducing GWP (eq.kg CO2). But the material and industrial facilities of the new production system may require more energy and different chemical processes, leading to an increase of GWP or other environmental impacts.

More complete social indicators developed by the United Nations UNEP-SETAC Life Cycle Initiative can be found in [15]. They are classified into 31 subcategories and concern different stakeholders, namely workers, local communities, societies, consumers and value-chain actors. In this chapter, the chosen indicators for the social category ($q = 8$) are slightly different from those of [15]. This subset of indicators is designed specifically for the textile supply chain in the background of delocalization, bearing in mind the limited scope of the assessment of alternative product designs and their corresponding suppliers. They include

- Impact of Low Employee's Income (LEI – I_1^{soc})
- Non-Community Liaison (NCL – I_2^{soc})
- Negative Effects for Worker's Health (NEWH – I_3^{soc})
- Negative Effects for Worker's Safety (NEWS – I_4^{soc})
- Noise Level (NL – I_5^{soc})
- Site Nuisance (SN - I_6^{soc})
- Product Nuisance (PN - I_7^{soc})
- Staff Non-Training Level (SNTL - I_8^{soc})

These indicators are mainly related to the industrial facilities and processes within an establishment, and the corresponding data are easily available while the database required for the guideline [15] is often incomplete. Different from the cardinal indicators of the environmental category, the indicators of the economic and social categories are linguistic or ordinal values, given by professional experts from their evaluation and analysis of internal data of the related company and its supply chain.

In this paper, these indicators take values from the following set: {very low, low, medium, high, very high} = { $\tilde{b}_1$, $\tilde{b}_2$, $\tilde{b}_3$, $\tilde{b}_4$, $\tilde{b}_5$ }. Moreover, all the indicators of the three categories have been consistently defined so that low values correspond to good textile materials and high values to bad textile materials. Thus, the best textile material corresponds to the lowest value of the overall evaluation criterion R.

These evaluation indicators take into account not only the criteria related to product design and production but also those of the related supply chain and human factors concerning this product.

3   Computing the Overall Criterion from the Multiple Evaluation Indicators

In this chapter, we assume that there exist n textile materials, denoted as $T = \{T_1, T_2, ..., T_n\}$. The main function of the proposed decision making system is the aggregation of all the risk indicators presented in Section 2, in order to select or rank the most appropriate textile material from T in terms of sustainable development. Based on the performance analysis of the existing methods, we apply a method of fuzzy multicriteria decision making for aggregating all the fuzzified evaluation indicators with linguistic weights.

3.1   Existing Methods for Multicriteria Decision Making

The authors of [12, 13] have summarized the existing methods for multicriteria decision making. These methods can be classified into three groups:

- **Multi-attribute utility theory** that aims at aggregating different sources of data or different points of view in a single function that is then optimized. These methods have been widely used in different applications. The main difficulty in this group is the construction of an appropriate utility function and determination of relevant weights for different evaluation criteria. The effects of compensation, i.e. the good effect of one criterion may be compensated by the bad effect of another criterion, and incomparability between criteria should be minimized. One representative method of this group is AHP [17].

- **Interactive methods** that consist of interactive and iterative exploration of all alternatives. These methods are suited to problems with an almost infinite number of alternatives and are then not adapted to the ranking of a set of limited textile materials.

- **Outranking methods** that consist of pairwise comparison of alternatives according to each criterion with introducing indifference and preference thresholds. In this group, the method PROMETHEE [18] is the most well-known and has been successfully applied to the environment-related decision making problems [19], including textile lifecycle assessment due to their high degree of pragmatism [12].

Compared with the multi-attribute utility theory, the outranking methods are easy for interpretation and can remove the effects of compensation and incomparability. However, they have the following drawbacks [20]: 1) they may only be applied if the decision maker can express his preference between two alternatives on any given criterion on a ratio scale. The preference between two alternatives can be changed by removing or adding one alternative. 2) They may be applied if the decision maker can express the importance of each criterion, which is often uncertain or imprecise for the decision maker. 3) The weights of the criteria express trade-offs between the criteria. 4) They may only be used with criteria where the differences between evaluations are meaningful. 5) It is impossible to take discordance into account when constructing outranking relations.

The outranking methods may be efficient for processing cardinal and normalized evaluation criteria. However, for processing human-evaluation-based ordinal criteria, we need to propose new methods in order to integrate all types of variables and further optimize the result of decision making with the extended evaluation criteria presented in Section 2.

In this context, fuzzy multicriteria decision making methods may be powerful tools for dealing with the subjectivity, imprecision and vagueness in the evaluation criteria and the related weights. The corresponding aggregation operations enable to combine the individual opinions on group decision making. Some representative methods of fuzzy multicriteria decision making can be found in [21, 22]. These methods have been successfully applied in tool steel materials selection under fuzzy environment [23] and risk decision-making in natural hazards [24].

3.2   Proposed Procedure for Aggregating Evaluation Indicators

This procedure was proposed in [25] and successfully applied in the design of fashion-oriented industrial products [26]. It is based on a number of hierarchical operations for aggregating fuzzy evaluation indicators and their corresponding fuzzy weights. First, we transform all the numerical and linguistic evaluation indicators into fuzzy variables for the following reasons:

1) Taking into account significant data ranges and removing the effects of scales and non significant data ranges according to professional knowledge of experts.

2) Obtaining physically interpretable results of evaluation and classifying all materials according to significant ranking results.

3) Maintaining conformity between all the evaluation indicators (environmental, economic and social) and with the knowledge of experts, usually expressed in linguistic variables and rules.

4) Removing the effects of compensation and incompatibility by using appropriate linguistic operations and introducing the knowledge of experts.

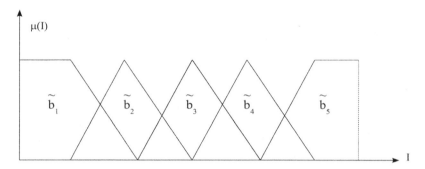

Fig. 2. Membership functions of each fuzzy evaluation indicator (I: any numerical evaluation indicator)

For each numerical indicator, we have two possibilities for transforming it into a fuzzy variable. If there exists a representative learning database for all materials, the corresponding measured data can cover the whole range of this indicator and we can apply the fuzzy c-means algorithm to regroup these measures into five classes, corresponding to $B=\{\tilde{b}_1, \tilde{b}_2, \tilde{b}_3, \tilde{b}_4, \tilde{b}_5\}$ defined in Section 2. If there does not exist any representative database for all materials, we invite a number of professional experts to analyze the numerical value of each indicator. From this analysis, each expert selects a linguistic value from the set B. The corresponding five membership functions take triangular and trapezoidal forms and their parameters are determined by aggregating the judgments of all the experts (see Fig. 2). If no expert knowledge is available for selecting linguistic values from the set B, one can uniformly divide the range of the corresponding evaluation indicator I into five equal subintervals, each corresponding to one fuzzy value of B. The range of the indicator I can be obtained by computing the minimum and maximum of the evaluation values for all the material samples.

For each linguistic indicator, the corresponding fuzzy variable can also be defined by aggregating the judgments of different experts or by uniformly dividing the range of evaluation values into five equal subintervals.

In general, the corresponding fuzzy evaluation indicator takes into account the distribution of evaluation scores given by all the experts. For a specific material, it is a vector of five components, each representing the membership degree of this material to one fuzzy value of the set B. The sum of these components is 1 since B represents a fuzzy 5-partition.

By following the same principle, we transform the weights of the evaluation indicators into fuzzy variables.

Theoretically, a relevant textile material in terms of sustainable development corresponds to the minimization of all the risk evaluation indicators. However, in practice, these indicators often conflict and can not reach their minimum values in the same time. Therefore, a relevant material should correspond to the minimum value of the aggregation of these indicators with relevant weights. These weights should be carefully determined according to each specific application. These applications are usually related to company strategies: economic benefit oriented product development; balance between environment, benefit and social impacts; human safety oriented product development; hard-condition application-oriented products; and other technical, economic and social strategies.

Next, we apply the procedure of fuzzy multicriteria decision making [25] for aggregating these fuzzy risk evaluation indicators with linguistic weights. The weights at Levels 3 and 2, denoted as $\{ W_i^j \}$ and $\{ W^j \}$ respectively, take values from the set {less important, important, very important, strongly important} in which the importance of the criteria are qualified with four linguistic values. Compared with the other fuzzy multicriteria aggregation methods, this method is more flexible and capable of treating both linguistic evaluation indicators and linguistic weights. The details of this procedure are given below.

For any textile material T_k, $k \in \{1, 2, ..., n\}$, we define its relevancy degrees (evaluation criteria) for the three categories using fuzzy number arithmetics $(+, \times)$, i.e.

$$R^{eco}(T_k) = \sum_{i=1}^{p}(W_i^{eco} \times I_i^{eco}(T_k)) \text{ for the economic category}$$

$$R^{soc}(T_k) = \sum_{i=1}^{q}(W_i^{soc} \times I_i^{soc}(T_k)) \text{ for the social category}$$

$$R^{env}(T_k) = \sum_{i=1}^{r}(W_i^{env} \times I_i^{env}(T_k)) \text{ for the environmental category}$$

Next, these three relevancy degrees can be further aggregated to generate the overall relevancy degree, i.e.

$$R(T_k) = W^{eco} \times R^{eco}(T_k) + W^{soc} \times R^{soc}(T_k) + W^{env} \times R^{env}(T_k)$$

Evidently, $R(T_k)$, $R^{eco}(T_k)$, $R^{soc}(T_k)$, $R^{env}(T_k)$ are also defined on $B = \{ \tilde{b}_1, \tilde{b}_2, \tilde{b}_3, \tilde{b}_4, \tilde{b}_5 \}$. After the normalization, we have

$$\overline{R}(T_k) = \frac{R(T_k)}{\sum_{i=1}^{n} R(T_i)_0^R}$$

where $\sum R(T_i)_0^R$ is the upper bound of 0-cut of the fuzzy number $\sum R(T_i)$.

Evidently, the values of $\overline{R}(T_k)$ varies between 0 and 1. The closer its value is to 0, the more relevant the corresponding material T_k is with respect to the predefined risk evaluation indicators. Therefore, we define a fuzzy positive ideal solution ($r^+ = 0$) and a fuzzy negative solution ($r^- = 1$) and compute the distances between the overall relevancy degree of the material T_k and these two ideal solutions [27, 28]. We then have

$$d_k^+ = d(\overline{R}(T_k), r^+) \text{ and } d_k^- = d(\overline{R}(T_k), r^-)$$

where $d(x, y)$ is a fuzzy distance between two fuzzy sets. Finally, we compute the ranking criterion as follows.

$$Rank(T_k) = \frac{1}{2}(d_k^- + (1 - d_k^+))$$

The material T_k that corresponds to the maximum value of the ranking criterion for all the materials is the most relevant to the risk evaluation criteria defined in Section 2. It is considered as the most appropriate material in terms of sustainable development.

4   Determination of the Weights

The weights of the three categories and the different indicators are first determined according to the sensitivity to evaluated or measured data and then optimized or adjusted using the professional knowledge of experts.

In Section 3, all the evaluation indicators were transformed into fuzzy values defined on the set B using the fuzzy c-means algorithm, equal partition or expert knowledge. The sensitivity criterion of each indicator I_i^j ($i \in \{1, 2, \ldots\}$ and $j \in \{$eco, soc, env$\}$) related to all the representative material samples $T_1, \ldots, T_n$ can be defined by

$$S_i^j = \frac{2}{n(n-1)} \sum_{k \neq l} \left| defuz(I_i^j(T_k)) - defuz(I_i^j(T_l)) \right|$$

S_i^j is the defuzzified averaged distance for all pairs of material samples for the evaluation indicator I_i^j. In the procedure of defuzzification, $\tilde{b}_1$, $\tilde{b}_2$, $\tilde{b}_3$, $\tilde{b}_4$, $\tilde{b}_5$, considered as weights of the averaging operation, are set to be 0, 0.25, 0.5, 0.75, and 1 respectively. As the difference between two defuzzified indicators is smaller than 1, then S_i^j varies between 0 and 1. A small value of S_i^j means that the indicator I_i^j is less sensitive to evaluated or measured data and then has less influence on the final decision or ranking result. In the same way, the indicator I_i^j is very sensitive and has more influence on the final decision if the corresponding S_i^j is close to 1.

By uniformly dividing the range of the sensitivity criteria for all the evaluation indicators into four equal subintervals {very low, low, high, very high}, we transform them into fuzzy sets and obtain the rules relating the data sensitivity to the linguistic weights of the evaluation indicators. The four fuzzy values of the data sensitivity criteria correspond to the following values of the weights: less important, important, very important, and strongly important. This correspondence is set up according to the following principles.

1)   If the data sensitivity of an evaluation indicator is very low, then its weight should be a little important.

2)   If the data sensitivity of an evaluation indicator is very high, then its weight should be strongly important.

3)   Medium values of data sensitivity criteria correspond to medium importance levels (important or very important) for the related evaluation indicators.

In the same way, we can also define the weights of the three categories according to their sensitivity criteria. The sensitivity criterion for each category is defined as the average of the sensitivity values calculated from all the evaluation indicators of this category, i.e.

$$S_j = \frac{1}{z} \sum_{i=1}^{z} S_i^j$$

where $z = p$ if $j = eco$, $z = q$ if $j = soc$ and $z = r$ if $j = env$.

In practice, the sensitivity to evaluated or measured data should not be the unique criterion for determining the weights of the evaluation indicators and the categories. The strategies of the company for its supply chain, marketing, applications of finished products as well as other technical, economic and social considerations should be integrated into the procedure of weight definition. In many cases, these elements play a more important role than the nature of measured data.

For simplicity, we design a questionnaire in which a number of questions related to the strategies of the company are asked. Several professional experts are invited to give the importance levels (the previously defined four levels) of these questions for each indicator and each category. Several examples of these questions are given below.

1)   Economic benefit oriented development
2)   Balance between environment, benefit and social impacts
3)   Coordination between environment and benefit
4)   Human safety oriented development
5)   Consumer preference oriented development
6)   Hard-condition application-oriented products
7)   Ecology oriented products

In contrast with the previous evaluation, the evaluation of the weights for the indicators and categories is mainly related to general strategies of the company instead of specific textile materials and specific processes. For each indicator I_i^j , by combining all the possibilities of the sensitivity criterion S_i^j (very low, low, high , very high) and the importance level L_{ik}^j related to each question Q_k, $k \in \{1, ..., w\}$, we obtain the corresponding weight W_{ik}^j by using the following rules extracted from the expert knowledge.

1)   If S_i^j is very high and L_{ik}^j is strongly important, then W_{ik}^j is strongly important

2)   If S_i^j is very high and L_{ik}^j is very important, then W_{ik}^j is very important

3) If S_i^j is very high and L_{ik}^j is important, then W_{ik}^j is important

4) If S_i^j is very high and L_{ik}^j is less important, then W_{ik}^j is less important

5) If S_i^j is high and L_{ik}^j is strongly important, then W_{ik}^j is strongly important

6) If S_i^j is high and L_{ik}^j is very important, then W_{ik}^j is very important

7) If S_i^j is high and L_{ik}^j is important, then W_{ik}^j is important

8) If S_i^j is high and L_{ik}^j is less important, then W_{ik}^j is less important

9) If S_i^j is low and L_{ik}^j is strongly important, then W_{ik}^j is very important

10) If S_i^j is low and L_{ik}^j is very important, then W_{ik}^j is very important

11) If S_i^j is low and L_{ik}^j is important, then W_{ik}^j is important

12) If S_i^j is low and L_{ik}^j is less important, then W_{ik}^j is less important

13) If S_i^j is very low and L_{ik}^j is strongly important, then W_{ik}^j is very important

14) If S_i^j is very low and L_{ik}^j is very important, then W_{ik}^j is important

15) If S_i^j is very low and L_{ik}^j is important, then W_{ik}^j is a important

16) If S_i^j is very low and L_{ik}^j is less important, then W_{ik}^j is less important

In these rules, the importance level of the indicator I_i^j related to the question Q_k is a little more important than the sensitivity criterion S_i^j in the determination of the weight W_{ik}^j. This is because the quantity of experimental data is often limited and the database can not be really representative in most situations. The final value of W_{ik}^j can be obtained by aggregating the judgments of all the experts. Formally, the aggregated result is a fuzzy set expressed by a 4-dimensional vector of which the sum of all components is 1. The final weight of the indicator I_i^j can be obtained by calculating the maximum value of the weights for all the questions (Q_1, ..., Q_w), i.e. $W_i^j = \max_k \{W_{ik}^j\}$. This means that the weight of the indicator I_i^j is important if it is important to at least one strategy of the company, and that this weight is not important if it is not important to any strategy of the company.

The weight of each category can be determined in the same way.

5   An Illustrative Example

This example aims at supporting decision makers in choosing the best scenario of a sustainable supply chain by considering risks in the aspects of environment, economics and social impacts. This case illustrates the comparison of three fabric samples for producing apparel items evaluated as the same functional unit. It includes not only the determination of raw materials and processes in terms of environment, but also the selection of fabric suppliers in the supply chain in terms of cost, quality and social impacts. The strategies of the company are integrated into the procedure of weight definition for the evaluation indicators.

The three fabric materials of interest are T_1, T_2 and T_3. The unified ranges of evaluation indicators for environment are defined by a group of experts. Next, each range is uniformly divided into 5 subintervals, each corresponding to one fuzzy value (see Fig. 2). The values of the environmental evaluation indicators for these three materials are given in Table 1.

Table 1. Numerical and fuzzy data of the environmental evaluation indicators

Materials – Indicators	Range	T_1	T_2	T_3
I_1^{env} (Kg phosphate eq.)	[0, 0.004]	Num: 0.002	Num: 0.0025	Num: 0.0038
		Fuz: (0 0 1 0 0)	Fuz: (0 0 0.5 0.5 0)	Fuz: (0 0 0 0.2 0.8)
I_2^{env} (Kg CO2 eq.)	[0, 10]	Num: 6.5	Num: 9.5	Num: 8
		Fuz: (0 0 0.4 0.6 0)	Fuz: (0 0 0 0.2 0.8)	Fuz: (0 0 0 0.8 0.2)
I_3^{env} (Kg R11 eq.)	[0, 1]	Num: 0.6	Num: 0.5	Num: 0.95
		Fuz: (0 0 0.6 0.4 0)	Fuz: (0 0 1 0 0)	Fuz: (0 0 0 0.2 0.8)
I_4^{env} (Kg SO2 eq.)	[0, 10]	Num: 7	Num: 9.5	Num: 6
		Fuz: (0 0 0.2 0.8 0)	Fuz: (0 0 0 0.2 0.8)	Fuz: (0 0 0.6 0.4 0)
I_5^{env} (Kg DCB eq.)	[0, 1.8]	Num: 0.1	Num: 0.1	Num: 1.8
		Fuz: (0.78 0.22 0 0 0)	Fuz: (0.78 0.22 0 0 0)	Fuz: (0 0 0 0 1)
I_6^{env} (Kg DCB eq.)	[0, 0.7]	Num: 0.68	Num: 0.6	Num: 0.68
		Fuz: (0 0 0 0.12 0.88)	Fuz: (0 0 0 0.57 0.43)	Fuz: (0 0 0 0.12 0.88)
I_7^{env} (Kg DCB eq.)	[0, 1932]	Num: 1800	Num: 1650	Num: 1800
		Fuz: (0 0 0 0.27 0.73)	Fuz: (0 0 0 0.6 0.4)	Fuz: (0 0 0 0.27 0.73)
I_8^{env} (Kg Ethene eq.)	[0, 0 .009]	Num: 0.008	Num: 0.005	Num: 0.003
		Fuz: (0 0 0 0.32 0.68)	Fuz: (0 0 0.78 0.22 0)	Fuz: (0 0.67 0.33 0 0)
I_9^{env} (Kg DCB eq.)	[0, 0.04]	Num: 0.01	Num: 0.015	Num: 0.038
		Fuz: (0 1 0 0 0)	Fuz: (0 0.5 0.5 0 0)	Fuz: (0 0 0 0.2 0.8)
I_{10}^{env} (Kg)	[0, 500]	Num: 150	Num: 120	Num: 450
		Fuz: (0 0.8 0.2 0 0)	Fuz: (0.04 0.96 0 0 0)	Fuz: (0 0 0 0.4 0.6)

Table 2. Aggregated evaluation data for the indicators of economics and social impacts

Materials-Indicators	T_1		T_2		T_3	
I_1^{eco}	Su_1	(0 0.4 0.6 0 0)	Su_3	(0.5 0.5 0 0 0)	Su_5	(0 0 0 0.5 0.5)
	Su_2	(0 0 0 0.8 0.2)	Su_4	(0 0.1 0.9 0 0)	Su_6	(0 0 0 0.1 0.9)
I_2^{eco}	Su_1	(0 0 0.6 0.4 0)	Su_3	(0.3 0.7 0 0 0)	Su_5	(0 1 0 0 0)
	Su_2	(0 0 0.5 0.5 0)	Su_4	(0.5 0.5 0 0 0)	Su_6	(0 1 0 0 0)
I_3^{eco}	Su_1	(0 0 0.2 0.8 0)	Su_3	(0.9 0.1 0 0 0)	Su_5	(0 0 0 0.5 0.5)
	Su_2	(0 0.3 0.7 0 0)	Su_4	(0.9 0.1 0 0 0)	Su_6	(0 0 0 0.1 0.9)
I_4^{eco}	Su_1	(0.7 0.3 0 0 0)	Su_3	(1 0 0 0 0)	Su_5	(0 0.6 0.4 0 0)
	Su_2	(0 0 0.6 0.4 0)	Su_4	(0.8 0.2 0 0 0)	Su_6	(0 0 0.5 0.5 0)
I_5^{eco}	Su_1	(0 0 0.6 0.4 0)	Su_3	(0.3 0.7 0 0 0)	Su_5	(0 0 0 0.7 0.3)
	Su_2	(0 0 0 0.8 0.2)	Su_4	(0.5 0.5 0 0 0)	Su_6	(0 0 0 0.9 0.1)
I_6^{eco}	Su_1	(0 0.4 0.6 0 0)	Su_3	(0 0.3 0.7 0 0)	Su_5	(0 0 0.3 0.7 0)
	Su_2	(0 0 0 0.2 0.8)	Su_4	(0 0 0.5 0.5 0)	Su_6	(0 0 0 0.8 0.2)
I_7^{eco}	Su_1	(0 0.5 0.5 0 0)	Su_3	(0 0 0.3 0.7 0)	Su_5	(0 0.5 0.5 0 0)
	Su_2	(0 0.5 0.5 0 0)	Su_4	(0 0 0.3 0.7 0)	Su_6	(0 0.5 0.5 0 0)
I_8^{eco}	Su_1	(0 0.4 0.6 0 0)	Su_3	(0.7 0.3 0 0 0)	Su_5	(0 0 0 0.5 0.5)
	Su_2	(0 0 0.8 0.2 0)	Su_4	(0.5 0.5 0 0 0)	Su_6	(0 0 0 0.1 0.9)
I_1^{soc}	Su_1	(0 0 0.5 0.5 0)	Su_3	(0 0.3 0.7 0 0)	Su_5	(0.1 0.9 0 0 0)
	Su_2	(0.8 0.2 0 0 0)	Su_4	(0.4 0.6 0 0 0)	Su_6	(1 0 0 0 0)
I_2^{soc}	Su_1	(0 0 0 0.6 0.4)	Su_3	(0.3 0.7 0 0 0)	Su_5	(0 0 0.5 0.5 0)
	Su_2	(0 0 0.5 0.5 0)	Su_4	(0.3 0.7 0 0 0)	Su_6	(0 0 0 0.2 0.8)
I_3^{soc}	Su_1	(0 0 0 0.5 0.5)	Su_3	(0 0 0 0.3 0.7)	Su_5	(0 0.4 0.6 0 0)
	Su_2	(0 0.4 0.6 0 0)	Su_4	(0 0 0.1 0.9 0)	Su_6	(0.1 0.9 0 0 0)
I_4^{soc}	Su_1	(0 0 0 0.8 0.2)	Su_3	(0 0 0 0.4 0.6)	Su_5	(0.2 0.8 0 0 0)
	Su_2	(0 0.5 0.5 0 0)	Su_4	(0 0 0.5 0.5 0)	Su_6	(0.5 0.5 0 0 0)
I_5^{soc}	Su_1	(0 0 0 1 0)	Su_3	(0 0 0 0.4 0.6)	Su_5	(0 0 0 0.3 0.7)
	Su_2	(0 0.5 0.5 0 0)	Su_4	(0 0.1 0.9 0 0)	Su_6	(0 0 0.5 0.5 0)
I_6^{soc}	Su_1	(0 0 0.7 0.3 0)	Su_3	(0 0 0.2 0.8 0)	Su_5	(0 0 0.3 0.7 0)
	Su_2	(0 0.5 0.5 0 0)	Su_4	(0 0.3 0.7 0 0)	Su_6	(0 0.5 0.5 0 0)
I_7^{soc}	Su_1	(0.8 0.2 0 0 0)	Su_3	(0 0.5 0.5 0 0)	Su_5	(0.5 0.5 0 0 0)
	Su_2	(0.8 0.2 0 0 0)	Su_4	(0 0.5 0.5 0 0)	Su_6	(0.5 0.5 0 0 0)
I_8^{soc}	Su_1	(0 0.4 0.6 0 0)	Su_3	(0 0 0.1 0.9 0)	Su_5	(0 0.4 0.6 0 0)
	Su_2	(0 0.7 0.3 0 0)	Su_4	(0 0 0.4 0.6 0)	Su_6	(0.9 0.1 0 0 0)

The linguistic indicators for economics and social impacts are evaluated by experts, and their aggregated results for the three textile materials T_1, T_2 and T_3 and six suppliers are given in Table. 2. Each material can be provided by two suppliers whose social environment and production strategies are quite different. In general, the suppliers Su_1, Su_3 and Su_5 aim at providing medium-quality products with a lower production cost, while the suppliers Su_2, Su_4 and Su_6 produce high-quality products with a relatively high production cost while maintaining a balance between environmental, economic and social impacts in their activities.

From the evaluation data provided in Table 1 and Table. 2, we can easily calculate the sensitivity criterion for each evaluation indicator and then transform it into a fuzzy value (a 4-dimensional vector). The corresponding results are given in Table. 3. From these results, we find that the indicators I_6^{env} and I_7^{env} are very insensitive to the evaluation data and they play a less important role in the final decision or the final ranking result. Therefore, low values should be given to the corresponding weights. In the same way, I_3^{eco}, I_5^{env} and I_9^{env} are very sensitive to the evaluation data and we need to assign high weights to them.

Table 3. Fuzzy sensitivity values for all the evaluation indicators

Indicators (env.)	Sensitivity S_i^{env}	Indicators (economic)	Sensitivity S_i^{eco}	Indicators (social)	Sensitivity S_i^{soc}
I_1^{env}	0.3	I_1^{eco}	0.4	I_1^{soc}	0.29
I_2^{env}	0.2	I_2^{eco}	0.25	I_2^{soc}	0.39
I_3^{env}	0.3	I_3^{eco}	0.51	I_3^{soc}	0.35
I_4^{env}	0.23	I_4^{eco}	0.34	I_4^{soc}	0.40
I_5^{env}	0.63	I_5^{eco}	0.37	I_5^{soc}	0.28
I_6^{env}	0.08	I_6^{eco}	0.26	I_6^{soc}	0.18
I_7^{env}	0.06	I_7^{eco}	0.15	I_7^{soc}	0.17
I_8^{env}	0.39	I_8^{eco}	0.46	I_8^{soc}	0.38
I_9^{env}	0.47				
I_{10}^{env}	0.44				

From the data in Table. 3, we obtain the sensitivity values for all three categories, i.e. S_{env} =0.31, S_{eco} =0.34, S_{soc} =0.31. These three categories have similar sensitivity values related to evaluation data.

Also, according to Table. 3, the range of the sensitivity values for all the indicators is between [0, 0.63]. Having uniformly divided it into four subintervals, we obtain the corresponding fuzzy values of the sensitivity criteria.

Moreover, for the company of interest, we define three strategies for management of the supply chain. The first strategy (St1) is the maximization of the profits of the company with minimal costs. It does not consider the environmental and social impacts in the production of materials and selection of suppliers. The second (St2) is the maximization of economic benefits with minimal environmental impacts. The third (St3) is to set up a balance between economic benefits, reduction of costs, environmental impacts and social impacts. According to these strategies, the experts define the importance levels for all the evaluation indicators and all the questions. Next, these importance levels are further combined with the sensitivity values and we obtain the final weights of the evaluation indicators (see Table. 4).

Table 4. The final aggregated weights for all the evaluation indicators

Weights (env.)	Strategies			Weights (eco.)	Strategies			Weights (social)	Strategies		
	St 1	St 2	St3		St 1	St 2	St 3		St 1	St2	St3
W^{env}_1	1	0	0	W^{eco}_1	0	0	0	W^{soc}_1	1	0	0
	0	0	0		0	0	0		0	1	0
	0	0	0		0	0	0		0	0	1
	0	1	1		1	1	1		0	0	0
W^{env}_2	1	0	0	W^{eco}_2	0	0	0	W^{soc}_2	1	1	0
	0	0	0		0	0	0		0	0	0
	0	1	1		1	1	1		0	0	1
	0	0	0		0	0	0		0	0	0
W^{env}_3	1	0	0	W^{eco}_3	0	0	0	W^{soc}_3	0	0	0
	0	0	0		0	0	0		1	0	0
	0	1	1		0	0	0		0	1	0
	0	0	0		1	1	1		0	0	1
W^{env}_4	1	0	0	W^{eco}_4	0	0	0	W^{soc}_4	0	0	0
	0	0	0		0	0	0		1	0	0
	0	1	1		0	0	0		0	1	0
	0	0	0		1	1	1		0	0	1
W^{env}_5	1	0	0	W^{eco}_5	0	0	0	W^{soc}_5	1	1	0
	0	0	0		0	0	0		0	0	0
	0	1	1		0	0	0		0	0	1
	0	0	0		1	1	1		0	0	0
W^{env}_6	1	0	0	W^{eco}_6	0	0	0	W^{soc}_6	0	0	0
	0	0	0		0	0	0		1	0	0
	0	1	1		1	1	1		0	1	1
	0	0	0		0	0	0		0	0	0
W^{env}_7	1	0	0	W^{eco}_7	0	0	0	W^{eco}_7	0	0	0
	0	0	0		0	0	0		0	0	0
	0	1	1		0.5	0.5	0.5		1	1	0
	0	0	0		0.5	0.5	0.5		0	0	1
W^{env}_8	1	0	0	W^{eco}_8	0	0	0	W^{soc}_8	0	0	0
	0	0	0		0	0	0		1	0	0
	0	1	1		1	1	1		0	1	0
	0	0	0		0	0	0		0	0	1
W^{env}_9	1	0	0								
	0	0	0								
	0	0.5	0.5								
	0	0.5	0.5								
W^{env}_{10}	1	0	0								
	0	0	0								
	0	1	1								
	0	0	0								

Next, we apply the method presented in Section 3.2 for aggregating evaluation data of each category and each strategy of the company. The corresponding results are given in Table. 5. The weights of the categories are given by experts because their sensitivity values are rather close each other. The total normalized relevancy degrees as well as the ranking values of all the materials and all the suppliers under different strategies are also given in this table.

Table 5. Aggregated relevancy degrees for the three categories and the three strategies

Relevancy degree	T_1-Su_1	T_1-Su_2	T_2-Su_3	T_2-Su_4	T_3-Su_5	T_3-Su_6
Strategy St1	W_{env}=(0 1 0 0), W_{eco}=(0 0 0 1), W_{soc}=(1 0 0 0)					
R^{env}	(1 0 0 0 0)	(1 0 0 0 0)	(1 0 0 0 0)	(1 0 0 0 0)	(1 0 0 0 0)	(1 0 0 0 0)
R^{eco}	(0.7 1.65 3.02 1.47 0)	(0 0.72 2.59 2.6 0.94)	(3.37 2.17 0.72 0.58 0)	(2.87 1.57 1.49 0.92 0)	(0 1.69 1.02 2.5 1.64)	(0 1.02 0.92 2.2 2.64)
R^{soc}	(0.54 0.3 0.4 0.5 0.2)	(0.54 0.83 0.63 0 0)	(0 0.33 0.4 0.53 0)	(0 0.43 0.89 0.66 0)	(0.4 0.96 0.5 0.23 0)	(0.83 0.99 0.17 0 0)
Normalized R	(0.14 0.24 0.42 0.2 0)	(0.05 0.1 0.36 0.36 0.13)	(0.52 0.3 0.1 0.08 0)	(0.44 0.22 0.21 0.13 0)	(0.05 0.23 0.14 0.35 0.23)	(0.05 0.14 0.13 0.31 0.37)
Rank	0.58	0.4	0.82	0.74	0.38	0.3
Strategy St2	W_{env}=(0 0 0 1), W_{eco}=(0 0 0 1), W_{soc}=(0 1 0 0)					
R^{env}	(0.5 1.5 1.8 1.68 1.5)	(0.5 1.5 1.8 1.68 1.5)	(0.55 1.2 2.1 1.7 1.6)	(0.55 1.2 2.1 1.7 1.6)	(0 0.45 0.62 3.5 4.28)	(0 0.45 0.62 3.5 4.28)
R^{eco}	(0.7 1.65 3.02 1.47 0)	(0 0.72 2.59 2.6 0.94)	(3.37 2.17 0.72 0.58 0)	(2.87 1.57 1.49 0.92 0)	(0 1.69 1.02 2.5 1.64)	(0 1.02 0.92 2.2 2.64)
R^{soc}	(0.54 0.4 1 1.22 0.47)	(0.8 1.6 1.27 0 0)	(0 0.43 0.69 1.07 1.47)	(1.32 0.73 1.47 1.34 0)	(0.5 1.68 1 0.47 0)	(1.67 1.68 0.33 0 0)
Normalized R	(0.09 0.22 0.34 0.24 0.11)	(0.05 0.18 0.32 0.29 0.16)	(0.26 0.23 0.2 0.17 0.14)	(0.25 0.19 0.26 0.2 0.1)	(0.01 0.16 0.12 0.36 0.35)	(0.03 0.12 0.1 0.34 0.41)
Rank	0.49	0.42	0.58	0.57	0.28	0.26
Strategy St3	W_{env}=(0 0 0 1), W_{eco}=(0 0 0 1), W_{soc}=(0 0 0 1)					
R^{env}	(0.5 1.5 1.8 1.68 1.5)	(0.5 1.5 1.8 1.68 1.5)	(0.55 1.2 2.1 1.7 1.6)	(0.55 1.2 2.1 1.7 1.6)	(0 0.45 0.62 3.5 4.28)	(0 0.45 0.62 3.5 4.28)
R^{eco}	(0.7 1.65 3.02 1.47 0)	(0 0.72 2.59 2.6 0.94)	(3.37 2.17 0.72 0.58 0)	(2.87 1.57 1.49 0.92 0)	(0 1.69 1.02 2.5 1.64)	(0 1.02 0.92 2.2 2.64)
R^{soc}	(0.8 0.6 1.4 2.9 0.97)	(1.34 2.6 2.4 0.33 0)	(0.2 1.37 1.1 1.6 2.6)	(0.47 1.64 2.6 2 0)	(0.77 2.7 1.74 1 0.47)	(3.6 2.3 0.7 0.5 0.54)
Normalized R	(0.1 0.18 0.3 0.3 0.12)	(0.09 0.24 0.33 0.22 0.12)	(0.2 0.22 0.19 0.19 0.2)	(0.19 0.21 0.3 0.22 0.08)	(0.03 0.22 0.15 0.31 0.29)	(0.15 0.16 0.1 0.27 0.32)
Rank	0.46	0.49	0.51	0.55	0.35	0.39

From Table. 5, we can find that the ranking values of different suppliers producing the same textile material are close to each other related to the other suppliers. The relevancy order of the three materials is T_2>T_1>T_3. However, the difference of the ranking values between these materials decreases after the integration of environmental and social indicators. Suppliers Su_1, Su_3 and Su_3 are more relevant to the strategy of profit maximization with minimal costs. However, suppliers Su_2, Su_4 and Su_6

are more relevant to the strategy of setting up a balance between economic development and reduction of environmental and social impacts. This conclusion is completely appropriate to the nature of these suppliers.

6   Conclusion

This chapter applied a method of data aggregation with multiple fuzzy criteria for selecting the most appropriate textile material and the most suitable supplier. In this process of data aggregation, the linguistic weights of the evaluation indicators are determined using a predefined sensitivity criterion and expert knowledge on enterprise strategy related to its supply chain, marketing, application contexts of finished products and other issues. In this way, the final aggregated evaluation index takes into account the criteria of sustainable development and the specific development strategy of the related company.

This work can be further improved by re-considering the social and economic indicators and their evaluation methods with the cooperation of researchers specialized in sociology and economy. The process of weight generation for the evaluation indicators and the categories can also be further optimized by systematically formalizing and re-organizing expert knowledge.

References

1. Andriantiatsaholiniaina, L.C., Kouikoglou, V.S., Phillis, Y.A.: Evaluating Strategies for Sustainable Development: Fuzzy Logic Reasoning and Sensitivity Analysis. Ecological Economics 48, 149–172 (2004)
2. Quaddus, M.A., Siddique, M.A.B.: Modelling Sustainable Development Planning: a Multicriteria Decision Conferencing Approach. Environment International 27, 89–95 (2001)
3. World Commission on Environment and Development. Our Common Future. Oxford University Press, Oxford (1987)
4. Howarth, G., Hadfield, M.: A Sustainable Product Design Model. Materials and Design 27, 1128–1133 (2007)
5. Tang, C.S.: Perspectives in supply chain risk management. International Journal of Production Economics 103, 451–488 (2006)
6. Tuncel, G., Alpan, G.: Risk assessment and management for supply chain networks: a case study. Computers in Industry 61, 250–259 (2010)
7. Cruz, J.M., Wakolbinger, T.: Multiperiod effects of corporate social responsibility on supply chain networks, transaction costs, emissions, and risk. International Journal of Production Economics 116, 61–74 (2008)
8. Cruz, J.M.: The impact of corporate social responsibility in supply chain management: multicriteria decision-making approach. Decision Support Systems 48, 224–236 (2009)
9. Linkov, I., Satterstrom, F.K., Kiker, G., Batchelor, C., Bridges, T., Ferguson, E.: From comparative risk assessment to multi-criteria decision analysis and adaptive management: recent development and applications. Environment International 32, 1072–1093 (2006)
10. Eduljee, G.H.: Trends in risk assessment and risk management. The Science of the Total Environment 249, 12–23 (2000)

11. Trudel, J.S.: Strategic Guide for Sustainable Development – Spider Thread: Textile of the Future. In: Report of the Canadian Textile Sustainable Development Working Committee (April 15, 2008)
12. Boufateh, I., Perwuelz, A., Rabenasolo, B., Jolly-Desolt, A.M.: Multiple Criteria Decision Making for Environmental Impact Optimization. In: Proceedings of the 39th International Conference on Computers and Industrial Engineering – CIE 39, Troyes, France, July 6-8 (2009)
13. Boufateh, I., Perwuelz, A., Rabenasolo, B., Jolly, A.M.: Optimization of environmental and social criteria in the textile supply chain: European state of the art and perspectives of research. Product Lifecycle Management, Special Publication 3, 667–676 (2007)
14. Guinée, J.B. (ed.): Life Cycle Assessment: an Operational Guide to the ISO Standards: Parts 1 and 2. Ministry of Housing, Spatial Planning and Environment and Centre of Environmental Science. Den Haag and Leiden, The Netherlands (2001)
15. Andrews, E.S., Barthel, L.-P., Beck, T., Benoit, C., Ciroth, A., Cucuzzella, C., Gensch, C.-O., Hebert, J., Leasge, P., Manhart, A., Mazeau, P., Mazijn, B., Methot, A.-L., Moberg, Å., Norris, G., Parent, J., Prakash, S., Reveret, J.-P., Spillemaeckers, S., Ugaya, C., Valdivia, S., Weidema, B., Benoit, C., Mazijn, B. (eds.): Guidelines for Social Life Cycle Assessment of Products. UNEP-SETAC Life-Cycle Initiative (2009)
16. Lo, J.L.W.: 'Benchmarking Performance: Supply Chains in the Textile and Garment Industries'. PhD thesis, Automatique, Productique, Communication et Ergonomie des Systèmes Homme-Machine à l'Université de Valenciennes et du Hainaut-Cambresis, France (2005)
17. Saaty, T.L.: The Analytical Hierarchy Process. McGraw-Hill, New York (1980)
18. Brans, J., Vincke, P., Mareschal, B.: How to Select and How to Rank Projects: the PROMETHEE Method. European Journal of Operational Research 24, 228–238 (1986)
19. Geldermann, J., Spengler, T., Rentz, O.: Fuzzy Outranking for Environmental Assessment – Casy Study: Iron and Steel Making Industry. Fuzzy Sets and System 115, 45–65 (2000)
20. Keyser, W.D., Peeters, P.: A Note on the Use of PROMETHEE Multicriteria Methods. European Journal of Operational Research 89, 457–461 (1996)
21. Chen, S.M., Tan, J.M.: Handling multicriteria fuzzy decision-making problems based on vague set theory. Fuzzy Sets and Systems 67, 163–172 (1994)
22. Chen, S.M., Wang, C.H.: A generalized model for prioritized multicriteria decision making systems. Expert Systems with Applications 36, 4773–4783 (2009)
23. Chen, S.M.: A new method for tool steel materials selection under fuzzy environment. Fuzzy Sets and Systems 92, 265–274 (1997)
24. Chen, K., Blong, R., Jacobson, C.: MCE-RISK: integrating multicriteria evaluation and GIS for risk decision-making in natural hazards, vol. 16, pp. 387–397 (2001)
25. Lu, J., Zhang, G., Ruan, D., Wu, F.: Multi-Objective Group Decision Making: Methods. In: Software and Applications with Fuzzy Set Technology. Imperial College Press, London (2007)
26. Zeng, X., Zhu, Y., Koehl, L., Camargo, M., Fonteix, C., Delemotte, F.: A Fuzzy Multi-Criteria Evaluation Method for Designing Fashion Oriented Industrial Products. Soft Computing 14, 1277–1285 (2010)
27. Liberatore, J.M., Stylianou, C.M.: Expert Support System for New Product Development Decision Making: a Modelling Framework and Applications. Management Science 41, 1296–1316 (1995)
28. Belliveau, P., Griffin, A., Somermeyer, S.: PDMA ToolBook 2 for New Product Development. Wiley, New York (2004)

Chapter 9
Fuzzy Decision System for Safety on Roads

Matilde Santos and Victoria López

Facultad de Informática, Universidad Complutense de Madrid,
c/ Profesor García Santesmases s/n. 28040-Madrid, Spain
msantos@dacya.ucm.es, vlopez@fdi.ucm.es

Abstract. The topic of road safety is a crucial problem because it has become nowadays one of the main causes of death, despite the efforts made by the countries trying to improve the roads conditions. When starting a journey, there are different factors, both objective and subjective, that influence on the driving safety. In this work, we apply fuzzy logic to model these subjective considerations. A committee machine that combines the information provided by three fuzzy systems has been generated. Each of these fuzzy systems gives a degree of risk when traveling taking into account fuzzy conditions of three variables: car (age, last check, the wear on brakes and wheels, etc.); driver (tiredness, sleeping time, sight, etc.); and characteristics of the trip (day or night, weather conditions, length, urban or country road, etc). The final system gives not only the degree of risk according to this fuzzy prediction but the degree in which this risk could be decreased if some of the conditions change according to the advice the fuzzy decision system provides, such as, for example, if the driver takes a rest, or if the tyres are changed.

1   Introduction

The topic of road safety is a crucial problem because it has become nowadays one of the main causes of death, despite the efforts made by the countries trying to improve the roads conditions. When starting a journey, there are different factors, both objective and subjective, that influence on the driving safe (Bedard et al. 2002; Chen 2007), and we should be able to take all of them into account to improve the safety on roads. This paper has a clear purpose, to present other ways to improve road traffic safety.

In the literature, we can find a large body of research on traffic accidents based on the analysis of statistical data. Most of them work with the probability approach of risk. Conventionally, researchers use linear or nonlinear regression models and probabilistic models (Kweon and Kockelman 2003; Chong et al. 2005). But these models are often limited in their capability to fully explain the process when underlying so many nonlinear processes and uncertain factors. That is why in this paper we will show an original approach that takes into account not only data but subjective factors. There are some authors who have also applied fuzzy logic to classify roads sections in order to identify the more dangerous elements that need interventions (Cafiso et al. 2004; Jayanth et al. 2008), or who have used it for predicting the risk of accidents, for example, on wet pavements (Xiao et al. 2000). But few of them refer to factors that

J. Lu, L.C. Jain, and G. Zhang: Handbook on Decision Making, ISRL 33, pp. 171–187.
springerlink.com   © Springer-Verlag Berlin Heidelberg 2012

are evaluated in a fuzzy way, i.e., linguistic characteristics that can be implemented by means of soft computing techniques to increase the safety on roads (Imkamon et al. 2008; Valverde et al. 2009).

On the other hand, conventional quantitative techniques are unsuitable for complex system analysis that requires approaches closer to human qualitative reasoning and decision making processes to manage uncertainty. In this framework fuzzy logic plays an essential role embedding the human ability to summarize information for decision making with incomplete, vague and imprecise data (Klir and Yuan 2005; Zadeh 1975). It has been proved as a useful tool when dealing with the uncertainty and the vagueness of real life (López et al. 2010; Farias et al. 2010).

In this work, a committee machine that combines the information provided by three fuzzy systems has been generated. Each of these fuzzy systems gives a degree of risk when traveling under certain fuzzy conditions that are related to the corresponding input variable: car (age, last check, the wear on brakes and wheels, etc.), driver (tiredness, sleeping time, sight, etc.), and characteristics of the trip (day or night, weather conditions, length, city or road, etc). The final system gives not only the degree of risk of traveling according to this fuzzy prediction but the degree in which this risk could be decreased if some of the conditions change such as, for example, if the driver takes a rest, or if the tyres are changed, according to the advice the fuzzy decision system provides.

The paper is organized as follows. In Section 2, the modeling of the reliability and risk of any system is analytically described. Section 3 is devoted to the description of the involved variables of the fuzzy systems. Section 4 shows the committee machine that has been implemented to obtain the risk of traveling under certain conditions and the advice the fuzzy decision system provides in order to enhance the risk. The paper ends with the conclusions.

2   Modeling Reliability and Risk

In this section, we introduce a classical way to define and model risk according to its definition in probability. The main reason for developing this section is to show another possibility in modeling the risk and then to compare both ways. This probabilistic approach is the most used in applications that involve any kind of vehicles and it is also very popular in modeling risk and risk reduction for verification and validation of engineering systems (Logan et al. 2005; Logan et al. 2003).

Reliability engineering is a discipline that deals with the study of the ability of a system or component to perform its required functions under stated conditions for a specified period of time (Liu 2010). In this section we report risk in terms of reliability. More specifically, reliability here is going to be defined as the probability of that a functional unit will perform its required functions for a specified interval under stated conditions.

Reliability engineers rely heavily on statistics, probability and reliability theories, etc. For example, some of the techniques that are used in reliability are reliability prediction, prognosis, fault diagnosis, Weibull analysis, reliability testing and accelerated life testing. Because of the large number of reliability techniques available, their expense, and the varying degrees of reliability required for different situations, most projects develop a reliability program plan to specify the tasks that will be performed in order to evaluate the risk expected by a specific system (López et al. 2009). This section provides an overview of some of the most common reliability and risk engineering tasks.

Automobile engineers deal with reliability requirements for the vehicles (and their components). They must design and test their products to fulfill hard specifications. Once this model of reliability of a specific system is developed, a risk analysis and evaluation are due.

For engineering purposes, reliability is mathematically defined as:

$$R(t) = 1 - F_T(t) = \Pr(T > t) = \int_t^\infty f(s)ds \qquad (1)$$

where T is the random variable 'Time before failure of the device', F is the function of distribution of T, $f(s)$ is the failure probability density function and t is the length of the period of time (which is assumed to start from zero).

Reliability engineering (and then risk analysis) is concerned with two key elements of this definition:

1. Reliability is a probability; failure is regarded as a random phenomenon, T is the random variable 'Time until next device's failure' and it is a recurring event without any information on individual failures, the causes of failures, or the relationships between failures, except that the likelihood for failures to occur varies over time according to a given probability function.
2. Reliability is applied to a specified period of time. Also other units may sometimes be used. For example, for this traffic application, reliability might be specified in terms of miles, age of the car, etc. A piece of mechanical equipment of a car may have a reliability rating value in terms of cycles of use.

Different methods and tools can be used to calculate reliability. Every system requires a different level of reliability and gets different level of risk as consequence. A car driver can operate under a wide range of conditions including the motivation and health condition of the driver, car status, and other environment variables. The consequences of failure could be serious, but there is a correspondingly higher budget. A stapler may be more reliable than a car, but has a very much different set of performing conditions, insignificant consequences of failure, and a much lower budget.

Reliability requirements are included in the system or subsystem specifications. Requirements are specified in terms of some reliability parameters such as the mean time between failures (MTBF) or, equivalently, the number of failures during a given period of time. It is very useful for systems that work frequently as vehicles and electronic equipment. Risk increases as the MTBF decreases and therefore reliability increases too. For systems with a clearly defined failure time the empirical distribution function of these failure times can be easily determined. Early failure rate studies determine the distribution with a decreasing failure rate over the first part of the bathtub curve. Although it is a general praxis to model the failure rate with an exponential distribution, the empirical failure distribution is often parameterised with a Weibull or a log-normal model which are more realistic.

Another point to take into account when evaluating the risk of complex systems is the combination of several components and subsystems that affects not only the cost but also the risk. In this case the risk evaluation may be performed at various levels, for each component, subsystem and for the whole system. Many factors of different nature must be addressed during this evaluation, such as temperature, speedy, shock, vibration, heat, and so on.

Reliability design starts with the development of a model by using blocks diagram and fault trees to provide a graphical way of evaluating the relationship between the different parts of the system. One of the most important design techniques is redundancy in which if one part of the system fails, there is an alternative path. For example, a vehicle might use two light bulbs for signaling the brake. If one bulb fails, the brake light still works using the other bulb. Redundancy increases the system reliability decreasing the risk of failure of the system; however, it is difficult and expensive and it is therefore limited to critical parts of the system.

Risk evaluation deals with failure rate that mathematically is defined in terms of probability and reliability by means of the risk function. This risk function $\lambda(t)$ is the frequency with which a system or component fails per unit of time t (second, hour or so). For example, an automobile's risk function or failure rate in its tenth year of service may be many times greater than its failure rate during the first year. It is also known in the literature as hazard function or rate and it can be computed as the quotient between the failure density function $f(t)$ and the reliability $R(t)$, although there are other useful relations as the following formula,

$$\lambda(t) = \frac{f(t)}{R(t)} = -\frac{R'(t)}{R(t)} = -\frac{\partial}{\partial t}(\ln R(t))$$

(2)

In practice, the mean time between failures (MTBF) is often used instead of failure rate. This is because for an exponential failure distribution, the risk function becomes constant $\lambda(t) = \lambda$; this property is the well known 'memoryless' characteristic of exponential distribution. However, for other distributions as Weibull or log-normal, the risk function may not be constant with respect to time and they can be monotonic increasing ('wearing out'), monotonic decreasing (Pareto distribution, 'burning in') or not monotonic.

Table 1 shows the formulation of two of the most popular distribution in reliability: exponential and Weibull. There are also other important distribution but their formulation is too complex for our purpose.

Table 1. Reliability distributions used to compute risk

Distribution Name	Reliability Function	Risk Function
Exponential λ	$R(t) = \exp(-\lambda t)$	$\lambda(t) = \dfrac{\lambda \exp(-\lambda t)}{\exp(-\lambda t)} = \lambda$
Weibull (β,γ,η)	$R(t) = \exp(-\left(\dfrac{t-\gamma}{\eta}\right)^{\beta})$	$\lambda(t) = \dfrac{\beta}{\eta}\left(\dfrac{t-\gamma}{\eta}\right)^{\beta-1}$

Modeling the system usually requires the combination of several subsystems or components that are connected in series, parallel or are 'K out of N' independent systems. In any of those cases, both reliability and risk functions are represented for more general formulas as it is shown in Table 2.

Table 2. Reliability and risk functions for compound systems

System Design	Reliability Function	Risk Function
Series	$R(t) = \prod_{i=1}^{n} R_i(t)$	$\lambda(t) = \sum_{i=1}^{n} \lambda_i(t)$
Parallel	$R(t) = 1 - \prod_{i=1}^{n} F_i(t)$	$\lambda(t) = \dfrac{f(t)}{R(t)}$
K out of N (independent components)	$R(t) = \sum_{j=k}^{n} \binom{n}{j} R(t)^j (1 - R(t))^{n-j}$	$\lambda(t) = \dfrac{f(t)}{R(t)}$

The complexity of the calculus increases in an environment with a lot of devices or components. The study of a system such as a human-machine interaction tool is too complex to assure the goodness of the method. These models incorporate predictions based on parts-count failure rates taken from historical data. Very often these predictions are not accurate in an absolute sense, so they are only useful in order to assess relative differences in design alternatives.

Figure 1 shows an example of a system with seven components of four different types (according to its color). It consists of four subsystems connected in series, where the second and the third ones are at the same time parallel subsystems (that is, compound subsystems). In a vehicle, for example, components can represent brakes, lights, tyres, or any other part of the car in this case.

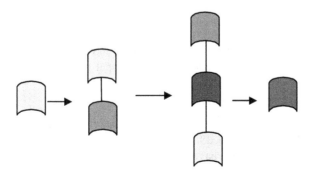

Fig. 1. System with for compound subsystems

3   Description of the Fuzzy System That Evaluates the Risk on Road

The whole fuzzy system that has been designed in this chapter represents the risk on roads when driving due to different factors. It is not easy to collect the right data needed to establish a correlation between, for example, human factors such as the fatigue or the drowsiness, and the accidents (Moore-Ede et al. 2004; Petridou and Moustaki 2000; Shinar 1978). The right choice of risk factors is a very key upon which success of failure of evaluation models depend. In this sense, a fuzzy model is a robust system that allows to represent the uncertainty of considerations on the weather, the status of the driver, etc.

Three main fuzzy subsystems have been defined for each of the variables under consideration: the environment, the driver and the car. For each of them, different fuzzy sets have been assigned to the selected input variables. The membership functions are adjusted manually by trial and error. The rule base for each one is designed based on the experience and understanding of the process, and are weighted accordingly to this knowledge. The antecedents of each rule are combined by the AND operator (that is implemented as the product or the minimum, depending on the subsystem).

The rules are evaluated in parallel using fuzzy reasoning. Fuzzy inference methods are classified in direct methods and indirect methods. Direct methods, such as Mamdani's and Sugeno's are the most commonly used. These two methods only differ in how they obtain the outputs. Indirect methods are more complex. In this work, Mamdani's is used except when the output is a binary one (then Sugeno's is applied).

The output of each subsystem is obtained by applying the Centre of Area (COA) or the MOM (Mean of Maximum) defuzzification methods. Defuzzification is the process of converting the degrees of membership of output linguistic variables into numerical values.

The combination of the fuzzy outputs of those subsystems gives the final degree of risk when making the trip under those conditions. This final output is again the result obtained by a fuzzy inference.

All the fuzzy systems have been implemented in Matlab©.

3.1   Environment Fuzzy Subsystem

The first fuzzy system refers to the Environment and all the factors related to this aspect of the driving. It is a Mamdani one (the output of the rule base is a linguistic value).

The fuzzy input variables of this subsystem are:

-      Day/night: range [0 10]. This represents the level of luminosity. Two fuzzy trapezoidal sets have been defined for Dark and Clear labels (Figure 2). Complementary membership functions have been considered as this factor depends on the location, the season of the year, etc.

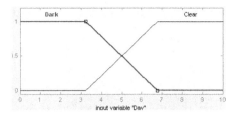

Fig. 2. Fuzzy sets assigned to the input variable luminosity

- Weather: range [0 10]. Three triangular membership functions represent Bad, Normal and Good weather (Figure 3). In this case, the fuzzy sets assigned are not symmetric.

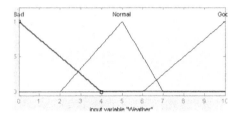

Fig. 3. Fuzzy sets assigned to the input variable weather

- Length: range [0 500] Km. This variable is represented by three asymmetric trapezoidal fuzzy sets meaning Short, Medium and Long (Figure 4). They represent the distance of the tour. It could be assume that for long distances, the risk is higher because it influences on other factors such as driver's fatigue, for example.

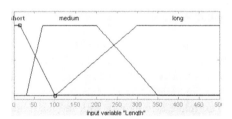

Fig. 4. Fuzzy sets assigned to the input variable length

- Road type: range [0 5] where 0 means rural or secondary road, and 5 motorway. Three Gaussian functions represent the fuzzy sets Rural, Normal and Highway (Figure 5).

These environmental factors come from external and independent sources. They result from measurements of some parameters such as time of the day, the type of roadway, etc.

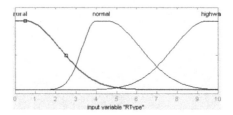

Fig. 5. Fuzzy sets assigned to the input variable Road type

Nine fuzzy rules have been defined to link those conditions and they are weighted according to the impact of each factor on the risk. The defuzzyfication method applied in this system is mom (mean of maximum).

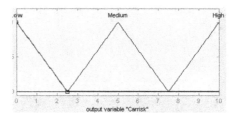

Fig. 6. Fuzzy sets assigned to the output variable risk for any of the three fuzzy subsystems (car, driver, and environment)

The fuzzy output is the risk that entails to travel under those conditions. The range, between 0 and 10, is covered by three triangular membership functions meaning Low, Medium and High (Figure 6). As it is possible to see, the fuzzy sets do not overlap, therefore the output is linear.

3.2   Driver Fuzzy Subsystem

The second fuzzy system refers to the Driver (Miyajima et al. 2007; Evans 1996). There are many behavioral issues that could be considered. For example, fatigue is one of the most pervasive yet under-investigated causes of human error related driving accidents. But there is not simple tool or objective way for investigators to collect the right data needed to correlate driving risk with fatigue in an analytical way. Behavioral factor are interrelated and any attempt of categorization can only be arbitrary.

Having said this, in this application, the primary factors that have been taking into account because they affect the driver behaviour and therefore have been selected as input variables for the fuzzy logic model are:

- Visual capacity: range [0 10], with labels Low, Medium and High, defined by asymmetric triangular fuzzy sets (Figure 7). It has been considered as an independent input but it is not, as it depends for example on the weather, luminosity, etc.

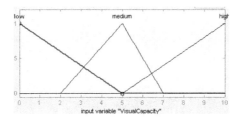

Fig. 7. Fuzzy sets assigned to the input variable visual capacity

- Tiredness or fatigue (Brown 1994): range [0 10]. It is represented by three triangular membership functions: Little, Medium and Much (Figure 8). This is quite a crucial factor but difficult to model.

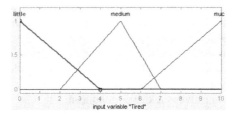

Fig. 8. Fuzzy sets assigned to the input variable tiredness

- Sleep (Horne and Reyner 1995): [0 10], number of hours that the driver has slept the previous night. Trapezoidal fuzzy sets represent the values of Little, Medium and Much (Figure 9). For example, slept hours fewer than 4 are considered Little, with high risk of crashes as the fatigue of the driver may make him slow in the reactions.

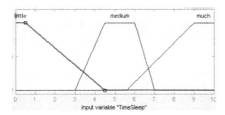

Fig. 9. Fuzzy sets assigned to the input variable time slept

- Motivation: range [0 10]. Two fuzzy sets, Low and High, are defined for this variable with trapezoidal functions (Figure 10).

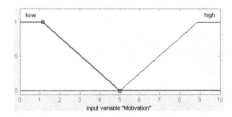

Fig. 10. Fuzzy sets assigned to the input variable motivation

These factors, among others, reduce the capability for adjustment to suddenly changing conditions that may culminate in an accident.

The output variable is again a fuzzy measurement of the risk that entails the trip when the driver is under those conditions, defined as in the environment system (Figure 6).

At the same time, some of these inputs and other environment inputs are related to another fuzzy system, Rest. This fuzzy system, which has a strong influence on the driver, has the following inputs that have been previously defined: Tiredness, Sleep, Motivation, Length, and Road type.

The output of this fuzzy system, time_relax, is the fuzzy necessity of rest. The defuzzyfication of this variable (three trapezoidal sets for Little, Medium and High) gives the minutes, between 0 and 60, which the driver should rest. This rest decreases the risk of driving as it improves the driver condition.

3.3   Car Fuzzy Subsystem

The third system refers to the Car status. There are different fuzzy inputs related to the car conditions and different ways to deal with them. In fact, although the car system is the main element of this block, two more fuzzy systems have been defined that are closely related, **Tyres/brakes** and **Light**. They have some inputs in common.

On the one hand, the fuzzy subsystem **Car** has the following inputs:

-    Lights: range [0 10]. Three Gaussian fuzzy sets have been defined for Bad, Normal and Good conditions of the lights (Figure 11).
-    Tyres: range [0 10]. The same as for lights (Figure 11). They stand for the status of the tyres, if they should be replace or can be kept.
-    Brakes: [0 10]. The same as for lights (Figure 11). Acceptable brakes conditions (Normal or Good) are considered when normalized wear is lower than 6. Otherwise they will not supply enough friction to stop the car at the required distance.

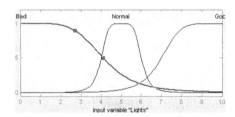

Fig. 11. Fuzzy sets assigned to the input variable lights, tyres and brakes

- Check: range [0 24] months. It can be Recent (less than 10 years), Medium or Far, all of them represented by symmetric triangular membership functions, as it is possible to see in Figure 12.

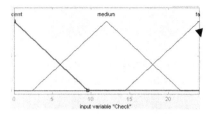

Fig. 12. Fuzzy sets assigned to the input variable check

- Kilometers: range [0 300000] Km. The fuzzy sets defined for this input are Few, Medium and Many, represented by asymmetric triangular membership functions (Figure 13).

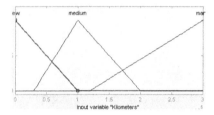

Fig. 13. Fuzzy sets assigned to the input variable kilometers

- Car Age: [0 20]: Little, Medium and Big (triangular), see Figure 14. Car age between 0-3 years stands for almost new, where more than 8 years is considered old.

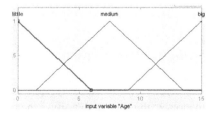

Fig. 14. Fuzzy sets assigned to the input variable age of the car

The fuzzy output of the system is the risk regarding the car that will be the third input to the final **Risk** fuzzy system. At the same time, Lights is also an input to **Light** system. This system receives inputs that are in common with other systems, such as Day/night, Weather, Lights and Visual Capacity. These inputs have been already defined.

In the same way, the **Tyres/brakes** system has the following inputs described before: Weather, Length and Road type, and two new ones:

-    Tyres: that can take values between 0 and 10 for Bad, Normal and Good conditions (Figure 11)
-    Brakes: again three fuzzy sets for [0 10] (Figure 12).

These two systems, **Light** and **Tyres/Breaks** are Sugeno-fuzzy type. They provide two possible outputs: OK or Change, suggesting lights change or brakes or tyres replacement if necessary.

As an example of the formulation of the rules of the any of the fuzzy inference systems, Figure 15 shows the rules for the car subsystem. In this case, the rules have six variables in the antecedent, and the output is the level of risk provided for that subsystem.

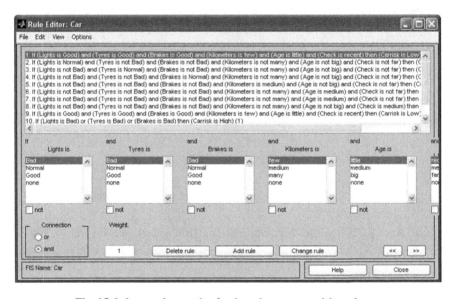

Fig. 15. Inference fuzzy rules for the subsystem car risk evaluator

4   Risk Fuzzy System Implementation

The committee machine combines all of the fuzzy systems described in section 3 (see Figure 17). As it is possible to see, the final output gives the risk of driving taking into account all the factors that have been considered in each fuzzy subsystem. The inputs of this system are: the Risk_driver, Risk_environment and Risk_car, which come from each main subsystem. The output is defined by seven triangular fuzzy sets in a universe of discourse [0 100] (Figure 16) and it is defuzzified by the centroide method. In this case, fifteen rules are considered.

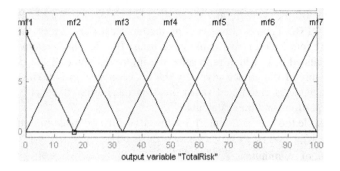

Fig. 16. Fuzzy sets assigned to the final output variable risk

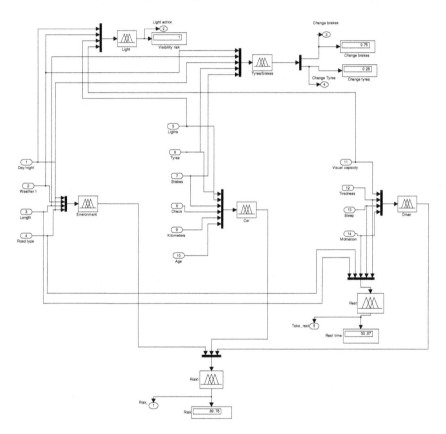

Fig. 17. Fuzzy committee machine that evaluates the risk of driving

To interact with the system, an interface has been developed (Figure 18). By means of slides, the inputs variables can be set to different values by the user. Information about the range of possible values for each factor is provided by the corresponding buttons (Info).

The outputs of the system are given in two ways: the REAL RISK and the ENHANCED RISK. The risk level is computed as Real risk, that is, the value that is obtained taking into account the actual conditions. Besides, the system gives another value, the Enhanced risk, which represents the improved value of the risk that would be obtained if any of the advice given by the decision system is taking into account. This new risk value is calculated when pressing the button "Accept" after marking any of the suggestions and then "calculate".

As it is possible too see in the interface, the advice may consist in four actions:

- Take a rest of … minutes
- Have your tyres changed
- Have your brakes changed
- Light action: do something to improve the visibility

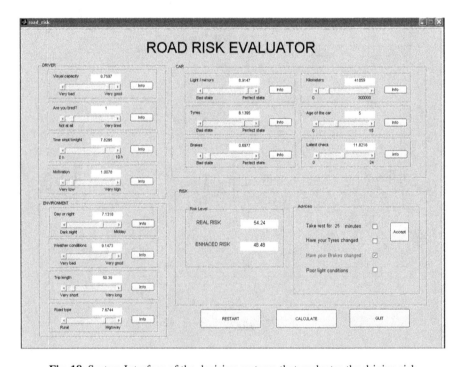

Fig. 18. System Interface of the decision systems that evaluates the driving risk

The value of the final risk is then recalculated by considering only the pieces of advice that are marked, that is, that have been followed. Then, this part of the system is set to the corresponding optimum condition once it has been applied (for example, tiredness = 0, brakes are now perfect, etc.).

The implementation of the correctness actions is presented in Fig. 19. The acceptation of a specific piece of advice acts as a switch (action_accepted) that is propagated to the corresponding fuzzy subsystem.

The output of each fuzzy subsystem is shown by means of graphics (scopes in Figure 19). Therefore it is possible to see if there has been any step because the conditions have improved respect to a previous situation.

As it is also possible to see in Figure 19, there are other inputs that could have been included, as the status of the mirrors, which were taken into account in some versions of this tool. This input is equivalent to the input lights, and it does not add any significant information. The list of possible factors to be included could be countless.

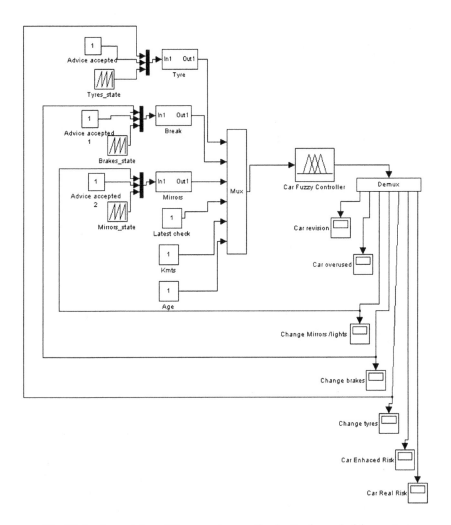

Fig. 19. Implementation of the correctness action by the fuzzy decision system

The risk level calculated by this fuzzy system does not mean probability of accident but influence of some factors on the safety of the trip.

5   Conclusions

Risk taking, on a long term basis, is in many ways equivalent to accident proneness. This paper intends to be a contribution to reducing or avoiding those car accidents on roads.

This work describes a simple analysis tool that investigates the relationship between some subjective factors that affect the driving and the risk of travelling under those conditions.

Three main issues have been considered: the driver, the car and the environment. For each of them, a fuzzy subsystem that calculates the risk of driving regarding some subjective factors has been designed. A combination of all of them in a more complex system gives the final value of risk.

The presented results indicate that, in addition to calculating the risk of driving under certain conditions, the fuzzy decision system can be applied to determine specific corrective actions that should be undertaken to improve safety.

References

Bedard, M., Guyatt, G.H., Stones, M.J., Hireds, J.P.: The independent contribution of driver, crash, and vehicle characteristics to driver fatalities. Accident Analysis and Prevention 34, 717–727 (2002)

Brown, I.D.: Driver fatigue. Human Factors 36, 298–314 (1994)

Cafiso, S., Lamm, R., La Cava, G.: Fuzzy model for safety evaluation process of new and old roads. J. Transportation Research 1881, 54–62 (2004)

Chong, M., Abraham, A., Paprzyckil, M.: Traffic Accident Analysis Using Machine Learning Paradigms. Informatica 29, 89–98 (2005)

Chen, Y.L.: Driver personality characteristics related to self reported accident involvement and mobile phone use while driving. Safety Science 45, 823–831 (2007)

Evans, L.: The dominant role of drive behaviour in traffic safety. American Journal Public Health 86, 784–786 (1996)

Farias, G., Santos, M., López, V.: Making decisions on brain tumour diagnosis by soft computing techniques. Soft Computing 14, 1287–1296 (2010)

Horne, J.A., Reyner, L.A.: Sleep related vehicle accidents. British Medical Journal 310, 565–567 (1995)

Imkamon, T., Saensom, P., Tangamchit, P., Pongpaibool, P.: Detection of hazardous driving behaviour using fuzzy logic. In: Proc. of the IEEE ECTI-CON II, pp. 657–660 (2008)

Jayanth, J., Hariharakrishnan, C.V., Suganthi, L.: Fuzzy Clustering of Locations for Degree of Accident Proneness based on Vehicle User Perceptions. Proceedings of World Academy of Science, Engineering and Technology 33, 182–185 (2008)

Klir, G.J., Yuan, B.: Fuzzy Sets and Fuzzy Logic: Theory and Applications, ch. 4. Prentice Hall, India (2005)

Kweon, Y.J., Kockelman, D.M.: Overall injury risk to different drivers: combining exposure, frequency and severity models. Accident Analysis and Preventions 35, 441–450 (2003)

Liu, B.: Uncertain Risk Analysis and Uncertain Reliability Analysis. Journal of Uncertain Systems 4(3), 163–170 (2010)

Logan, R.W., Nitta, C.K., Chidester, S.K.: Risk Reduction as the Product of Model Assessed Reliability, Confidence, and Consequence. Journal of Defence Modeling and Simulation 2(4), 191–207 (2005)

Logan, R.W., Nitta, C.K.: Validation, Uncertainty, and Quantitative Reliability at Confidence (QRC). In: AIAA-2003-1337 (2003)

López, V., Santos, M., Montero, J.: Improving Reliability and Performance in Computer Systems by means of Fuzzy Specifications. In: Intelligent Decision Making Systems, pp. 351–356. World Scientific (2009)

López, V., Santos, M., Montero, J.: Fuzzy Specification in Real Estate Market Decision Making. International Journal of Computational Intelligence Systems 3(1), 8–20 (2010)

Miyajima, C., Nishiwaki, Y., Ozawa, K., Wakita, T., Itou, K., Takeda, K., Itakura, F.: Driver modeling based on driving behaviour and its evaluation in driver identification. Proc. IEEE 95(2), 427–437 (2007)

Moore-Ede, M., Heitmann, A., Guttkuhn, R., Trutschel, U., Aguirre, A., Crok, D.: Circadian Alertness Simulator for Fatigue Risk Assessment in Transportation: Application to Reduce Frequency and Severity of Truck Accidents. Aviation, Space, and Environmental Medicine 75(1), A107–A111 (2004)

Petridou, E., Moustaki, M.: Human factors in the causation of road traffic crashes. European Journal of Epidemiology 16, 819–826 (2000)

Shinar, D.: Psychology on the raoad. In: The human factor in traffic safety, pp. 29–40. John Wiley & Sons, USA (1978)

Valverde, L., Santos, M., López, V.: Fuzzy decision system for safety on roads. In: Intelligent Decision Making Systems, World Scientific, pp. 326–331 (2009)

Xiao, J., Kulakowsk, B.T., Ei-Gindy, N.: Prediction of risk of wet-pavement accidents: fuzzy logic model. J. Transportation Research 1717, 28–36 (2000)

Zadeh, L.: The concept of linguistic variable and its application to approximate reasoning, parts I, II, III. Inform. Sci. 8, 199–249; 8, 301–357; 9, 43–80 (1975)

Part II

Risk Management in Business Decision Making

Chapter 10
A New Latex Price Forecasting Model to Reduce the Risk of Rubber Overproduction in Thailand

Jitian Xiao[1] and Panida Subsorn[1,2]

[1] School of Computer and Security Science,
Edith Cowan University, 2 Bradford St, Mt Lawley, WA 6050, Australia
j.xiao@ecu.edu.au
[2] Suan Dusit Rajabhat University, Bangkok, Thailand

Abstract. One of the key areas in risk management in the public rubber industry in Thailand (PARIT) is to accurately forecast rubber latex prices thus to adjust rubber production in a timely manner. Accurately forecasting rubber latex price may not only reduce risks of overproduction and costs of over stocking, but also respond promptly and directly to global market thus improve in gaining higher sales in the competitive rubber marketing environment. This chapter presents a rubber latex price forecasting model, with three variations, i.e., one-year prediction, 6-month prediction and 4-month prediction, each embedding with either non-neural or neural network training techniques. The model is validated using actual rubber latex prices trend data, which in turn compared with experimental forecasting results to determine forecasting accuracy and the best-fitting model for policy makers in PARIT.

Keywords: Risk management, latex price, forecast, neural network training techniques.

1   Introduction

The last several years have witnessed a tremendous increase and volatility in commodity prices, such as natural rubber prices. The most recent episode with such an increase in commodity prices is the period from 2002 and 2008, which has been labeled a "supercycle" by many observers. After a sharp sell-off during the 2001 recession, the rubber prices rose to unprecedented levels between 2006 and 2008 only to collapse in late 2008 due to the global economic downturn [10]. At the initial stage of the supercycle, the demand ran ahead of available supply, reducing inventories. The supply-side of market then began to respond by raising prices. Higher prices then affected demand, with resultant slowing down of consumption growth. At the same time, investments were made, which, after a period of time, began to yield higher production. With production growth first catching and then surpassing consumption growth, market began to shift from deficit to equilibrium, and then to surplus. The

J. Lu, L.C. Jain, and G. Zhang: Handbook on Decision Making, ISRL 33, pp. 191–203.
springerlink.com   © Springer-Verlag Berlin Heidelberg 2012

transformation in markets was completed by 2008, although prices kept escalating for a period of time before it dived sharply. Even so, the behavior of markets during the last 5 years was normal in the broadest sense [10].

In addition to raising commodity prices, the last few years also saw a very high level of volatility in commodity prices [10]. The various players in the rubber supply chain have evolved different responses to the volatility [6]. The increased volatility has raised the need for reliable and timely market intelligence. While buyers and purchasing agents need to have a good feel of where prices are heading before entering a new contract and negotiation, suppliers and production planners also need a concise yet efficient way to forecast the market demand and price movement.

Southeast Asia (Thailand, Indonesia and Malaysia) produced over 70 percent of global natural rubber, with Thailand being the largest of the three producers [17, 10]. Thailand, as the world's largest rubber exporter, sells rubber and raw rubber products to countries around the world, and possesses about 39 percent share of the world market [24, 15, 18]. Their exported rubber products include rubber latex, rubber sheet, rubber block and other primary rubber products which contribute to production of tyres, gloves and shoes, etc. To monitor global rubber production and market trends, Thai Rubber Association offers world's natural rubber production data on its website. Although the data is more than 4 months old, it does give an accurate breakdown of production data by the major natural rubber producing countries.

Rubber latex prices in the public agricultural rubber industry in Thailand (PARIT) are affected by many factors, such as global rubber demand, changes in supply, input costs, government policies, economic and political factors in local and global markets, etc. A small change in these factors may cause large rubber price fluctuations and cause difficulties for policy development and planning. It is also traditionally accepted that it is more difficult to pinpoint an exact cause for upward or downward pricing pressure on rubber prices at any point in time because the markets are globally more competitive and rubber production costs typically do not influence rubber market prices. In addition, exchange rates and interest costs influence prices of all commodities.

According to Dana and Gilbert [6], price management techniques have the potential to improve the functioning of the rubber supply chain in developing economies. They pointed out that many developing countries still lack expertise on market-based approach to managing risk, a problem which is exacerbated by cumbersome decision-making processes that permit insufficient delegations for quick response to market signals. Dana and Gilbert suggested some important steps to the agricultural commodity risk management for the developing countries, including those of applying modern financial techniques for identifying and quantifying risk, and monitoring price exposure throughout the course of the season, and establishing the type of risk management monitoring and reporting functions, and so on.

Risk management is an important step to sustain the natural rubber industry in Thailand. One of the key areas in risk management in PARIT is to accurately forecast rubber latex prices thus to adjust rubber production accordingly and in a timely manner. Accuracy in forecasting rubber latex price not only reduces risks of overproduction and costs of over stocking, but also allows for a prompt and direct response to the global market, ultimately improves in gaining higher sales in the competitive rubber marketing environment. Results from forecasting directly affect PARIT in the areas of risk management, planning, production, sales and prices. Taking rubber plant planning as an example, too much planting may result in rubber overproduction which increases rubber inventory, leading to rubber latex price cut; while too little planting may lead to losing opportunity for higher economic income from the rubber exports in the Thai rubber industry sector.

Forecasting, as a significant capability of decision support systems, provides useful information in supporting organizations' decision making processes. It is one of the critical steps in facilitating desired management and improving various performances of organizations because it enables prediction of future events and conditions by statistically analyzing and utilizing data or information from the past [12, 25]. While the forecasting results may directly affect many areas such as planning, production, sales and prices [9, 16] in PARIT, the success of performing forecasting activities depends on a trustworthy tool to enhance accuracy of the forecasting results.

This chapter investigates a possible rubber latex price forecasting model, with three variations, i.e., one-year prediction, 6-month prediction and 4-month prediction, each embedding with either non-neural or neural network training techniques. The model is validated using actual rubber latex prices trend data, and its outcomes are compared with experimental forecasting results to determine forecasting accuracy and the best-fitting model for policy makers in PARIT.

The rest of the chapter is organized as follows: Section 2 briefly introduces the rubber industry in Thailand. Section 3 presents the methods used in this study. Experiments are developed by employing non-neural network training and neural network training techniques in Section 4. Section 5 concludes the chapter.

2   The Public Agricultural Rubber Industry in Thailand

Thailand is the world's largest rubber exporter. It exports rubber or raw rubber products to many countries over the world, such as Japan, China and the United States of America, etc., with a 39 percent share of the world market [15, 18]. The exported products include rubber sheet, rubber block, rubber latex and other primary rubber products, whereas only a small fraction of rubber products is reserved for manufacturing within Thailand [24, 15, 18].

The agricultural sector is one of the significant growth sectors in Thailand. This sector is responsible for rubber production planning and selling price policy development. However, little attention has been paid to enhancing price forecasting models or improving the accuracy in forecasts in this sector even though several studies in PARIT have focused on management, control and forecasting in the recent years [24, 11]. A focus on improving forecasting is exceptional as the agricultural sector has been using similar models for a long period of time. Their existing models apply traditional statistical techniques only.

The Office of Agricultural Economics (OAE) in Thailand makes its rubber latex price trend data available in its website, which are utilized as information references for non-government agencies to create their own production planning. The data also assists Thai agriculturists to decide upon quantities of rubber to produce for the next period of time. In addition, the data is also employed to conduct research for a rubber latex price index every month to provide information about the current situation and assist planning of emergency policy for the future [1, 2].

We attempt to derive a feasible forecasting model for rubber latex prices in PARIT using artificial intelligence (AI) techniques, such as non-neural and neural network training techniques. We use the monthly rubber price data from PARIT's websites to create an experimental environment for testing the model and verifying the proposed techniques in this study. The outcomes of this study may be used to assist policy makers in the agricultural sector in forecasting future rubber price trends in a competitive environment through policy development and implementation. This may assist in advancing the agricultural sector's forecasting practice and thereby contribute to Thailand's economic development.

3   The Rubber Price Forecasting Model and Analysis Procedure

There has been a dramatic development in combining traditional forecasting techniques, such as the statistics-based forecasting techniques, to IT-related forecasting techniques, especially AI-based techniques. Examples of such techniques include fuzzy logic, neural network, genetic algorithms and hybrid ones [14]. While these modern forecasting techniques have been widely used around the world, especially in developed countries, it is not the case in some traditional areas and sections in Thailand, such as in PARIT. The traditional statistics-based forecasting models, and similar ones, have been used by PARIT for a long period of time [24, 11]. The adoption of modern forecasting techniques may not only improve the accuracy in forecasting results, but also enhance forecasting process and risk management practices in PARIT.

There have been some works on utilizing AI techniques for forecasting purposes in the past few years. For example, Co's work [4] compared the performance of artificial neural networks with exponential smoothing and ARIMA models in forecasting rice exports from Thailand. However, our work differs from his work in a few ways. First, our model is for forecasting rubber latex prices, while his work is for forecasting rice exports. Secondly, while we focus on creating forecast models using neural network and non-neural networking training techniques, his work compared the performance of artificial neural networks with traditional forecasting models.

Existing forecasting methods differ in objectives and application problems within the organizations. Basically, one forecasting technique may be efficient for particular application scenarios but it may be unsuitable or inaccurate for other situations. This study, based on a comparative method, attempts to gain a feasible forecasting model that supplies less forecasting errors for the Thai rubber industry. We first provide a description of non-neural and neural network tech-niques in the next subsection, and then present the main components and data analysis procedures of the new forecasting model.

3.1   The Training Techniques

Two types of training techniques, i.e., *neural network (NN)* training technique and *non- neural network (non-NN)* training one, are used in our latex price forecasting model. A brief description of each will be given in this subsection.

3.1.1 Non-NN Training Technique
Time series analysis is often applied in prediction for the component analysis of historical data sets to determine a forecasting model used to predict the future [19].

There are several well-known time series forecasting techniques such as simple moving average (SMA), weight moving average (WMA), exponential smoothing (ES) and seasonal autoregressive integrated moving average (SARIMA) [12, 19]. Among these techniques, ES has the capability to create trend and seasonal analysis efficiently for time series forecasting, while SARIMA has the capability and efficiency of creating seasonal time series forecasting based on a moving average (MA). This study deployed ES and SARIMA as base techniques for rubber latex price forecasting.

The formula for the ES technique, particularly simple seasonal exponential smoothing, is shown below:

$$L(t) = \alpha(Y(t) - S(t-s)) + (1-\alpha)L(t-1)$$
$$S(t) = \delta(Y(t)) + (1-\delta)S(t-s)$$
$$\hat{Y}_t(k) = L(t) + S(t+k-s)$$

where $\hat{Y}_t(k)$ is the model-estimated k-step ahead forecasting at time t for series Y, t the trend, L the length of time, S the seasonal length, α the level smoothing weight and δ is the season smoothing weight [21].

Time series data from several consecutive periods of time are added and divided to obtain mean values to create the prediction [19]. The formula of the SARIMA technique or equivalent to autoregressive integrated moving average (ARIMA) (0, 1, (1, s, +1)) (0, 1, 0) with restrictions among MA parameters is as follows:

$$\Phi(B)[\Delta y - \mu] = \Theta(B)a_t \qquad t = 1,..., N$$

where

$$\Phi(B) = \varphi_p(B)\Phi_P(B)$$
$$\Theta(B) = \theta_q(B)\Theta_Q(B)$$

and N is the total number of observations, a_t $(t = 1, 2, ..., N)$ is the white noise series which normally distributed with mean zero and variance σ_a^2, p is the order of the non-seasonal autoregressive element, q is the order of the non-seasonal moving average element, P is the order of the seasonal autoregressive element, Q is the order of the seasonal moving average element, s it the seasonality or period of the model, $\varphi_p(B)$ is the autoregressive (AR) polynomial of B of order p, $\theta_q(B)$ is the MA polynomial of B of order q, $\Phi_P(B)$ is the seasonal AR polynomial of BS of order P, $\Theta_Q(B)$ is the seasonal MA polynomial of BS of order Q, Δ is the differencing operator, B is the backward shift operator, and μ is the optional model constant or the stationary series mean. Independent variables $x_1, x_2, ..., x_m$ may be included in the model as the formula is shown below.

$$\Phi(B)\left[\Delta\left(y_t - \sum_{i=1}^{m} c_i x_{it}\right) - \mu\right] = \Theta(B)a_t$$

where c_i, $i = 1, 2, ..., m$, is the regression coefficients for the independent variables.

3.1.2 NN Training Technique

Neural network is a well-known predictive technique that claims to provide more reliable results than other forecasting techniques [19, 22, 23]. This technique creates a relationship between dependent and independent variables from several training data sets during the learning process. The results from this learning process are called *neurons*. Neurons arrange themselves in a level form and have connection lines to transfer or process data from the input, hidden to output layers. Each connection line presents weights between each layer connection. Moreover, neurons adjust their weights via an activation function, which is the processing function to create results that are used to calculate the desirable results [19, 7]. The activation function deployed in this study is the sigmoid function, which is a non-linear function. Its formula is

$$\gamma(c) = \frac{1}{1 + \exp(-c)}$$

To improve the reliability and accuracy of the forecasting results, we employed a supervised learning technique, i.e., feed-forward back propagation neural network (BPN), which was considered to be a suitable learning technique for rubber latex price forecasting. The supervised learning technique is used to adjust weights for producing forecasting with fewer errors between an output from NN and a desirable output. A feed-forward architecture is a one way connection from the input, hidden to output layers within the network in the model [22]. The input layer consists of independent variables or predictors. The hidden layer consists of unobservable nodes or units, presenting a function to be utilized for independent variables or predictors. The output layer consists of dependent variable(s). Additionally, a feed-forward BPN has the capability to simulate the complicated relationship of the function correctly, which is called a universal approximator, without having previous knowledge of the function relationships [19, 22, 8].

However, overtraining of data sets may cause low efficiency of non-training data sets in the forecasting model. Thus data separation is introduced to solve this problem by dividing the same data set into two groups, namely a training data set and a test data set [19, 8]. Based on the Crowther and Cox's principle [5, 20], this study partitioned the data set at 70 percent for the training data set and 30 percent for the testing data set.

The multilayer perceptron (MLP) was used in this study to facilitate the use of a supervised learning network and a feed-forward BPN. The MLP network is a function of one or many independent variables or predictors in the input layer which may reduce forecasting errors of one or many dependent variables in the output layer [26]. The MLP formulae are shown below.

Input layer: $J_0 = P$ units, $a_{0:1} \cdots a_{0:J_0}$ with $a_{0:j} = x_j$

Hidden layer: J_i units, $a_{i:1} \cdots a_{i:J_i}$ with $a_{i:k} = \gamma(c_{i:k})$ and $C_{i:k} = \sum_{j=0}^{J_{i-1}} w_{i:j,k} a_{i-1:j}$

where $a_{i-1:0} = 1$

Output layer: $J_l = R$ units, $a_{l:1}, \ldots, a_{l:J_l}$ with $a_{l:k} = \gamma(c_{l:k})$ and $c_{l:k} = \sum_{j=0}^{J_l} w_{l:j,k} a_{l-1:j}$

where $a_{i-1:0} = 1$

Additionally, error measurements in the forecasting model used in this study are the root mean square error (RMSE) and mean absolute percentage error (MAPE). The formula of this error measurement is

$$RMSE = \sqrt{\frac{\sum (Y(t) - \hat{Y}(t))^2}{n-k}}$$

where *RMSE* is root mean square error value of the forecasting model. *Y* is the original series of time (*t*), *t* is time series, *n* is the number of non-missing residuals and *k* is the number of parameters in the model.

MAPE is an average of the differences between the actual and forecasted data of the absolute percentage [3]. It is one of the commonly used measurements for assessing accuracy in quantitative forecasting because it is easy to compute and understand as it reported the results in typical percentage format [13].

The formula of this error measurement is represented as

$$\sum |PE|/N$$

where PE is the absolute value of the percent error and N the number of periods for the percent error [13], or

$$[(\sum | \text{Actual data} - \text{forecasted data} | / \text{Actual data}) \times 100 / N]$$

3.2 Components of the Forecasting Model

The new forecasting model for latex prices in PARIT has three main components: input, processing and output components. The input component consists of two main subcomponents, one for individual and the other for group policy makers to input data for forecasting. Policy makers may input new rubber data sets to create forecasts, or retrieve data from the existing rubber price database (and/or forecasting results database) to form dataset/s for forecasting.

The processing component consists of three main subcomponents, namely time-series forecasting, AI and error measurement. This study analyzes rubber data sets into non-neural and neural network training date sets. Each data set is separately processed using the Statistical Package for the Social Sciences (SPSS) to create forecasts. Forecasting errors for accuracy and reliability purposes can be measured with each data set.

The output component consisted of two main subcomponents, namely the forecasting results and the database. The results are produced during the processing components. Both forecasting data sets (non-neural and neural network training) are then compared to the actual corresponding data set. The forecast data is stored in the database for retrieval and modification for policy makers to make decisions.

3.3   Data Analysis Procedures

This study deployed rubber latex price data sets from the website of OAE in Thailand to examine a newly refined price forecasting model. These data sets were collected on a monthly and yearly basis from the period January 2004 to April 2009. We focus on rubber latex prices in three provinces: Rayong, Nakornsrithammarat and Pattani, Thailand. The data analysis process involved six procedures, as described below.

Data Preparation: Time series data from January 2004 to April 2009 are used to prepare data sets for four months, six months and one year for January to December 2007 and January to April 2009 forecasting. The reason for selecting forecasting for three different time periods is to enhance validity and reliability of the newly refined forecasting model.

Sequence Chart Creation: A sequence chart to consider rubber price trends from January 2004 to April 2009 is plotted before the forecasts are created. This chart displays rubber latex price trends to examine a seasonality factor within the trends, so that suitable forecasting techniques could be selected and utilized.

Data processing: Non-neural and neural network training are used for analysis with application of forecasting techniques provided in SPSS, namely ES and SARIMA.

Error Measurement: Errors in the forecasts are examined while the SPSS application created the forecasts for non-neural and neural network training. Forecasting accuracy is thereby strengthened.

Results comparison: Forecasting results from non-neural and neural network training are compared with the actual price data sets for these provinces to determine the possibility of using this new forecasting model in this industry.

Figures Creation: Figures are created to present forecasting results, comparisons and errors to policy makers and/or decision makers, so that they may have a better understanding or visual summary when planning and/or making decisions.

The following section presents experimental results based on the data from the three selected provinces where the market prices for rubber latex in Thailand are most significant. The results are compared between non-neural and neural network training. The results are then classified and presented in four months, six months and one year subsets, respectively.

4   Experimental Results

This section presents the rubber latex price trends and forecasting results. The trends data from three selected provinces, Rayong, Nakornsrithammarat and Pattani in Thailand are used as testing data sets. The data span from January 2004 to April 2009, as shown in Fig. 1. It can be seen that the rubber latex price data from the abovementioned three provinces have similar trends. Generally, the rubber latex prices increase in the middle of the year as demand for rubber rises while there is less rubber production. Prices decrease at the end of the year as there is high rubber production and normal rubber demand. In Thailand, the rainy season starts from June and ends in October, and the winter or mild season starts from November and ends in February the next year. It is apparent that rubber latex prices are time series data (with seasonality factors).

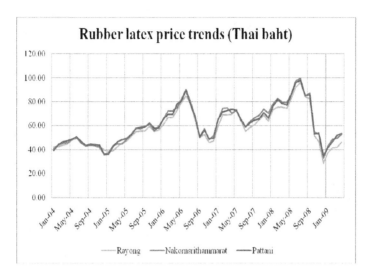

Fig. 1. Summary of rubber latex price trends

Rubber latex price forecasting results are used with a hold-out sample method to perform forecasting result validation and evaluation of forecasting accuracy. Rubber latex prices data collected are based upon a hold-out sample method in this section. A hold-out sample method was also utilized to prevent data overtraining. As mentioned early, the data set was separated into two groups, called training and testing data sets at the rates of 70 and 30 percent, respectively. The confidence interval was set to 95 and 99 percent for RMSE and MAPE in the SPSS software for each experiment.

While we have conducted a range of forecasting trends, we show only a part of the experimental results to demonstrate the feasibility of our method. For example, we present the forecasting results in four months trends, six months trends and one year trends, respectively. We selected the experimental results for Rayong province as a representative exemplar to present the forecasting results as those from other provinces also have the similar forecasting results.

4.1   One Year Rubber Latex Price Forecasts

Fig. 2 shows the actual Rayong price forecasts for January to December 2007. Two types of forecasts, using either non-neural or neural network training approach, are shown in the figure. For the 4-month prediction, three sets of forecasting results (i.e., January-to-April curve, May-to-August curve, and September-to-December curve) are concatenated to form a single one-year forecast curve. Similarly, the 6-month prediction concatenated two half-year predictions in the figure. As a comparison, the actual rubber latex prices curve is also shown in the figure. It demonstrates that the one year prediction is the most accurate one among the three predictions. Actually, the one year prediction has RMSE and MAPE values which were 3.6328 or 5.13 percent for non-neural training and 4.3943 for neural network training, respectively. In contrast, the RMSE and MAPE values for the four month prediction and the six month prediction are much higher. In addition, the 6-month prediction is more accurate than the

4-month prediction. This is because that the 6-month prediction has a RMSE and MAPE of 3.9404 or 5.52 percent for non-neural network training and 4.5526 for neural network training, and the 4-month prediction has a RMSE and MAPE of 4.0245 or 5.65 percent for non-neural network training and 4.5796 for neural network training.

The 2007 forecast by the NN training technique for Rayong province was shown in Fig. 3. Similar to the forecasting results displayed in Fig. 2, the one year forecast created lower RMSE and MAPE values, which were 3.8957 or 5.41 percent, than the other two forecasts. The RMSE and MAPE values were 4.0053 or 5.58 percent for the six month forecast and 4.0951 or 5.71 percent for the four month forecast, both of which were much higher than the one year forecast. It can be summarized that the one year forecast performed better than the four month and the six month forecasts.

Comparing the actual rubber latex price curve with the forecasting curves demonstrated that the one-step ahead forecast, January to December, was the most accurate one among the three periods of forecasts for both non-NN and NN training techniques. Based on the forecasting results, RMSE and MAPE values, the one year forecast provided the best-fitting forecasting model for adoption in PARIT, followed by the six month and the four month forecasts. Furthermore, the forecasting results with the use of the non-NN training technique, particularly the ES technique for three different time intervals, were better than the NN training technique. Therefore, it can be summarized that the one year forecast with the non-NN training technique provided a better performance than the other 2007 forecasts for Rayong province. Both the non-NN and NN training techniques created similar forecasts and none was found to be close to the actual data. This occurred because of a short-term fluctuation in the input data. Additionally, the major reason that the one year forecast performed better than the four month and the six month forecasts is the seasonality which can generate fluctuations in data. This suggests that further research is needed to follow in this direction in the future.

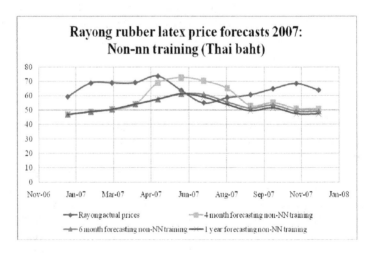

Fig. 2. Summary of Rayong rubber latex price forecasts by the non-NN training for January to December 2007

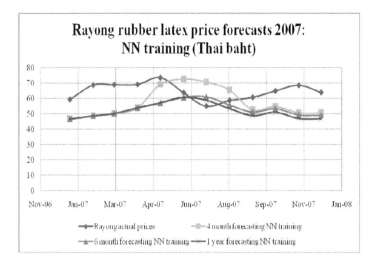

Fig. 3. Summary of Rayong rubber latex price forecasts by the NN training for January to December 2007

4.2 Four Month Rubber Latex Price Forecasts

Apart from Rayong rubber latex price forecasts for January to December 2007, another experiment for January to April 2009 was conducted in order to examine the proposed forecasting model further. Non-NN and NN training forecasting results were, once again, compared with actual rubber latex prices for January to April 2009, and are shown in Figure 4. It demonstrated that the results produced using the NN technique was marginally better than those of the non-NN training technique. The RMSE and MAPE values were 6.2279 or 7.70 percent for the non-NN training and 6.2211 or 7.53 percent for the neural network training.

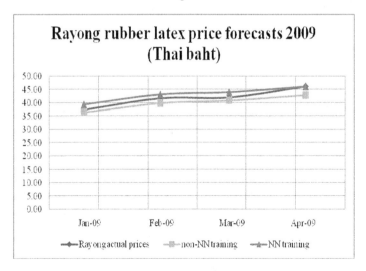

Fig. 4. Summary of Rayong rubber latex price forecasts for January to April 2009

For the four month prediction, the RMSE and MAPE values suggest that the model that includes neural network training prediction provides the best-fitting forecasting model. Based on this analysis, these results demonstrate that the neural network training method generated more accurate forecast results than that of the non-neural network training method. Our experiments also showed that the forecasting with neural network training support is more useful when there is not much data fluctuation to cause forecasting noise or errors. Similar to the case of one year forecasting, it does require independent variables in forecasts, unlike non-neural network training. Again, both non-neural network training and neural network training showed similar trends, as displayed in the Fig. 4.

5   Conclusions

This chapter described a new latex price forecasting model for PARIT, in the aim of improving the accuracy of the latex price forecasting thus to reduce the risk of rubber overproduction in PARIT.

We investigated the best-fitting forecasting model for rubber latex prices forecasting. The method was based on non-neural and neural network training techniques. We compared the forecasting results with the actual rubber latex price data to determine the best-fitting forecasting model. Experiments have shown that the model with non-neural network training generates more accurate forecasts for one year prediction, and the model with neural network training generates more accurate forecasting result for the four month prediction.

To our knowledge, this preliminarily study brings a new perspective to policy makers in PARIT in creating forecasting with AI techniques. This method may be considered as a possible decision support tool in rubber latex price forecasting in Thailand. Further research on longer time-span rubber price forecasting is needed to judge more clearly how effective this forecasting model may be applied to rubber price forecasting in PARIT.

References

1. Center of Agricultural Information, Crop forecasting in Thailand, C.o.A. Information. The Office of Agricultural Economics, OAE, Bangkok (2002)
2. Center of Agricultural Information, Agricultural prices information management, Center of Agricultural Information. The Office of Agricultural Economic, OAE, Bangkok (2009)
3. Chiulli, R.M.: Quantitative analysis: An introduction Automation and Production Systems. In: Methodologies and Applications. CRC Press, Singapore (1999)
4. Co, H.C.: Forecasting Thailand's rice export: Statistical techniques vs. artificial neural networks. Computers & Industrial Engineering 53, 1140–1154 (2007)
5. Crowther, P.S., Cox, R.J.: A Method for Optimal Division of Data Sets for Use in Neural Networks. In: Khosla, R., Howlett, R.J., Jain, L.C. (eds.) KES 2005. LNCS (LNAI), vol. 3684, pp. 1–7. Springer, Heidelberg (2005)
6. Dana, J., Gilbert, C.L.: Managing Agricultural Price Risk in Developing Countries. In: Geman, H. (ed.) Risk Management in Commodity Markets: from Shipping to agricultural and Energy, pp. 207–237. John Wiley & Sons, Ltd. (2008)

7. Fausett, L.: Fundamentals of neural networks: architectures, algo-rithms, and application. Prentice-Hall, USA (1994)
8. Funahashi, K.: On the approximate realization of continuous mappings by neural networks. Neural Networks 2, 183–192 (1989)
9. Geurts, M., Lawrence, K.D., Guerard, J.: Forecasting sales. JAI PRESS INC., Greenwich (1994)
10. Hodge, L., Mothersole, J.: Pricing Risks – Be prepared for the Next Round of Commodity Price Swing. In: 95th ISM Annual International Supply Management Conference (April 2010)
11. Leechawengwong, M., Prathummintra, S., Thamsiri, Y.: The Devel-opment of Basic Rubber Database. Rubber Research Institute of Thailand, Bangkok (2002)
12. Markland, R.E., Sweigart, J.R.: Quantitative methods: Applications to managerial decision making. John Wiley & Sons Inc., Canada (1987)
13. Mentzer, J.T., Moon, M.A.: Sales forecasting management: A de-mand management approach, 2nd edn. SAGE Publications, California (2005)
14. Negnevitsky, M.: Artificial intelligence: a guide to intelligent systems, 2nd edn. Addison-Wesley, New York (2005)
15. Office of Industrial Economics & Economic Research and Training Center. In: Study of Industrial Structure for Enhancing Competitiveness of Rubber Product Industries. Ministry of Industry, Bangkok (1998)
16. Olson, D.L., Courtney, J.F.: Decision support models and expert systems. Macmillan Publishing Company, New York (1992)
17. The Rubber Economist Ltd Risk Management in the Rubber Industry (2010), http://www.therubbereconomist.com/The_Rubber_Economist/ Risk_management_in_the_rubber_industry.html (searched on November 20, 2010)
18. Rubber Research Institute of Thailand (1997); Academic rubber information. Rubber Research Institute of Thailand, Bangkok (2007)
19. Sangpong, S., Chaveesuk, R.: Time series analysis techniques for forecasting pineapple yield in various sizes. In: การประชุมวิชาการด้านการวิจัยดำเนินงานแห่งชาติ ประจำปี ๒ ๕๕๐ (2007)
20. Soyhan, H.S., Kilic, M.E., Gokalp, B., Taymaz, I.: Performance comparison of MATLAB and neuro solution software on estimation of fuel economy by using artificial neural network. In: The Second International Conference "Intelligent Information and Engineering Systems" (INFOS 2009), Varna, Bulgaria, pp. 71–76 (2009)
21. SPSS Inc. TSMODEL algorithms, Statistical Package for the Social Sciences (2009)
22. SPSS Inc. Introduction to neural networks. Statistical Package for the Social Sciences (SPSS) (2009-1)
23. SPSS Inc. MLP algorithms. Statistical Package for the Social Sciences (2009-2)
24. Subsorn, P.: Enhancing rubber forecasting: The case of the Thai rub-ber industry. In: The Ninth Postgraduate Electrical Engineering & Computing Symposium (PEECS 2008). The University of Western Australia, Perth (2008)
25. Tomita, Y.: Introduction to forecasting, http://www.math.jmu.edu/~tomitayx/math328/Ch1Slide.pdf (retrieved January 8, 2008)
26. Twomey, J.M., Smith, A.E.: Validation and verification. In: Artificial Neural Networks for Civil Engineers: Fundamentals and Applications. ASCE Press, New York (1996)

Chapter 11
An Agent-Based Modeling
for Pandemic Influenza in Egypt

Khaled M. Khalil, M. Abdel-Aziz, Taymour T. Nazmy, and Abdel-Badeeh M. Salem

Faculty of Computer and Information Science Ain Shams University Cairo, Egypt
{kmkmohamed,mhaziz67}@gmail.com, ntaymoor@yahoo.com,
absalem@asunet.shams.edu.eg

Abstract. Pandemic influenza has great potential to cause large and rapid increases in deaths and serious illness. The objective of this paper is to develop an agent-based model to simulate the spread of pandemic influenza (novel H1N1) in Egypt. The proposed multi-agent model is based on the modeling of individuals' interactions in a space-time context. The proposed model involves different types of parameters such as: social agent attributes, distribution of Egypt population, and patterns of agents' interactions. Analysis of modeling results leads to understanding the characteristics of the modeled pandemic, transmission patterns, and the conditions under which an outbreak might occur. In addition, the proposed model is used to measure the effectiveness of different control strategies to intervene the pandemic spread.

Keywords: Pandemic Influenza, Epidemiology, Agent-Based Model, Biological Surveillance, Health Informatics.

1   Introduction

The first major pandemic influenza H1N1 is recorded in 1918-1919, which killed 20-40 million people and is thought to be one of the most deadly pandemics in human history. In 1957, a H2N2 virus originated in China, quickly spread throughout the world and caused 1-4 million deaths world wide. In 1968, an H3N2 virus emerged in Hong Kong for which the fatalities were 1-4 million [16]. In recent years, novel H1N1 influenza has appeared. Novel H1N1 influenza is a swine-origin flue and is often called swine flue by the public media. The novel H1N1 outbreak began in Mexico, with evidence that there had been an ongoing epidemic for months before it was officially recognized as such. It is not known when the epidemic will occur or how sever it will be. Such an outbreak would cause a large number of people to fall ill and possibly die.

In the absence of reliable pandemic detection systems, computer models and systems have become important information tools for both policy-makers and the general public [15]. Computer models can help in providing a global insight of the infectious disease outbreaks' behavior by analyzing the spread of infectious diseases in a given population, with varied geographic and demographic features [12]. Computer models promise an improvement in representing and understanding the complex social

J. Lu, L.C. Jain, and G. Zhang: Handbook on Decision Making, ISRL 33, pp. 205–218.

structure as well as the heterogeneous patterns in the contact networks of real-world populations determining the transmission dynamics [4]. One of the most recent approaches of such sophisticated modeling is agent-based modeling [3]. Agent-based modeling of pandemics recreates the entire populations and their dynamics through incorporating social structures, heterogeneous connectivity patterns, and meta-population grouping at the scale of the single individual [3].

In this paper we propose a stochastic multi-agent model to mimic the daily person-to-person contact of people in a large scale community affected by a pandemic influenza in Egypt. The proposed model is used to: (i) assess the understanding of transmission dynamics of pandemic influenza, (ii) assess the potential range of consequences of pandemic influenza in Egypt, and (iii) assess the effectiveness of different pandemic control strategies on the spread of the pandemic. We adopt disease parameters and the recommended control strategies from WHO [16]. While, we use Egypt census data of 2006 [7] to create the population structure, and the distribution of social agent attributes. Section 2 reviews different epidemiological modeling approaches: mathematical modeling, cellular automata based modeling, and agent based modeling. While, section 3 reviews related multi-agent models in literature. Section 4 discusses the proposed model, and section 5 validates the proposed model. Section 6 discusses the pandemic control strategies and their effect on the spread of the pandemic. Section 7 presents the modeling experiments and analysis of results, and then we conclude in Section 8.

2   Epidemiological Modeling Approaches

The search for an understanding of the behavior of infectious diseases spread has resulted in several attempts to model and predict the pattern of many different communicable diseases through a population [5]. The earliest account was carried out in 1927 by Kermack and McKendrick [9]. Kermack and McKendrick created a mathematical model named SIR (Susceptible-Infectious-Recovered) based on ordinary differential equations. Kermack and McKendrick started with the assumption that all members of the community are initially equally susceptible to the disease, and that a complete immunity is conferred after the infection. The population is divided into three distinct classes (see Fig 1): the susceptible (S) healthy individuals who can catch the disease; the infectious (I) those who have the disease and can transmit it; and the recovered (R) individuals who have had the disease and are now immune to the infection (or removed from further propagation of the disease by some other means).

Fig. 1. SIR (Susceptible–Infectious–Recovered) Model

Let $S(t), I(t)$, and $R(t)$ be the number of susceptible, infected and recovered individuals, respectively, at time t, and N is the size of the fixed population, so we have:

$$N = S(t) + I(t) + R(t)$$

(1)

Upon contact with an infected a susceptible individual contracts the disease with probability β, at which time he immediately becomes infected and infectious (no incubation period); infectious recover at an individual rate γ per unit time. Based on mentioned assumptions; Kermack and McKendrick derived the classic epidemic SIR model as follows:

$$\frac{dS}{dt} = -\beta SI$$

$$\frac{dI}{dt} = \beta SI - \gamma I \tag{2}$$

$$\frac{dR}{dt} = \gamma I$$

From equations (1) and (2), we found that SIR model is deterministic and doesn't study the nature of population vital dynamics (handling newborns and deaths). Following Kermack and McKendrick, other physicians contributed to modern mathematical epidemiology; extending SIR model with more classes and supporting vital dynamics such as: SEIR (Susceptible–Exposed–Infectious–Recovered), and MSEIR (Immunized–Susceptible–Exposed–Infectious–Recovered) models [9]. However, mathematical models had not taken into account spatial and temporal factors such as variable population structure, and dynamics of daily individuals' interactions which drive more realistic modeling results [1].

The second type of developed models is cellular automata based models, which incorporate spatial parameters to better reflect the heterogeneous environment found in nature [13]. Cellular automata based models are an alternative to using deterministic differential equations, which use a two-dimensional cellular automaton to model location specific characteristics of the susceptible population together with stochastic parameters which captures the probabilistic nature of disease transmission [2]. Typically a cellular automaton consists of a graph where each node is a cell. This graph is usually in the form of a two-dimensional lattice whose cells evolve according to a global update function applied uniformly over all the cells [13]. Cell state takes one of the SIR model states, and is calculated based on cell present state and the states of the cells in its interaction neighborhood. As the Cellular automata based model evolves, cells states will determine the overall behavior of a complex system [2]. However, cellular automata based models neglect the social behavior and dynamics interactions among individuals in the modeled community. Therefore, cellular automata gave the way to a new approach; Agent-based models.

Agent-based models (ABM) are similar to cellular automata based models, but leverage extra tracking of the effect of the social interactions of individual entities [1]. Agent-based model consists of a population of agents, an environment, and set of rules managing agents' behavior [12]. Each agent has two components: a state and a step function. Agent state describes every agent attributes values at the current state. The step function creates a new state (usually stochastically) representing the agent attributes at the next time step. The great benefit of agent-based models is that these models allow epidemiological researchers to do a preliminary "what-if" analysis with the purpose of assessing systems' behavior under various conditions and evaluating which alternative control strategies to adopt in order to fight epidemics [12].

3   Multi-agent Related Models

This section discusses the risk-based decision making process which includes the main risk-based decision making activities, the types of decision making process and the decision support technology for risk-based decision making. The decision support technology discussed briefly as this chapter focusing on the five main elements of risk-based decision making framework for investment in real estate industry.

3.1   Main Risk-Based Decision Making Activity

Related agent-based models are Perez-Dragicevi model, BIOWar, and EpiSims. Perez-Dragicevi [12] had developed a multi-agent model to simulate the spread of a communicable disease in an urban environment using measles outbreak in an urban environment as a case study. The model uses SEIR (Susceptible–Exposed–Infectious–Recovered) model and makes use of census data of Canada. The goal of this model is to depict the disease progression based on individuals' interactions through calculation of ratios of susceptible/infected in specific time frames and urban environments. BIOWar [10] is a computer model that combines computational models of social networks, communication media, and disease transmission with demographically resolved agent models, urban spatial models, weather models, and a diagnostic error model to produce a single integrated model of the impact of a bioterrorist attack on an urban area. BIOWar models the population of individual agents as they go about their lives. BIOWar allows the study of various attacks and containment policies as revealed through indicators such as medical phone calls, insurance claims, death rates, over-the counter pharmacy purchases, and hospital visit rates, among others. EpiSims [8] is an agent-based model, which combines realistic estimates of population mobility, based on census and land-use data of USA, with configurable parameters for modeling the progress of a disease. EpiSims involves a way to generate synthetic realistic social contact networks in a large urban region.

However, the proposed agent-based model will differ from the above models for: (i) usage of census data of Egypt, (ii) proposed extension to SIR model, (iii) and studying the effect of different control strategies on the spread of the disease. This study is considered very important which incorporate Egypt population structure in modeling process. We have adopted Egypt census data of 2006 for creating realistic social contact networks such as home, work and school networks. While, the proposed extension to SIR model encompasses new classes; modeling the real pandemic behavior and control states such as (in contact, quarantine, not quarantined, dead, and immunized). In addition, we involve the study of the effect of different control strategies on the spread of the pandemic influenza. We plan in future work to integrate the proposed model with different simulation tools and models such as weather models, transportation models, and decision support models to build a complete system for pandemic management in Egypt.

4   Proposed Model

In what follows, we propose an extension to SIR model. Then we propose the multi-agent model based on the proposed extension of SIR model states. Finally, we validate the proposed multi-agent model by aligning with the classical SIR model.

4.1   Proposed Extension to SIR Model

We propose an extension to SIR model by adding extra classes to represent more realistic agent states (see Fig 2). In addition, we adopt stochastic approach to traverse among agent states using normal distribution. Agents are grouped based on the proposed extension to SIR model into nine classes. The first class is the (S) Susceptible agents, who are not in contact with infectious agents and are subject to be infected. At the start of the modeling, all agents fall in the (S) Susceptible class. The second class is the (C) in Contact agents, who are in direct contact with other infectious agents. The third class is the (E) Exposed agents, who are infected agents during the incubation time (latent) of the disease. The fourth class is the (I) Infectious agents, who are contagious. The fifth class is the (Q) Quarantined agents, who are infected agents quarantined by the health care authorities. The sixth class is the (NQ) Not Quarantined agents, who are infected agent but not quarantined. The seventh class is the (D) Dead agents. The eighth class is the (R) Recovered agents. The ninth class is the (M) Immunized agents, who are immunized against the disease infection.

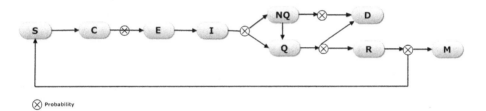

⊗ Probability

Fig. 2. State chart of proposed extension to SIR model. (S) Susceptible, (C) in Contact, (E) Exposed, (I) Infectious, (Q) Quarantined, (NQ) Not Quarantined, (D) Dead, (R) Recovered, and (M) Immunized.

Fig 3 presents flow chart which explains in details the sequence of the state chart of the proposed extension to SIR model. All population members are born susceptible then may contact contagious agents (move into in-contact class). In-contact agents may acquire the infection (move into the exposed class) based on given distribution. Exposed agents remain non-contagious for given latent time. At the end of the latent time, agents will become contagious (move into the infectious class). Infected agents may ask for doctor help and thus become quarantined by health care authorities (move into quarantined class), or ignore disease symptoms (move to non-quarantined class) based on given distribution. Non-quarantined agents are the main source of disease in this model. Non-quarantined agents may ask for doctor help and thus become quarantined, or die (move to dead class) based on given distribution. Quarantined agents may response to disease drugs and become recovered (move to recovered class) or die (move to dead class) based on given distribution. Recovered agents may become immunized (move to immunized class) based on given distribution, or become susceptible again.

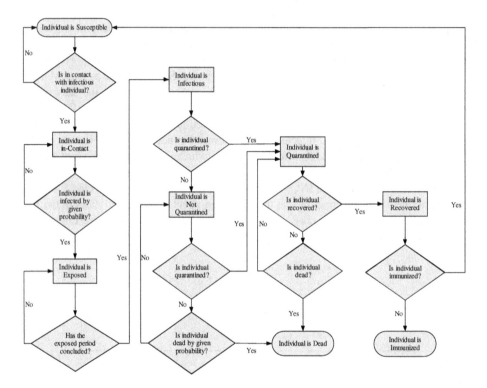

Fig. 3. Flow chart of the proposed extension to SIR model

4.2   Proposed Multi-agent Model

The proposed multi-agent based model attempts to realistically represent the behavior of individuals' daily activities, and the natural biological process of the pandemic influenza spread among individuals as a result of individuals' interactions. Proposed agent-based model involves (i) population agents, (ii) agents' rules which govern the behavior of the agents, (iii) and the infection transmission patterns following the proposed extension to SIR model. Agents represent human population, in which each agent is involved in a sequence of daily basis activities according to the agent social type. These daily activities allow agents to interact themselves in groups or even travel and join other groups. The daily activities of working, travelling, and public gathering are modeled, while agents' states are calculated on discrete time steps during agent life time.

Proposed multi-agent model has several parameters such as: simulation parameters, disease model parameters, agents' attributes, and population distribution based on census data. First: simulation parameters which include (i) number of days to be simulated, (ii) random seed for the gaussian random number generator, (iii) population size, (iv) and initial agents. Second: disease model parameters which include: (i) incubation time which is the average time of infected agent before being contagious, (ii) percentage of recovered infected agents after treatment, (iii) percentage of immunized agents after the recovery, (iv) percentage of dead agents, (v) average minimum and maximum time required to recover infected agent, and finally (vi) percentage of quarantined infected agents (see Table 1 for parameters default values). Third: agent attributes which are crucial for describing the nature of the pandemic and control the behavior of agent among time and space. Agent attributes includes: (i) health state (based on proposed extension to SIR Model states), (ii) social activities level (High, Moderate, Low), (iii) daily movement, (iv) spatial location, (v) infection time, (vi) social type (SPOUSE, PARENT, SIBLING, CHILD, OTHERFAMILY, COWORKER, GROUPMEMBER, NEIGHBOR, FRIEND, ADVISOR, SCHOOLMATE, OTHER), and (vii) agent social networks. Social activities level controls the number of daily contacts of the agent, which proportionally affect the number of in-contact agents interacting with the infected agent. Increasing number of in-contact agents adds more chance to the reproduction number to increase which means epidemic outbreak [4]. Reproduction number (R) can be defined as the average number of secondary agents infected by a primary agent case [16]. We have distributed number of daily infected cases based on the social activities level as following: Low: 2 agents, Moderate: 3 agents, and High 4 agents. Social type distribution is based on Egypt Census data of 2006 [7]. Egypt census data has classified population into five classes (see Table 2). The distribution of social types based on census data leads to different social network structure. All involved distributions are assumed to be Gaussian distribution with mean of 0 and standard deviation of 1. Distributions are subject to be replaced in future work according to the availability of more information about Egypt population.

At the beginning of the system startup, configurations are loaded and user is required to set up initial agents. The simulation runs in a loop for a pre-specified number of days. When the simulation starts, the first step is to create the initial agents instances. Next, agents start practicing their natural daily activities. During each day, every member of the agent community moves around, and communicates with their social network agents or with public agents. See Fig 4 for the spatial representation of agents moving and contacting each other, and their health states represented with different icons. Daily moved distance by agents and the number of contacted agents are randomly determined by each agent attributes. During the day activities, simulation keeps track of the social networks of each agent which can be at work, home, or school.

Selection of contact agents is random and most likely results in contacting new agents who are created at runtime. Newly created agents are initially susceptible, and are placed randomly across the landscape. Agent social types are drawn from normal distribution based on the population structure of census data of Egypt 2006. Agent health state and disease clock are changed according to the proposed extension to SIR model. Infected agent will affect all agents in his social networks to be exposed. Thus, probability of agent to be infected increases according to the number of infected agents in his social networks.

Table 1. Default values of model parameters

Parameter	Default Value
Number of simulated days	50
Random Seed	0
Population Size	72798031*
Agents	Initial agents
Incubation Time (latent)	2 days**
Percentage of immunized agents	0.95**
Percentage of recovered agents	0.9**
Percentage of dead agents	0.14**
Min. time to be recovered	5 day*s*
Max. time to be recovered	14 days**
Percentage of quarantined agents	0.1

* Egypt census data of 2006 [7].
** WHO [16] values for novel H1N1 pandemic.

Table 2. Age distribution of Egypt

Age (years)	Percentage	Possible social types
< 4	10.60 %	SIBLING, CHILD, OTHER
05–14	21.10 %	SIBLING, CHILD, OTHERFAMILY, COWORKER, GROUPMEMBER, NEIGHBOR, FRIEND, SCHOOLMATE, OTHER
15–44	49.85 %	SPOUSE, PARENT, SIBLING, OTHERFAMILY, COWORKER, GROUPMEMBER, NEIGHBOR, FRIEND, ADVISOR, SCHOOLMATE, OTHER
45–59	12.36 %	SPOUSE, PARENT, SIBLING, OTHERFAMILY, COWORKER, GROUPMEMBER, NEIGHBOR, FRIEND, ADVISOR, OTHER
> 59	6.08 %	SPOUSE, PARENT, SIBLING, OTHERFAMILY, GROUPMEMBER, NEIGHBOR, FRIEND, OTHER

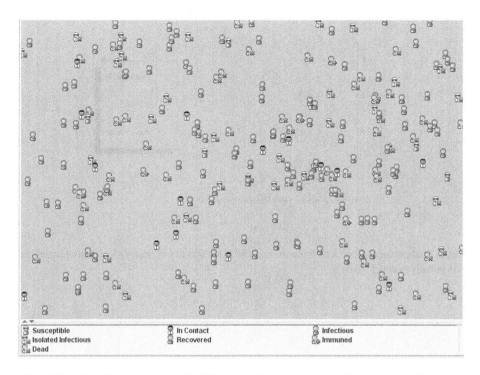

Susceptible	In Contact	Infectious
Isolated Infectious	Recovered	Immuned
Dead		

Fig. 4. Snapshot displays agents with different health states moving & contacting each others

5   Proposed Model Validation

It is often very difficult to validate epidemiological simulation models due to the lack of reliable field data, and the lack of real geographical location of the individual cases occurred. We have to validate the proposed multi-agent model against other available models that have been validated such as SIR model [14]. SIR Model has a long history and has proved to be a plausible model for real epidemics. The proposed model should be aligned with the SIR model at least for some simplified scenarios. In order to align the proposed model to SIR model, we have evaluated SIR model using Mathematica [11] based on given parameters (basic reproduction number R0 =3, duration of Infection = 9.5, initially immunized = 0, initially infected = 0.01% - see Fig 5) and we have evaluated the same model parameters in the proposed multi-agent model (see Fig 6). Two graphs are not a perfect match, but the proposed multi-agent model graph match the general behavior of SIR model graph. Two graphs differ by the magnitude and the smoothness of the curves. The source of difference of curves behavior is confined in the following factors: (i) the heterogeneous structure of the population, (ii) different reproduction numbers which are calculated for each agent independently, (iii) and the usage of random variable for infection time instead of deterministic values in SIR model.

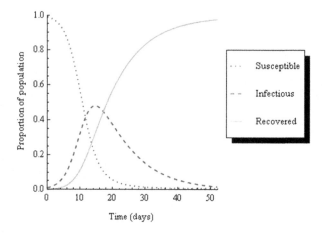

Fig. 5. Mathematica SIR Model. Parameters: R0=3, Duration of Infection=9.5, initially immunized = 0, initially infected = 0.01%.

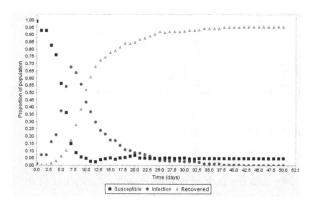

Fig. 6. Proposed multi-agent model. Parameters: 3 infected agents – Population Size: 300 agents – Duration: 50 days.

6   Pandemic Control Strategies

Control strategies are useful for the development of an action plan to control disease outbreak. Controlling outbreak is related to the peak of infectious and the time required reaching the peak. Time required reaching the peak is helpful for giving time for different control strategies to be effective. Without a properly planned strategy, the pandemic chaos might be disastrous causing large-scale fatalities and substantial economic damage. A proven control strategy would incorporate increasing awareness of population, vaccination, social distancing, and quarantining decisions [16] [6]. The proposed multi-agent model permits injection of control strategies to study different scenarios for controlling the outbreak. User can determine control strategies and the coverage percentage applied to the population. Control strategies will affect the agent health states, and the agent daily activities. Increasing awareness will increase the number of doctors' visit and the number of quarantined infected agents. Vaccination moves susceptible and in-contact agents to be immunized. Social distancing and quarantining reduce the number of possible contacts among individual agents. In experiments we will run different scenarios of applied control strategies.

7   Experiments and Analysis of Results

The proposed multi-agent model was programmed using Java programming language and run on AMD Athlon 64 X2 Dual 2.01 GHz processor with 1GB memory. To demonstrate the model behavior, we have run five scenarios of pandemic influenza in a closed population of 1000 agents, and initially three infected agents. Each run takes about eight minutes to be completed. Simulated scenarios are: (i) population with no deployed control strategies, (ii) population with deployed 50% of increasing aware-ness control strategy, (iii) population with deployed 50% of vaccination control strat-egy, (iv) population with deployed 50% of social distancing control strategy, and (v) population with deployed 50% of quarantining control strategy. We display suscepti-ble, infectious and removed curves of each scenario to be compared with each other. In the first scenario with no deployed control strategies, we have found that epidemic has a steep infection curve which reaches its peak (608 infected agents – 60.8% of the population is infected) on day 10 (see Fig 7). At the end of the simulation, we have analyzed the distribution of health states among social types. Mortality rate is high among schoolmates, neighbors, and advisors. Numbers of immunized agents are close for schoolmates, and parents, while equals zero with child agents (see Fig 8).

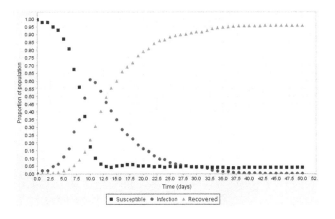

Fig. 7. Scenario 1: pandemic peak is at day 10

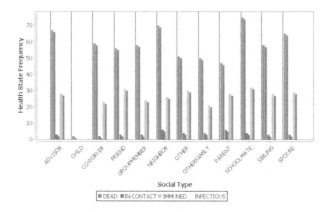

Fig. 8. Scenario 1: distribution of health states to social types

We have a pandemic peak at day 10. Thus, we choose to apply control strategies from day 8 to day 12 in the rest of the scenarios to study the effect of the control strategy on the pandemic peak. In the second scenario, we found that the number of infected agents is reduced during the deployment period of increase awareness control strategy to reach it is minimum value of 67 infected agents (6.7 % of the population) at day 12 (see Fig 9.a). This is because agents are asking for doctor help when they have influenza symptoms.

In the third scenario, we found that the number of infected agents is reduced to 320 agents (32% of the population) at day 9 (see Fig 9.b). In the fourth and the fifth scenarios, we found that the number of infected agents is increased to 614 (61.4% of the population) at day 10 (see Fig 9.c - d). This means that there are control strategies which not affect the pandemic spread when applied on given outbreak duration such as: social distancing and quarantining through the outbreak peak. Finally, we conclude that the aggregate attack rate is exponentially increasing without any deployed control strategies. Attack rate is controlled by different factors such as: the daily travelling distance of agents, and the percentage of vaccinated agents. As a result, proper combination of deployed control strategies can be effective to decrease the pandemic damage.

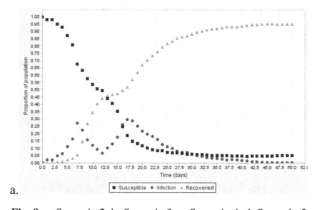

a.

Fig. 9. a: Scenario 2, b: Scenario 3, c: Scenario 4, d: Scenario 5

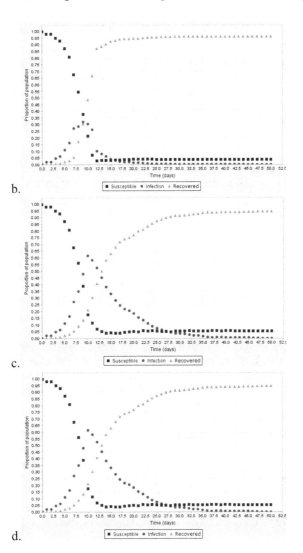

b.

c.

d.

Fig. 9. (*continued*)

8   Conclusion

The field of computational epidemiology has arisen as a new branch of epidemiology to understand epidemic transmission patterns, and to help in planning precautionary measure. For this reason a spatially explicit agent-based epidemiologic model of pandemic contagious disease is proposed in this paper. The methodology for this paper involves the development of a multi-agent model of pandemic influenza in Egypt. The proposed model simulates stochastic propagation of pandemic influenza outbreaks, and the impact of the decisions made by the healthcare authorities in population with millions of agents.

We have proposed extension to SIR model, in which we have investigated the agent attributes. The model can be easily customized to study the pandemic spread of any other communicable disease by simply adjusting the model parameters. We have simulated the spread of novel H1N1 pandemic in Egypt. Experiments are run in a closed population of 1000 agents, and initially 3 infected agents. Modeled novel H1N1 reaches infection peak (608 agents) with in 10 days without deployment of control strategies. Number of dead agents reaches its peak at the end of the simulation with mortality of 658 dead agents. Deployment of proper combination of control strategies can limit the pandemic chaos and reduce the fatalities and substantial economic damage. Further work on proposed model includes: agents with additional attributes that allow a better realistic model (e.g., ages, gender, etc), as well as finding optimal combination of control strategies to manage the pandemic outbreak waves.

References

1. Bonabeau, E.: Agent-based modeling: methods and techniques for simulating human systems. Proc. National Academy of Sciences 99(3), 7280–7287 (2002)
2. Ching Fu, S., Milne, G.: Epidemic Modeling Using Cellular Automata. In: Proceedings of the ACAL 2003: The First Australian Conference on Artificial Life, Canberra, Australia (2003)
3. Chowell, G., Hyman, J.M., Eubank, S., Castillo-Chavez, C.: Scaling laws for the movement of people between locations in a large city. Phys. Rev. E 68, 066102 (2003)
4. Colizza, V., Barrat, A., Barthélemy, M., Vespignani, A.: The modeling of global epidemics: stochastic dynamics and predictability. Bull. Math. Biol. 68, 1893–1921 (2006)
5. Daley, D.J., Gani, J.: Epidemic Modeling and Introduction. Cambridge University Press, NY (2005)
6. d'Onofrio, A., Manfredi, P., Salinelli, E.: Vaccinating behavior, information, and the dynamics of SIR vaccine preventable diseases. Th. Pop. Biol., 301–317 (2007)
7. Egypt Census Data (2006), http://www.capmas.gov.eg/
8. Eubank, Guclu, H., Marathe, M.V., et al.: Modeling disease outbreaks in realistic urban social networks. Nature 429(6988), 180–184 (2004)
9. Hethcote: The Mathematics of Infectious Diseases. SIAM Review 42(4), 599–653 (2000)
10. Carley, K.M., Fridsma, D.B., Casman, E., et al.: Biowar: Scalable agent-based model of bioattacks. IEEE Transactions on Systems, Man, and Cybernetics, Part A 36(2), 252–265 (2006)
11. Mathematica, http://www.wolfram.com/
12. Perez, L., Dragicevic, S.: An agent-based approach for modeling dynamics of contagious disease spread. International Journal of Health Geographics 8, 50 (2009)
13. Liu, Q.-X., Jin, Z., Liu, M.-X.: Spatial organization and evolution period of the epidemic model using cellular automata. Phys. Rev. E 74, 031110 (2006)
14. Skvortsov, Connell, R.B., Dawson, P., Gailis, R.: Epidemic Modeling: Validation of Agent-based Simulation by Using Simple Mathematical Models. In: Proceedings of Land Warfare Conference, pp. 221–227 (2007)
15. Timpka, T., Eriksson, H., Gursky, E.A., Nyce, J.M., Morin, M., Jenvald, J., Strömgren, M., Holm, E., Ekberg, J.: Population-based simulations of influenza pandemics: validity and significance for public health policy. Bull. World Health Organ. 87(4), 305–311 (2009)
16. World Health Organization, http://www.who.int/en/

Chapter 12
Supply Chain Risk Management:
Resilience and Business Continuity

Mauricio F. Blos[1,*], Hui Ming Wee[2], and Wen-Hsiung Yang[3]

[1] SONY do Brasil Ltda, Supply Chain Solutions, Manaus – AM, Brasil
blos31@yahoo.com.br
[2] Industrial and Systems Engineering Department,
Chung Yuan Christian University, Chungli, Taiwan
weehm@cycu.edu.tw
[3] University of Liverpool Management School, Liverpool L69 3BX, UK
jhs.why@gmail.com

Abstract. The last decade has been extraordinary, as it is marked by a succession of disasters. The risks in supply chain are critical because of increasing outsourcing, off-shoring, product variety, lean manufacturing and supply chain security. Moreover, organizations and societies are at a greater risk of system failure because of the massive interdependency throughout the global supply chains. Due to those facts, continuity is the main concern of any supplier. This chapter gives an overview on how a better supply chain decision making with risk can be made so as to achieve supply chain resilience and business continuity.

1   Introduction to Risk in Supply Chain

Risk in the supply chain is not a new phenomenon. Business organizations have always been exposed to the failure of a supplier to deliver the right quantity, at the right time, to the agreed quality and at the agreed price. One important characteristic to know is that even when the individual risk to each member of a chain is small, the cumulative effect over the hundreds or thousands of members in a large chain becomes very significant.

There are basically two kinds of risk to a supply chain: 1) internal risks that appear in normal operations, such as late deliveries, excess stock, poor forecasts, financial risks, minor accidents, human error, faults in information technology systems, etc; and 2) external risks that come from outside the supply chain, such as earthquakes, hurricanes, industrial action, wars, terrorist attacks, outbreaks of disease, price rises, problems with trading partners, shortage of raw materials, crime, financial irregularities, etc.

Leading companies take a broad, risk-based approach to securing their supply chains, based on comprehensive information on the operating environment and detailed analyses of the risks they need to address. Armed with this information they

---

* Corresponding author.

J. Lu, L.C. Jain, and G. Zhang: Handbook on Decision Making, ISRL 33, pp. 219–236.
springerlink.com   © Springer-Verlag Berlin Heidelberg 2012

develop and implement risk management strategies across the supply chain, and reduce their vulnerability to highly damaging disruptions.

As described in Table 1 and Fig. 1, the supply chain risks and vulnerabilities are raising more and more. Therefore, the awareness of supply chain risk management thinking is changing. Uncertainty seems to be a disturbing factor in organized life and continuous change feels like a treat to social life and economy.

One important aspect of risk management in supply chain is the timing of management actions. In principle, the company's or the network's risk management process should be continuous. Companies observe their operational environment and business processes and carry out decision and planning procedures, which have an effect of risks. This may be difficult in practice and it is advantageous to restrict the process to certain situations.

The activation of the risk management process is particularly associated with situations and events that are new and therefore, cause uncertainty. Such situations can be related to the establishment of a new network and changes in network relationships. Large purchases or projects and external signals, like customer or product specific investments and technology decisions also cause uncertainty. Identification of the company's internal signals may sometimes be a more convenient way to analyze uncertainty in the business environment.

According to [1] one view of risks that is common in the management literature is that a risk is thought of in terms of variability or uncertainty. In the insurance industry, the term risk is an accurate item to be insured. In this category, the writers seek to differentiate the two categories of risk, i.e. pure risk and speculative risk.

Table 1. The 11 Greatest Supply Chain Disasters

Rank	Company	Year(s)	Issue Problems	Impact Result
1	Foxmeyer Drug	1996	New order management and distribution systems don't work, and fulfilment cost targets built into contracts are unattainable	Huge sales losses; Foxmeyer files for bankrupcy, and is eventually bought by Mckesson
2	GM	1980s	CEO Robert Smith invests billions in robot technology that mostly doesn't work	Smith Fired; Low tech Toyota uses lean manufacturing to gain strong competitive advantage as GM's market share heads south
3	WebVan	2001	On-line grocer has many problems, including massive investment in automated warehouses that drain capital and aren't justified by demand	Company goes from billions in market cap to bankrupt in a matter of months
4	Adidas	1996	New warehouse system ~ actually first one then another ~ and DC automation just don't work	Company under~ships by 80% in January, incurs market share losses that persist by years
5	Denver Airport baggage handling system	1995	Complex, hugely expensive automated handing system never really works	Airport opens late, huge PR fiasco; system is only minimally used from start and shuttered totally in 2005
6	Toys R Us.com	1999	Can't fulfill thousands of orders for which it promises delivery by Christmas	Famous "we are sorry" e-mails 2 days before Christmas cause fire storm of negative PR; eventually oursources fulfillment to Amazon.com
7	Hershey Foods	1999	Order management and warehouse implementation issues cause Hershey to miss critical Halloween shipments	Company says at least $150 million in revenue lost; profit drops 19%, and stock goes from 57 to 38
8	Cisco	2001	Lacking adequate demand and inventory visibility, Cisco is caught with piles of product as demand slows	Company takes $2.2 billion inventory write~down; stock drops 50% and has stayed near that level since
9	Nike	2001	Trouble with new planning system causes inventory and orders woes	Stock drops 20%
10	Aris Isotoner	1994	Division of Sara Lee makes disastrous from Manila to even lower cost countries; cost rise instead as quality plummets	Sales are cut by 50%; company goes from strong profit to big losses; Sara Lee soon sells Isotoner unit to Totes
11	Apple	1995	Playing a conservative inventory strategy, Apple is swamped with damand for new Power Macs and can't deliver the goods	Apple takes PR black eye and loses PC market share which it never really recovers

Source: *SupplyChainDigest*

According to a survey report from [2], 74% of the organizations had experienced supply chain disruption in last 12 months; 35% had experienced increased supply chain disruption; and 88% expected to experience supply chain disruption in the next 12 months.

The main objective of any risk management practice is to increase risk adjusted returns, improve strategic judgment, and/or avoid extraordinary losses due to lawsuits, fines, operational failures, or negligence. The need for organizations to meet best practice in the management of supply chain risk has never been more important, or challenging, as they face potential economic downturn, increasing regulation and ever more complex supply chains that make it difficult to understand risk aggregation. Furthermore, supply chain risk management assumes importance in the wake of organizations understanding that their risk susceptibility is dependent on other constituents of their supply chain.

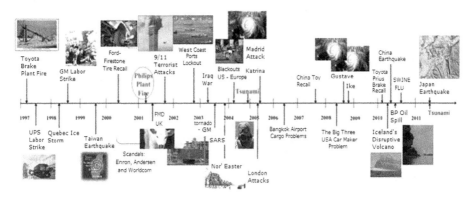

Fig. 1. Major Disasters from 1997 to 2011 (adapted from Dr. Debra Elkins, GM)

According to the [3], Risk Management Process involves applying logical and systematic methods for:

- Communication and consultation throughout the process;
- Establishing the context;
- Assessing risks (identification, analysis, evaluation) and treating risk associated with any activity, process, function, project, product, service or asset;
- Monitoring and reviewing risk;
- Recording and reporting the results appropriately.

In this context, the approach by using the ISO31000, the ISO28002 and the BS 25999 standards will be essential to explore the effective practices of two important and necessary variables of SCRM: resilience and business continuity.

2   Resilience in Supply Chain

In the supply chain literature, the topic of resilience emerged a few years ago and has recently become more widely recognized (Christopher and Peck, 2004 [4]; Craighead et al., 2007 [5]; Sheffi and Rice, 2005 [6]). There are many papers published on supply chain resilience, with focus on qualitative insights into the problem. The researchers tend to focus on supply chain disruptions from the point-of-view of either mitigation measures or response measures, but typically do not look at both stages simultaneously as Falasca et al. (2008) [7] proposed.

Tomlin (2006) [8] developed a model to determine the best contingency and mitigation strategies for a firm with a single product and two alternative suppliers. The parameters used by the author include supplier reliability, capacity and costs, transit lead time, volume and response time flexibility, as well as different inventory and demand considerations. Those parameters are then used to determine appropriate tactics for dealing with supply chain disruptions. The resulting model is most relevant to individual members of a supply chain and for tactical decision-making, and it does not explicitly take supply chain design decisions into consideration.

Lodree Jr, and Taskin (2007) [9] introduce an insurance risk management framework for disaster relief and supply chain inventory planning. This framework provides an approach by which decision makers can quantify the risks and benefits of stocking decisions related to supply chain disruptions and disaster relief efforts. The authors introduce different newsvendor problem variants that take into consideration demand uncertainty as well as the uncertainty related to the occurrence of extreme events. Optimal inventory levels are determined and the insurance premium associated with disaster-relief planning is calculated. Once again, the parameters used by the authors are tactical in nature and are most appropriate for individual supply chain members.

Huang et al. (2007) developed a dynamic system model for supply chains and applied it to the management of disruptive events in full-load states of manufacturing chains. Their model is used to analyze demand shocks and to determine the level of contingent resources that must be synchronously activated by the members of the chain. The authors describe how the model can be used to reduce the impact of disruptive events and thus to enhance risk management. In this case, although the parameters used by the authors are tactical in nature, they are appropriate for coordinating an entire supply chain by assessing the impacts of disruptive events on the entire system and by activating contingency plans for mitigating these impacts.

Datta, et al. (2007) presented an agent-based framework for studying multi-product, multi-country supply chains subject to demand variability, production, and distribution capacity constraints, with the aim of improving supply chain flexibility and resilience. The model developed by the authors showed the advantages of using a decentralized information structure and flexible decision rules, monitoring key performance indicators at regular intervals, and sharing information across members of the supply chain network. One key limitation of this study, however, is that it does not incorporate cost data into the agents' decision-making functions, limiting the possibilities of performing relevant trade-off analyses. The parameters used by the authors are tactical in nature, since different fundamental structures of the supply chain network are not discussed, but they are appropriate for coordinating an entire supply chain.

Tang (2006) [12] reviewed various models for managing supply chain risks and related various supply chain risk management strategies examined in the research literature with actual practices.

Kleindorfer and Saad (2005) [13] developed a conceptual framework for managing supply chain disruption risk that includes the tasks of specification, assessment, and mitigation. The authors analyze the relationship among diversification (extended to facility locations, sourcing options, and logistics), weakest link identification, leanness and efficiency, backup systems, contingency plans, information sharing, modularity of process and product designs, and other elements of agility and flexibility in relation to supply chain disruption drivers. Sheffi (2005) [14] discussed the use of safety stocks, extra capacity, redundant suppliers, standardized components and simultaneous processes and analyzed their effect on the resilience of a firm's supply chain in order to derive a series of qualitative insights into the problem. The considerations in both of those papers can be categorized as strategic, and those characteristics are most relevant at the level of individual members of a supply chain.

Peck (2005) [15] identified different sources and drivers of supply chain vulnerability and developed a multi-level conceptual framework for the analysis of the nature of supply chain vulnerabilities. The drivers identified by the author include products and processes, assets and infrastructure dependencies, organizations and inter-organizational networks, as well as the environment.

Before Falasca et al. (2008) [7], Craighead et al. (2007) [5] used an empirical research design to derive different insights and propositions that relate the severity of supply chain disruptions to three specific supply chain design characteristics (density, complexity, and node criticality), as well as to the supply chain mitigation capabilities of recovery and warning. The chosen design characteristics can be categorized as strategic in nature, and as appropriate to the study of an entire supply chain, however the authors do not explicitly derive any quantitative relationships between them.

The idea of resilience suggests the speed with which a chain can return to normal working after some kind of damage. So in different circumstances supply chain risk management might either try to prevent risky events from occurring (reducing vulnerability) or accept that they will occur and then return the chain to normal working as quickly as possible (increasing resilience).

2.1   Resilience Management System Requirements

The organization shall establish, document, implement, maintain, and continually improve a resilience management system in accordance with the requirements of ISO 28002 [16], and determine how it will fulfill these requirements. The requirements to build a resilient management system are: Understanding the Organization and its Context; Policy Management Review; Planning; Implementation and Operation; Checking and Corrective Action; and Management Review (Fig. 2).

Here we summarize the requirements:

1) Understanding the Organization and its Context

- The organization shall define and document the internal and external context of the organization.
- The organization shall identify and document some procedures related to risk protection and resilience in order to define the context for the management

system and its commitment to the management of risk and resilience within specific internal and external contexts of the organization.

- The organization shall define and document the objectives and scope of its resilience management system within specific internal and external context of the organization.
- The organization shall define the scope consistent with protecting and preserving the integrity of the organization and its supply chain including relationships with stakeholders, interactions with key suppliers, outsourcing partners, and other stakeholders.

2) Policy Management Review

Top management shall define, document, and provide resources for the organization's resilience management policy reflecting a commitment to the protection of human, environmental, and physical assets; anticipating and preparing for potential adverse events; and business and operational resiliency.

The policy statement of an organization shall ensure that within the defined scope of the resilience management system several responsibilities should be taking in action, including: a commitment to the enhanced organizational and supply chain sustainability and resilience; a commitment to risk avoidance, prevention, reduction, and mitigation; to provide a framework for setting and reviewing resilience management objectives and targets.

3) Planning

The main proposal of this requirement is to ensure that applicable legal, regulatory, and other requirements to which the organization subscribes are considered in developing, implementing, and maintaining its resilience management system. Furthermore, this requirement follows the risk assessment exigencies.

4) Implementation and Operation

This requirement which proposal is to implement and operate resilience in supply chain has the following procedures: resources, roles, responsibility, and authority for resilience management; competence, training, and awareness; communication and warning; documentation; control of documents: operational control; and incident prevention, preparedness, and response.

5) Checking and Corrective Action

The organization shall evaluate resilience management plans, procedures, and capabilities through periodic assessments, testing, post-incident reports, lessons learned, performance evaluations, and exercises. Significant changes in these factors should be reflected immediately in the procedures.

The organization shall keep records of the results of the periodic evaluations.

6) Management Review

Management shall review the organization's resilience management system at planned intervals to ensure its continuing suitability, adequacy, and effectiveness. This review shall include assessing opportunities for improvement and the need for changes to the resilience management system, including the resilience management system policy and objectives. The results of the reviews shall be clearly documented and records shall be maintained.

7) Resilience Management Policy

The resilience management policy is the driver for implementing and improving an organization's resilience management system, so that it can maintain and enhance its sustainability and resilience. This policy should therefore reflect the commitment of top management to:

a) Comply with applicable legal requirements and other requirements
b) Prevention, preparedness, and mitigation of disruptive incidents
c) Continual improvement

The resilience management policy is the framework that forms the basis upon which the organization sets its objectives and targets. The resilience management policy should be sufficiently clear to be capable of being understood by internal and external interest parties (particularly its supply chain partners) and should be periodically reviewed and revised to reflect changing conditions and information. Its area of application (i.e., scope) should be clearly identifiable and should reflect the unique nature, scale, and impacts of the risks of its activities, functions, products, and services.

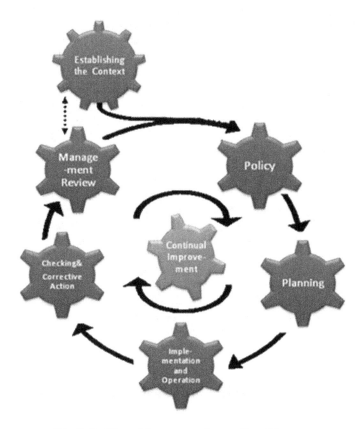

Fig. 2. Resilience Management System Flow Diagram

2.1.1   Best Practices of a Resilient Organization

Innovative organizations are fielding new ideas and deploying new solutions that increase both their risk intelligence and capacity for resilience. Three case studies reported in the literature are the following.

(1) Nokia changed product configurations in the nick of time to meet customer demand during a supply disruption. Both Ericsson and Nokia were facing a shortage of a critical cellular phone component (radio frequency chips) after a key supplier in New Mexico (Philips's semiconductor plant) caught fire during March 2000. Ericsson was slow in reacting to this crisis and lost €400 million in sales. In contrast, Nokia had the foresight to design their mobile phones based on the modular product design concept and to source their chips from multiple suppliers. After learning about Philips's supply disruption, Nokia responded immediately by reconfiguring the design of their basic phones so that the modified phones could accept slightly different chips from Philips's other plants and other suppliers. Consequently, Nokia satisfied customer demand smoothly and obtained a stronger market position. Please refer to Hopking [17] for details.

(2) Li and Fung Limited changed its supply plan in a flash to meet customer demand during a currency crisis. When the Indonesian Rupiah devalued by more than 50% in 1997, many Indonesian suppliers were unable to pay for the imported components or materials and, hence, were unable to produce the finished items for their US customers. This event sent a shock wave to many US customers who had outsourced their manufacturing operations to Indonesia. In contrast, The Limited and Warner Bros. continued to receive their shipments of clothes and toys from their Indonesian suppliers without noticing any problem during the currency crisis in Indonesia. They were unaffected because they had outsourced their sourcing and production operations to Li and Fung Limited, the largest trading company in Hong Kong for durable goods such as textiles and toys. Instead of passing the problems back to their US customers, Li and Fung shifted some production to other suppliers in Asia and provided financial assistance such as line of credit, loans, etc. to those affected suppliers in Indonesia to ensure that their US customers would receive their orders as planned. With a supply network of 4,000 suppliers throughout Asia, Li and Fung were able to serve their customers in a cost-effective and time-efficient manner. Despite the economic crisis in Asia, this special capability enabled Li and Fung to earn its reputation in Asia and enjoy continuous growth in sales from 5 billion to HK$17 billion from 1993 to 1999. Please refer to [18] and [19] for details.

(3) Dell changed its pricing strategy just in time to satisfy customers during supply shortage. After an earthquake hit Taiwan in 1999, several Taiwanese factories informed Apple and Dell that they were unable to deliver computer components for a few weeks. When Apple faced component shortages for its iBook and G4 computers, Apple encountered major complaints from customers after trying to convince its customers to accept a slower version of G4 computers. In contrast, Dell's customers continued to receive Dell computers without even noticing any component shortage problem. Instead of alerting their customers regarding shortages of certain components, Dell offered special price incentives to entice their online customers to buy computers that utilized components from other countries. The capability to influence customer choice enabled Dell to improve its earnings in 1999 by 41% even during a supply crunch [20], [21].

Creating supply chain resiliency requires more than utilizing the framework to identify, assess, and mitigate the enterprise risks. The long-term success of a resilient supply chain depends heavily on the organization's ability to foster a culture of reliability that stretches across departmental borders.

2.2   Business Continuity in Supply Chain

Over the years the need for disaster recovery, business recovery and contingency planning has increased. Realizing that most business processes today extend beyond the boundaries of a single entity, awareness of critical supply chain interdependencies has risen sharply. (Fig. 3)

Nearly all organizations rely in some way or another on supply chains to ensure business continuity, and they are vulnerable if supplies are interrupted. Business Continuity is the concern of the entire organization, not just the IT department. For supply chains, the basic requirement of business continuity is that the flow of materials is not interrupted by any disaster hitting the chain, or that it is able to return to normal as quickly as possible. Also, we should pinpoint the changing nature of business, such as:

- Increasing dependency on information and communications technology(ICT).
- Increasing system interdependency.
- Increasing commercial complexity.
- Increasing technical complexity.
- Increasing expectations from customer.
- Increasing Threat.

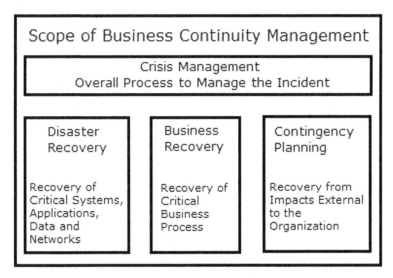

Fig. 3. Scope of Business Continuity Management

Source: Chartered Management Institute (2002)

In order to keep an organization's client base and business partners, business continuity standard [22] was created. BS 25999 offers the framework to develop, implement, maintain, and improve a dynamic Business Continuity Management (BCM) System that combats business interruption. The professional practices that this standard offers are based on 10 principles to know:

1. Project Initiation and Management.

2. Risk Evaluation and Control.

3. Business Impact Analysis.

4. Developing Business Continuity Management Strategies.

5. Emergency Response and Operations.

6. Developing and Implementing Business Continuity Plans.

7. Awareness and Training Programs.

8. Exercising and Maintaining Business Continuity Plans.

9. Crisis Communications.

10. Coordination with External Agencies.

Supply chain continuity is one aspect of a broader category of issues called Supply Chain Risk Management (SCRM). Supply chain continuity is an emerging discipline. One of the greatest threats to business continuity faced by manufacturers is loss of critical raw materials. Most organizations rely in some way on their supply chain to ensure business continuity and they are vulnerable if supplies are interrupted. BS 25999 ensures that organization will develop a network of supply chains so that when one link breaks, another link is ready to take its place. Furthermore, the flow of materials is not interrupted by any disaster hitting the chain or that it is able to return to normal as quickly as possible.

The recommendations to ensure business continuity in supply chain are [23]:

- Maintaining safety stock.
- Monitoring transportation entities and planning for contingent shipping arrangements.
- Observing product transportation paths, looking for potential bottlenecks.
- Developing plans to allow for acceleration of shipments.
- Implementing a crisis communications process with key vendors.
- Developing continuity plans to address in-house product receipt, inventory management and product shipment processes, with an emphasis on labor interruption and other single points of failure.

Business continuity efforts are enhanced with management system-oriented models that avoid professional jargon and focus on business outcomes. Putting the business continuity program into key organizational outputs is meant to focus on the objectives of management and to speak in the language of the organization. Because of this, the business continuity program is able to communicate real capability and output, rather than focusing on micro business continuity-specific projects. Then strategies and plans for how to recover should be developed that could be implemented, both before the incident (similar to risk management strategies) and after the incident, post-incident strategies are implemented to maintain partial or total product supply.

2.2.1 Supply Chain Continuity Framework

Various authors have proposed different development cycles for business continuity management, each of which places an emphasis on particular aspects of business continuity plan (BCP) such as CCTA (1995), Barnes (2001), Hiles and Barnes (2001), Starr, Newfrock, & Delurey (2002), and Smith (2002). In this section, a framework is created to allow a company to create supply chain visibility and better handle supply chain disruptions, improving the organization's performance. The framework is based in a BCP process life cycle with six stages (risk mitigation management, business impact analysis, supply continuity strategy development, supply continuity plan development, supply continuity plan testing and supply continuity plan maintenance) with five supply chain operational constructs (inventory management, quality, ordering cycle-time, flexibility and costumers) with the purpose to keep supply chain more resilience.

In creating a mitigation program the goal is to eliminate risks where possible and to lessen the negative impact of disasters that cannot be prevented. Mitigation is as simple as fastening down computer terminals in earthquake-prone areas or moving computers to a higher floor where floods are a potential threat to dispersing critical business operations among multiple locations, to relocating an operation from an area where the risks are extremely great or perhaps contractually transferring some operations. However, there is the need to have a BCP to maintain continuity of business during a destructive disruption.

According to Fig. 4 (Blos, Wee, & Yang, 2010), the BCP process life cycle has the business continuity best practices for developing and maintaining a supply chain continuity plan. Therefore, a comprehensive business continuity program that considers all internal and external links in the supply chain is essential if the business is to survive following a major disaster. With this affirmation, we developed a supply chain continuity framework, where the BCP process life cycle, together with the operational constructs, has the objective of keeping the supply chain more resilient:

1. Stage 1: Risk Mitigation Management:
   This first stage will assess the threats of disaster, existing vulnerabilities, potential disaster impacts, and identifying mitigation control are needed to prevent or reduce the risks of disaster related to the operational constructs (O.C).
2. Stage 2: Business Impact Analysis (BIA):
   This stage identifies mission-critical processes of the OC, and analyzes the impacts to business if these processes are interrupted as a result of disaster. Furthermore, the BIA is the foundation on which a comprehensive business continuity program is based. Its goal is to determine the most to the least time-critical business functions throughout the organization. For each of these functions, determine a related recovery time objective (RTO) and the target time to recover from disruption. For a supply management, a BIA will include a review of the manufacturing operations, transportation, distribution services, support technology, warehouses, and service centers, as well as a tradeoff in between them.

3. Stage 3: Supply Continuity Strategy Development:
    The third stage assesses the requirements and identifies the options for recovery of critical processes and resources related to the O.C. in the event they are disrupted by a disaster
4. Stage 4: Supply Continuity Plan Development:
    This stage develops a plan for maintaining business continuity based on the results of previous stages.
5. Stage 5: Supply Continuity Plan Testing:
    This stage will test the supply continuity plan document to ensure its currency, viability, and completeness.
6. Stage 6: Supply Continuity Plan Maintenance:
    The final stage will maintain the supply continuity plan in a constant ready-state for execution.

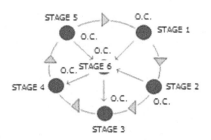

Fig. 4. Supply Chain Continuity Framework

2.2.2   Business Assessment Process

The business assessment is divided into two components, risk assessment and business impact analysis (BIA). Risk assessment is described to evaluate existing exposures from the organization's environment, whereas the BIA assesses potential loss that could be caused by a disaster.

2.2.2.1   Risk Assessment. The purpose of risk assessment is the generation of knowledge linking specific risk agents with uncertain but possible consequences [30], [31]. Furthermore, risk assessment is an evaluation of the exposures present in an organization's external and internal environment. It identifies whether or not the facility housing the organization is susceptible to floods, hurricanes, tornadoes, sabotage, etc. It then documents what mitigation steps have been taken to address these threats. The basis of risk assessment is the systematic use of analytical – largely probability-based – methods which have been constantly improved over the past years.

The basis of risk assessment is the systematic use of analytical – largely probability-based – methods which have been constantly improved over the past years. Probabilistic risk assessments for large technological systems, for instance, include tools such as fault and event trees, scenario techniques, distribution models based on Geographic Information Systems (GIS), transportation modeling and empirically driven human-machine interface simulations [32], [33]. With respect to human health, improved methods of modeling individual variation [34], dose-response relationships [35] and exposure assessments [36] have been developed and successfully applied. The processing of data is often guided by inferential statistics and organized in line with decision analytic procedures. These tools have been developed to generate knowledge about cause effect relationships, estimate the strength of these relationships, characterize remaining uncertainties and ambiguities and describe, in quantitative or qualitative form, other risk or hazard related properties that are important for risk management [37], [38]. In short, risk assessments specify what is at stake, calculate the probabilities for (un)wanted consequences, and aggregate both components into a single dimension [39].

According to [3], Risk Assessment is the overall of "Risk Identification, Risk Analysis and Risk Evaluation".

2.2.2.2   *Risk Identification.* The organization should identify sources of risk, areas of impacts, events and their causes and their potential consequences. The aim of this step is to generate a comprehensive list of risks based on those events that might enhance, prevent, degrade or delay the achievement of the objectives. It is also important to identify the risks associated with not pursuing an opportunity. Comprehensive identification is critical, because a risk that is not identified at this stage will not be included in further analysis. Identification should include risks whether or not their source in under control of the organization.

The organization should apply risk identification tools and techniques which are suited to its objectives and capabilities, and to the risks faced.

Relevant and up-to-date information is important in identifying risks. This should include suitable background information where possible. People with appropriate knowledge should be involved in identifying risks. After identifying what might happen, it is necessary to consider possible causes and scenarios that show what consequences can occur. All significant causes should be considered.

2.2.2.3   *Risk Analysis.* Risk analysis is about developing an understanding of the risk. Risk analysis provides and input to risk evaluation and to decisions on whether risks need to be treated and on the most appropriate risk treatment strategies and methods.

Risk analysis involves consideration of the causes and sources of risk, their positive and negative consequences, and the likelihood that those consequences can occur. Factors that affect consequences and likelihood should be identified. Risk is analyzed by determining consequences and their likelihood, and other attributes of the risk. An event can have multiple consequences and can affect multiple objectives. Existing risk controls and their effectiveness should be taken into account.

The way in which consequences and likelihood are expressed and the way in which they are combined to determine a level of risk will vary according to the type of risk, the information available and the purpose for which the risk assessment output is to be used. These should all be consistent with the risk criteria. It is also important to consider the interdependence of different risks and their sources.

The confidence in determination of risk and their sensitivity to preconditions and assumptions should be considered in the analysis, and communicated effectively to decision makers and other stakeholders if required. Factors such as divergence of opinion among experts or limitations on modeling should be stated and may be highlighted.

Risk analysis can be undertaken with varying degrees of detail depending on the risk, the purpose of the analysis, and the information, data and resources available. Analysis can be qualitative, semi-quantitative or quantitative, or a combination of these, depending of the circumstances. In practice, qualitative analysis is often used first to obtain a general indication of the level of risk and to reveal the major risks. When possible and appropriate, one should undertake more specific and quantitative analysis of the risks as a following step.

Consequences can be determined by modeling the outcomes of an event or set of events, or by extrapolation from experimental studies or from available data. Consequences can be expressed in terms of tangible and intangible impacts. In some cases, more than one numerical value or description is required to specify consequences for different times, places, groups or situations.

2.2.2.4   Risk Evaluation. The purpose of risk evaluation is to assist in making decisions, based on the outcomes of risk analysis, about which risks need treatment to prioritize implementation.

Risk evaluation involves comparing the level of risk found during the analysis process with risk criteria established when the context was considered. If the level of risk does not meet risk criteria, the risk should be treated.

Decisions should take account of the wider context of the risk and include consideration of the tolerance of the risks borne by parties other than the organization that benefit from the risk. Decisions should be made in accordance with legal, regulatory and other requirements.

In some circumstances, the risk evaluation can lead to ad decision to undertake further analysis. The risk evaluation can also lead to a decision not to treat the risk in any way other than maintaining existing risk controls. This decision will be influenced by the organization's risk appetite or risk attitude and the risk criteria that have been established.

2.2.2.5   Business Impact Analysis (BIA). A BIA is an assessment of an organization's business functions to develop an understanding of their criticality, recovery time objectives, and resource needs all based on the [25]. Moreover, BIA's primarily objective is to identify impact of business disruption and time sensitivity for recovery

During the BIA process it is evaluated the risk of business process failures and it is identified the critical and the necessary business functions and their resource dependencies [25].   Also, it is estimated the financial and operational impacts of a disruption; it is identified regulatory/compliance exposure; and it is determined the impact upon the client's market share and corporate image.

The BIA process is a crucial link between the risk management stage and the business continuity plan development stage. The BIA identifies the mission-critical areas of business and continuity requirements which become the main focus of the business continuity.

The benefits of BIA are:

1. Identify the requirements for recovering disrupted mission-critical areas.
2. Guarantee the business stability.
3. Guarantee an ordinate recovery.
4. Reduce the dependency of key personal.
5. Increment the active protection.
6. Guarantee the safety of people, customers and partners.

Business Continuity is a concern of the entire organization, not just the IT department. Therefore, all area of the organization should be involved in a comprehensive BIA. At least one representative from each business area of the organization should participate.

Recovery time requirements consist of several components. Collectively, these components refer to the length of time available to recover from a disruption. An understanding of these components is a prerequisite for conducting the BIA. The components of recovery time are: Maximum Tolerable Downtime (MTD), Recovery Time Objective (RTO), Recovery Point Objective (RPO), and Work Recovery Time (WRT). (Fig. 5)

MTD represents the maximum downtime the organization can tolerate for a business process (RTO+RPO); RTO indicates the time available to recover disrupted system resources; RPO extent of data loss measured in terms of a time period that can be tolerated by a business process. WRT is the time period to recover the lost data, work backlog, and manually captured data once the systems/resources are recovered or repaired.

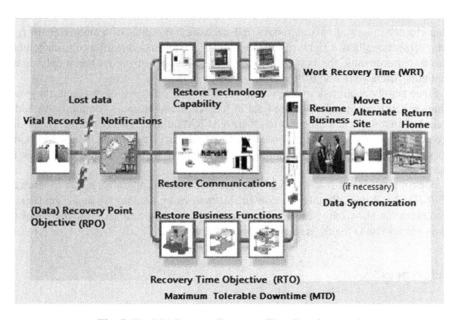

Fig. 5. The BIA Process (Recovery Time Requirements)

One result of the analysis is a determination of each business function's RTO. The RTO is the time within which business functions or application systems must be restored to acceptable levels of operational capability to minimize the impact of an outage. The RTO describes the time between the point of disruption and the point at which business functions or application systems must be operational and updated to current status. The RTO is associated with the recovery of resources such as computer systems, manufacturing equipment, communication equipment, facilities, etc.

RPO expresses the tolerance to a loss of data as a result of disruptive event. It is measured as the time between the last data backup and the disruptive event.

The BIA determines RPO for each application by asking participants the question "What is the tolerance, in terms of length of time, to loss of data that may occur between any two backup periods?" The response to this question indicates the values of RPO.

The BIA process consists of a sequence of 11 steps that interact together to identify the impacts of a business disruption and determine the requirements to restore disrupted critical business process: 1. Define BIA objectives, scope, and assumptions; 2. Identify business functions and process; 3. Assess financial and operational impacts; 4. Identify critical processes; 5. Assess MTDs and prioritize critical process; 6. Identify critical IT systems and applications; 7. Identify critical non-IT resources; 8. Determine RTO; 9. Determine RPO; 10. Identify work-around procedures; and 11. Generate BIA information summary.

At the end of the BIA process a report is generated that includes details of the key output from the steps of the BIA process and summarize its findings.

3   Conclusion and Future Research Proposal

This chapter reviewed two important and necessary variables of a supply chain risk management: resilience and business continuity. We discussed resilience management system requirements, the best practices of a resilient organization, the supply chain continuity framework and the business assessment process.

Since supply chain risk management is a relatively new area of research, there are many research opportunities in the field of resilience and business continuity. In addition, the most important aspect is to create a SCRM culture. This is not just risk management but performance improvement, where companies will make money by taking smart risks or lose money by failing to re-tool their legacy business processes to assess and mitigate risk effectively.

Focusing the worst risk scenario, our future research proposal suggests to manage PPPI (Public-Private Partnerships on Infrastructure) by creating strategic geological centers to monitor the weather risk events (*floods, hurricanes, volcano smokes, tsunamis, etc*) and creating strategic evacuation plans.

References

1. Hertz, D., Thomas, H.: Risk Analysis and its Applications. John Wiley & Sons, Chischester (1983)
2. Survey Report BCI, Experience of Supply Chain Disruption (2009)
3. ISO/DIS 31000: Risk Management – Principles and Guidelines on Implementation (2009)
4. Christopher, M., Peck, H.: Building the Resilient Supply Chain. International Journal of Logistics Management 15(2), 1 (2004)

5. Craighead, C.W., Blackhurst, J., Rungtusanatham, M.J., Handfield, R.B.: The Severity of Supply Chain Disruptions: Design Characteristics and Mitigation Capabilities. Decision Sciences 38(1), 131 (2007)
6. Sheffi, Y., Rice, J.B.: A Supply Chain View of the Resilient Enterprise. MIT Sloan Management Review 47(1), 41 (2005)
7. Falasca, M., Zobel, C.W., Cook, D.: A Decision Support Framework to As-sess Supply Chain Resilience. In: Proceedings of the 5th International ISCRAM Conference, pp. 596–605 (2008)
8. Tomlin, B.: On the Value of Mitigation and Contingency Strategies for Managing Supply Chain Disruption Risks. Management Science 52(5), 639 (2006)
9. Lodree Jr., E.J., Taskin, S.: An insurance risk management framework for disaster relief and supply chain disruption inventory planning. Journal of the Operational Research Society, 1–11 (2007)
10. Huang, H.Y., Chou, Y.C., Chang, S.: A dynamic system model for proactive control of dynamic events in full-load states of manufacturing chains. International Journal of Production Research 1, 1–22 (2007)
11. Datta, P.P., Christopher, M., Allen, P.: Agent-based modelling of complex production/distribution systems to improve resilience. International Journal of Logistics Research and Applications 10(3), 187–203 (2007)
12. Tang, C.S.: Perspectives in supply chain risk management. International Journal of Production Economics 103(2), 451–488 (2006)
13. Kleindorfer, P.R., Saad, G.H.: Managing Disruption Risks in Supply Chains. Production and Operations Management 14(1), 53–68 (2005)
14. Sheffi, Y.: Building a resilient supply Chain. Harvard Business Review Supply Chain Strategy 1(8), 1 (2005)
15. Peck, H.: Drivers of supply chain vulnerability: an integrated framework. International Journal of Physical Distribution & Logistics Management 35(3/4), 210 (2005)
16. ISO 28002: Resilience in the Supply Chain: requirements with Guidance for Use (2009)
17. Hopking, K.: Value opportunity three: improving the ability to fulfill demand. Business Week (2005)
18. George, A.S.: Li and Fung: beyond "filling in the mosaic". Harvard Business School, case 9-398-092 (1998)
19. McFarlan, W.: Li and Fung (A): Internet issues. Harvard Business School, case 9-301-009 (2002)
20. Veverka, M.: A DRAM shame, Barron's, October 15, p. 15 (1999)
21. Martha, J., Subbakrishna, S.: Targeting a just-in-case supply chain for the inevitable next disaster. In: Supply Chain Mgmnt Rev. (2002)
22. BS 25999: Business Continuity Management (BCM) Standard (2007)
23. Parier, M., Zawada, B.: Risky Business – Failing to Assess Supply Chain Continuity. In: A Study Conducted by the Council of Logistics Management States (2002)
24. CCTA, An introduction to business continuity management. HMSO, London (1995)
25. Barnes, James, C.: A guide to Business Continuity Planning. John Wiley & Sons Ltd. (2001)
26. Hiles, A., Barnes, P. (eds.): The Definitive Handbook of Business Continuity Management. J. Wiley and Sons, Chichester (2001)
27. Starr, R., Newfrock, J., Delurey, M.: Enterprise resilience: managing risk in the networked economy. Strategy and Business 30, 73–79 (2002)
28. Smith, D.J. (ed.): Business continuity management: Good practice guidelines. Business Continuity Institute, Caversham (2002)

29. Blos, M.F., Wee, H.M., Yang, J.: Analysing the external supply chain risk driver competitiveness: A risk mitigation framework and business continuity Plan. Journal of Business Continuity & Emergency Planning 4, 368–374 (2010)
30. Brindley, C.: Supply Chain Risk. Ashgate Publishing Company (2004)
31. Lave, L.: Health and Safety Risk Analyses: Information for Better Decisions. Science 236, 291–295 (1987)
32. Grahan, J.D., Rhomberg, L.: How Risks are Identified and Assessed (1996)
33. IAEA: Guidelines for Integrated Risk Assessment and Management in Large Industrial Areas (1995)
34. Stricaff, R.S.: Safety Risk Analysis and Process Safety Managesment: Principles and Practices (1995)
35. Hattis, D.: The conception of variability in risk analyses: Developments since 1980 (2004)
36. Olin, S., Farland, W., Park, C., Rhomberg, L., Scheuplein, R., Starr, T., Wilson, J.: Low Dose Extrapolation of Cancer Risks: Issues and Perspectives (1995)
37. US-EPA Environmental Protection Agency: Exposure Factors Handbook (1997)
38. IEC: Guidelines for Risk Analysis of Technological Systems, Report IEC-CD (Sec) 381 issued by the Technical Committee QMS/23. (European Community: Brussels) (1993)
39. Kolluru, R.V.: Risk Assessment and Management: A Unified Approach, pp. 1.3–1.41. Mc-Graw-Hill, New York (1995)

Chapter 13
A Fuzzy Decision System
for an Autonomous Car Parking

Carlos Martín Sánchez[1], Matilde Santos Peñas[1], and Luis Garmendia Salvador[2]

[1] Dpto. Arquitectura de Computadores y Automática, Facultad de Informática,
Universidad Complutense de Madrid
28040-Madrid, Spain
cmartins@fdi.ucm.es, msantos@dacya.ucm.es
[2] Dpto. Inteligencia Artificial e Ingeniería del Software, Facultad de Informática,
Universidad Complutense de Madrid
28040-Madrid, Spain
lgarmend@fdi.ucm.es

Abstract. In this paper, the design of a fuzzy decision system for autonomous parallel car parking is presented. The modeling of the problem, the physics involved, the fuzzy sets assigned to the linguistic variables and the inference rules are explained. Different fuzzy operators have been tested in the generated simulation environment and the results have been compared. The decision system is proposed as a benchmark to show the influence of the different fuzzy strategies on the final decision. The comparison is made in terms of number of maneuvers for parking the vehicle. It also depends on the range of measurements to environmental objects around the car and on the starting position. The final implementation of the parking system in a Java Applet can be tested in the web using any browser.

1   Introduction

Fuzzy logic has been proved as a useful approach to implement decision systems when risk is involved. When uncertainties and inaccuracies exist and they are unavoidable, fuzzy strategy provides a robust methodology to deal with that vagueness [1, 2].

On the other hand, the autonomous parking problem attracts a great deal of attention from research community and the automobile industry because of the complexity of this problem for non-holonomic vehicles (cars) and the possibility of numerous applications [3, 4].

The impact of different ways of parking on environmental effects, mainly vehicle emissions and air pollution, needs attention. Vehicle energy consumption and specially the urban air quality at street level, related to location and design of parking procedures, need to be assessed and quantified. The number of maneuvers when parking process is under way can have a strong influence on some environmental issues [5]. At the same time, the parking procedure can potentially disrupt traffic flow and

J. Lu, L.C. Jain, and G. Zhang: Handbook on Decision Making, ISRL 33, pp. 237–258.
springerlink.com   © Springer-Verlag Berlin Heidelberg 2012

require crowd management, for example in special events. In this sense, decision making on the best sequence of parking steps is a way of managing traffic problems and the potential risks this procedure entails.

The motion control for autonomously parking a vehicle is within the general problem of steering the car from an arbitrary point to a specified one. There are some approaches reported in the literature on the parking problem [6, 7, 8]. Most of them refer to street parking detection using different visual techniques [9, 10, 11]. Many others are focused on the path planning approach [12], i.e., planning a feasible path to reach a point and subsequently following the path, such as the developed by [13, 14, 15, 16, 17]. More recent works on autonomous driving do not deal with the parking problem but with more complex issues that include uncertainty and risk [18, 19] or control of vehicles [20].

For these approaches, accurate localization of the vehicle has to be provided during the motion. They usually require accurate kinematic and dynamic models of the vehicle and its environment. However, within these models some of the vehicle's parameters are uncertain or unknown, or an analytical model of the system is not available [21]. In this case, the fuzzy approach can be very useful as it allows to generate the parking maneuvers based on approximate reasoning [22, 23, 24, 25, 26]. That is, the exact position of the vehicle neither the obstacle are needed, but some linguistic rules to describe the motion of the car when parking.

This paper describes the design and implementation of a fuzzy decision system that shows different fuzzy strategies for parking a car. The problem specification, the physics involved, the fuzzy sets assigned to the linguistic variables and the inference rules are explained. For each of the different alternatives that are provided by the system, some criterions are calculated that allows us to decide the best one. That is, the best sequence of motions to park the vehicle with fewer maneuvers, more smoothly, etc.

The main goals of this parking system are:

- To park the car from any start location in any available space (given by the user in the implemented interface).
- To maintain the fuzzy rules as simple as possible.
- To develop a tool where it is possible to see the influence of different fuzzy strategies on the behaviour of that decision system.

So the resulting system is a good balance between simplicity, correct modeling and effectiveness. The system is proposed as a benchmark to show the influence of the logic operators on the decision, as in [27].

The final implementation of the parking system in a Java Applet can be interactive tested in the internet using any web browser at http://www.fdi.ucm.es/profesor/lgarmend/SC/aparca/index.html.

The paper is organized as follows. Section 2 presents an analysis and comparison of different intelligent approaches to the parking problem. In section 3 the problem specification and the model of the vehicle motion are presented. In section 4 the fuzzy simulation environment that has been generated to implement the autonomous car parking process is described. Section 5 shows the results of applying different fuzzy strategies; they are compared and discussed in terms of efficiency. Conclusions end the paper.

2   Analysis and Comparison of Different Approaches to the Parking Problem

The automatic parallel parking problem has previously attracted a great deal of attention among researchers. The research in car parking problem is derived from the general motion planning problem and is usually defined as finding a path that connects the initial configuration to the final one with collision-free capabilities and by considering the non-holonomic constraints [28].

Current approaches to solving this problem can be classified into two main categories: (1) the path planning approach, where a feasible geometry path is planned in advance, taking into account the environmental model as well as the vehicle's dynamics and constraints, and then control commands are generated to follow the reference path; (2) the skill-based approach, where fuzzy logic or neural networks are used to acquire and transfer an experienced human driver's parking skill to an automatic parking controller. There is no reference path to follow and the control command is generated by considering the orientation and position of the vehicle relative to the parking space [29].

The most common approaches to solve this problem are generally path tracking ones. Path planning is an open-loop approach. The accuracy of the resulting path depends on the accuracy of actuators. However, the effect of control actions is not completely reliable, for example, wheels may slip. The position error caused by actuators cannot be compensated by an open-loop strategy. In the case of open-loop parking, even though position errors may be compensated by subsequent iterative motions (i.e., backward or forward motions), this compensating strategy increases the time and cost of implementation. Some researches [28, 30] have proposed various intelligent control strategies on this topic for the path planning approach.

In contrast, a skill based intelligent approach (let say fuzzy or neuro-fuzzy) generates control commands at each sampling period based on the current position of the vehicle relative to the parking space.

Both the path planning approach and the intelligent approach rely on the environmental modeling and sensor information which are in general approximate, having uncertainties. In this sense, fuzzy systems employ a mode of approximate reasoning which enables them to make robust and meaningful decisions under uncertainty and partial knowledge. That is why the fuzzy approach seems to be useful when dealing with this problem, and therefore we have focused on it for this application.

In the fuzzy logic approaches of [31, 32, 33], the control command (i.e., the steering angle of the vehicle) was generated based on the relative longitudinal and lateral distance of the vehicle to the parking space and the orientation of the vehicle. Fuzzy rules were built for each of the parking steps [31] or for different parking positions [32, 33].

More sophisticated solutions have been tried, taking advantage of the synergy of soft computing techniques. For example, in [29] a genetic fuzzy system is used to tune a fuzzy logic controller for both a skid steering AGV and a front-wheel steering AGV. In this case the car-like mobile robot is always navigated to reach a ready-to-reverse position with the vehicle orientation parallel to the parking space. So, the vehicle is first reversed into the maneuvering space and then is moved forward to adjust its position inside the space, and these steps can be repeated several times until the desired final position is reached within some tolerance. Some promising results are shown but, in contrast with our paper, no information about the number of maneuvers during the parking process or the influence of different fuzzy logics is presented.

In [34], a neuro-fuzzy model has been developed for autonomous parallel parking of a car-like mobile robot. The fuzzy model has been identified by subtractive clustering algorithm and trained by ANFIS (Adpative Neural Fuzzy Inference System). The simulation results show that the model can successfully decide about the motion direction at each sampling time without knowing the parking space width, based on the direct sonar readings which serve as inputs. But then the input data to identify the neuro-fuzzy model are needed, therefore previous simulation or experimental data are to be generated. This information is not always available.

In [28], an automatic fuzzy behavior-based controller for diagonal parking is proposed where the fuzzy controller consists of four main modules in charge of four different tasks: deciding the driving direction to meet the constraints of the parking lot, selecting the speed magnitude, and planning the short paths when driving forward and backward. While the global structure of the fuzzy controls system has been obtained by emulating expert knowledge as drivers, the design of the different constituent modules has mixed heuristic and geometric-based knowledge. So some modules have been designed by translating heuristic expert knowledge into fuzzy rules. On the other hand, the modules which should perform forward and backward maneuvers are designed based on the fuzzy rules that have been extracted by identification and tuning process from numerical data corresponding to geometrically optimal paths.

To summarize, the parallel parking problem with car-like vehicles has attracted a great interest in the research community. Due to the strong nonlinearities inherent in the problem, and the complexity of the required sensor and vehicle models, the task is hard to attack by conventional methods. But is has been tackled with different intelligent techniques. Between them, fuzzy logic has been proved as a useful strategy to translate the expert knowledge of a human driver into rules and to deal with the uncertainty. The originality of our proposal is that we not only solve the problem by a fuzzy approximation but we have generated a simulated environment where different types of fuzzy logic can be tested and compared. We include linguistic constraints for the process of parking a car and show the influence of the selection of different approaches on the final number of maneuvers of the parking process.

3   Problem Description

The parking structures for vehicles in the urban environment may be classified as "lane", "diagonal" and "row" [35].

The problem description is a simplified model of the real world parallel parking strategy, with the following items (Figure 1):

- A vehicle (car-like)
- The border of the parking lane
- The parking bay
- Front and rear obstacles

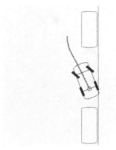

Fig. 1. Parking scenario

The goal is to park the car with the minimum number of maneuvers. The parking maneuver is considered complete when the necessary location with respect to the environmental objects is reached. That is, when the vehicle is in the parking bay, with an angle respect to the border of the parking lane very close to zero, and it keeps certain distance to the rest of obstacles.

3.1   The Car Model

The car is modeled as a four-wheeled vehicle with the steering wheels at the front. It is able to move forward or backward at a constant speed on the plane. No consideration has been taken into account about acceleration or braking in this simplified model because their influence on the topic of this work can be neglected.

Figure 2 describes the physical model of the car. The body frame of the car places the origin (x,y) at the center of rear axle (x and y are the coordinates of the midpoint of the vehicle's rear wheel axle). The x-axis points along the main axis of the car. Obviously, dealing with these car-like vehicles the variable z, height, is not considered in the motion of the car. θ is the orientation angle of the vehicle's frame, as shown in Figure 2. The next state of the position variables, x, y and z, is given by equation (1), where the dynamics of the vehicle's is defined. In this equation, variable s stands for the longitudinal velocity of the midpoint of the front wheel axle, and Φ is the steering angle (it is negative for the wheel orientation shown in Figure 2). In this application, the vehicle runs at constant speed. The vehicle can steer the front wheels in the interval [-45°, 45°]. The wheel base, that is, the distance between the front and rear axles, is represented as L. If the steering angle is fixed at Φ, the car travels in a circular motion, in which the radius of the circle is ρ. Note that ρ can be determined from the intersection of the two axles shown in Figure 2 (the angle between these axes is $|\Phi|$).

$$\dot{x} = f_1(x, y, \theta, s, \phi)$$
$$\dot{y} = f_2(x, y, \theta, s, \phi) \qquad (1)$$
$$\dot{z} = f_3(x, y, \theta, s, \phi)$$

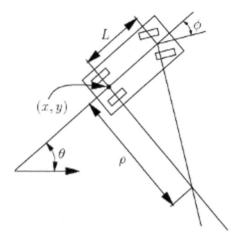

Fig. 2. Car model

The vehicle is controlled by the steering angle. This simplified model allows us to deal with the vehicle in order to apply fuzzy logic to parking it; an analytical model is not needed.

3.2   Modeling the Scenario

As it was said, the parking structures for vehicles in the urban environment may be classified as "lane", "diagonal" and "row". In the case of the lane structure, the parking bays are oriented parallel to the traffic lane, as in Figure 3. This structure is mainly used for parking along the streets. In this paper the lane structure for parallel parking is considered. The parallel parking maneuver results in lateral displacement of the vehicle into the parking bay.

A parallel parking maneuver consists of N iteratively repeated low-speed motions (backwards-forwards) with the coordinated lateral and longitudinal control aimed at obtaining the lateral displacement of the vehicle. The word "parallel" indicates that the start and end orientations of the car are the same for each iteration. The vehicle's orientation varies during the iterative motion. The number of such motions depends on the longitudinal spacing available within the parking bay and the necessary parking "depth" which depends on the width of the vehicle. As we will prove, for the same conditions, the number of motions will also depend on the type of fuzzy logic applied.

Different scenarios are created to test the car, depending on the start location of the vehicle. The start position for the parallel parking maneuver must ensure that the subsequent motion will be collision-free with the borders of a parking bay. Typically, the parking space is structured into bays of a rectangular form.

In all of them, the relevant details to describe the situation in a certain instant of time are:

- The angle formed between the car and the parking bay.
- The distance from the car to the border of the parking lane. This distance is measured from the most right wheel x value (see Figure 2)

- The distance to the front and back obstacles. These distances are measured from the front and rear bumpers of the car. The corners of the parking bay must be considered. A desired minimal safety distance between the vehicle and the nearest corner of the bay during the motion must be ensured.

In any case, the vehicle must be oriented almost parallel to the parking bay, and it must also reach a suitable start position in front of the bay.

One important aspect to remark in this decision system is that the objects of the environment are not avoided as just something that the car cannot go through or something not to collide with. Instead of doing so, the distance to the obstacles is used as information that helps the car to take a decision, i.e. if the car is going forward and the front object is quite near, it will decide to go backward and so on.

In order to obtain the rules of the decision system, three clearly different strategies, or maneuvers, are specified. They constitute the possible scenarios that are going to be considered. These three strategies are shown in Figure 3, depending on the initial location of the vehicle.

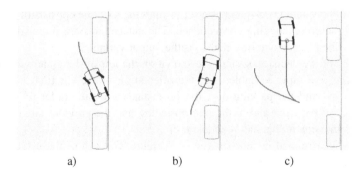

a)   b)   c)

Fig. 3. Three parking strategies considered, a), b) and c), respectively

a) In real life you would try to park in a single backward maneuver if the car is in front of the parking bay, taking into account that the dimensions of the bay are compared with those of the vehicle and the free bay is suitable for the vehicle's size. That is what the car does as well in the applet; it goes backward while steering right, and then at the right distance from the border of the parking lane, it starts to steer left.

b) If the car is behind the parking bay, it will have to move forward. Depending on the distance to the border of the parking lane, it will approach the border while moving forward, or it will move away from the rear obstacle, so it gets some leeway to try the backward strategy again.

c) If the car is far from the parking space, it will be driven to a position near the parking lane, and then the vehicle moves forward in order to start a backward movement into the parking bay.

Unlike a real driver would do, the car will not try to find a good start location, but it will try to park from wherever the user puts the car with the mouse in the graphic interface of the application.

4   The Car Parking Fuzzy Decision System

Fuzzy logic is a useful tool to represent approximate reasoning [2]. It allows us to deal with vagueness and uncertain environments. In this paper we have used the software tool Xfuzzy [36] to implement the fuzzy-inference-based system for car parking. It consists of several tools that cover the different stages in the design process of a fuzzy inference system. Xfuzzy has been programmed in Java and has GNU license. One of its most suitable characteristics is the capacity for developing complex systems and the flexibility to extend the set of available fuzzy functions.

The different tools of Xfuzzy can be executed independently. The simulation environment integrates them under a graphical interface that facilitates the design procedure to the users.

The decision system is designed as a set of rules each of one consists of some antecedents and the corresponding consequence. The antecedents are linguistic variables that are described by fuzzy terms, and so is the output variable.

The inputs of the decision system are shown on the left of the system displayed in Figure 4. The five input variables are: the orientation of the car (that is, the angle between the car and the parking bay, θ), the distances to the border of the parking lane, distance to the front and to the rear obstacles, and the previous strategy, that is, the previous movement that has been taken.

The rules are grouped in four subsystems (Figure 3). Each of these fuzzy subsystems computes an output by the defuzzification method of the center of area. The names of these four fuzzy systems indicate the strategy to be applied; they are: backward, forward approaching to the parking lane, forward moving away from the bay, and the last block is called strategy selector. It will be explained later.

The first three blocks are used to decide the angle in which the steering wheel should be steered, depending on the strategy to be implemented: backward, forward approaching, or forward moving away.

The outputs are also shown on the right hand of the image (Figure 4), they are named: strategy, steering_backward, steering_forward_approach, and steering_forward_moveaway. The three fuzzy subsystems calculate a steering output, but only one is going to be applied, the corresponding to the value of the output of the fourth subsystem, new strategy.

The last group of rules decides the best strategy to be applied. The defuzzification method used in this system is the maximum value of the membership functions of the other three, in order to choose only one of the strategies.

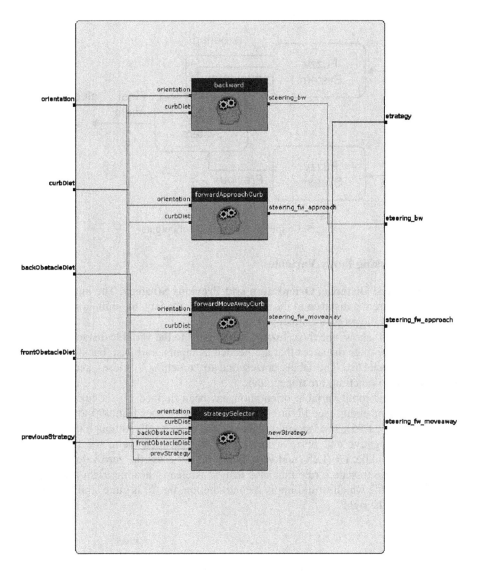

Fig. 4. The fuzzy car parking system implemented in Xfuzzy

This procedure can be called a coupling configuration (Figure 5) in the following sense. After computing the output of one of the first three fuzzy subsystems, this is combined with the information of the fourth (previous strategy) to calculate the final decision.

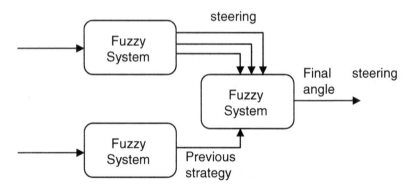

Fig. 5. Coupling configuration of outputs

4.1   Car Parking Fuzzy Variables

Input variables: Distance, Orientation and Previous Strategy. The input variables of the system are the distance to the obstacles and to the bay, the initial orientation of the vehicle, and the previous strategy (Fig. 4).

Figure 6 (left) shows the three fuzzy sets assigned to the variable distance, which is the same for the three distances that are needed as inputs: curbDist, backObstacleDist and frontObstacleDist. The labels correspond to "touching", "close" and "far", and the membership functions are trapezoidal.

The linguistic input variable orientation has been defined by six fuzzy sets with trapezoidal membership (see Figure 6, right). The range goes from -90 to 90 degrees, scaled to [-1.75, 1.75]. The "straight" label means close to 0°. Although the functions representing negative and positive angles overlap between them, the two that are nearer to the middle, L1 (Left) and R1 (Right), do not cross the zero value. This way it is not possible to state a rule that says that "it is clearly heading right and more or less heading left", which would imply a contradiction, but "it is quite straight and also a bit turned to the right".

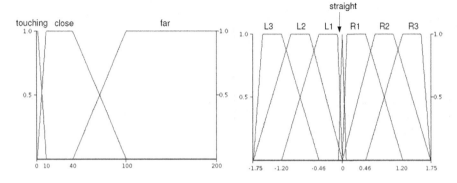

Fig. 6. Fuzzy input variables: distance (left) and orientation (right)

The other input variable, called previous_strategy, is modeled by classical or "crisp" sets because only one strategy will be applied at a time. The four singleton values are: "stop", "fwApproach", "fwMoveAway" and "backward" (see Figure 7, right).

Output Variables: Steering and Strategy. The fuzzy sets modeling the linguistic variables on the steering wheel are quite similar to the input variable orientation but the interval is now [-45°, 45°], scaled in this case to [-0.8, 0.8]. Besides, the overlap between them is a bit lower in order to produce a more precise output (Figure 7, left).

This linguistic variable corresponds to three different outputs, depending on the strategy that is going to be applied. So, there are steering_fw_approach, steering_bw, and steering_fw_away.

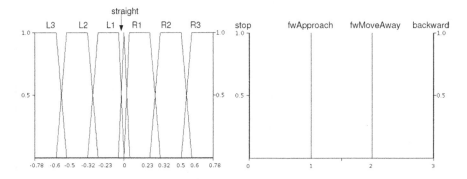

Fig. 7. Fuzzy variables: steering (left) and strategy (right)

The variable strategy is modeled as explained above for previous_strategy variable (Figure 7, right).

4.2   Car Parking Fuzzy Rules

Keeping in mind that one of the goals of this system is to maintain the rule base as simple as possible while designing the three strategies, the following notation is used:

Distance to the curb → Desired orientation

That is, given a certain distance to the border of the parking lane, the output should be the angle of the steering wheel that will make the car to be closer to the desired orientation.

Again, an indirect scheme is used to define the rules. Several fuzzy subsystems are applied and the output of each of them is used as input of the following one. The sequence is the following (Figure 8). This is applied for each strategy.

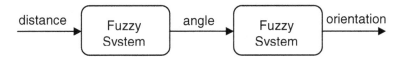

Fig. 8. Indirect rule inference

Usually a premise of a rule may contain the conjunction (AND or &), implemented by a t-norm [37]. It may combine more than one linguistic variable, i.e.,

"if x is X and y is Y"

or

"x==X & y==Y",

where x and y are fuzzy variables, such as Back obstacle distance, and X, Y are fuzzy sets (close, far, …).

Equal to X	Less than X	Less or equal than X	Greater than X	Greater or equal than X
x = X	x < X	x <= X	x > X	x >= X

Fig. 9. Interpretation of some linguistic labels used in the premises of the rules

In this application we also use expressions such as "x < X", that stands for the fuzzy premise "x is less that X". For example, given a fuzzy set X, the fuzzy sets "less than X", "less or equal than X", "more than X", "more or equal than X", are defined in figure 9 [19].

In this case, for the car parking, "less than touching" in the variable that gives the distance to the border of the parking lane means that the car is on the sidewalk.

The rules for the four fuzzy subsystems that have been implemented are detailed in the following sections.

4.2.1 Backward Strategy

According to the formulation of the rules stated above (Figure 8), the rules to compute the steering_backward angle can be expressed in a general way as (Table 1), where far, close, touching corresponds to Figure 4, left.

Table 1. Fuzzy general rules for the backward strategy

Distance	Desired_Orientation
Far	L3
Close	L1
Touching	Straight
Less than touching	R2

That means that if the car is far from the parking bay, in order to get it closer, it should be orientated or turned quite to the left (first row). But if the car is near the bay, the approaching angle should be only a bit to the left (second row). Finally, if the car is very close to the border of the parking lane, or even touching it, it should move with an angle next to zero (straight). But if the distance to the parking bay is less than touching, then the car is on the sidewalk. In that case, it should be approaching the border of the lane with an angle quite to the right to correct the position.

Let us take the first rule as an example. When the distance of the car to the border of the parking lane is big (far in Table 1), the resulting output tries to drive the steering wheel so the car reaches that L3 (quite to the left) desired angle.

To get this desired orientation to the left L3, if the car's angle is less than L3 -more to the left-, the steering output must be L3. If the angle is more than L3 - more to the right-, then the steering output should be R3. And when the car is at the desired angle, the steering should be straight (0 °).

Therefore, the first row of Table 1 is implemented in Xfuzzy by the following set of rules:

```
if(curbDist == far & orientation >  L3)
    -> steering_backward = R3;
if(curbDist == far & orientation <  L3)
    -> steering_backward = L3;
if(curbDist == far & orientation == L3)
    -> steering_backward = straight;
```

The same reasoning is done for the rest of the rules in Table 1, resulting in the following rules:

```
if(curbDist == close & orientation <  L1)
    -> steering_backward = L3;
if(curbDist == close & orientation >  L1)
    -> steering_backward = R3;
if(curbDist == close & orientation == L1)
    -> steering_backward = straight;

if(curbDist == touching & orientation <  straight)
    -> steering_backward = L3;
if(curbDist == touching & orientation >  straight)
    -> steering_backward = R3;
if(curbDist == touching & orientation == straight)
    -> steering_backward = straight;

if(curbDist < touching & orientation < R2)
    -> steering_backward = L3;
```

4.2.2 Forward Approaching the Bay

The rules for implementing the steering angle when the forward strategy has been applied are the following (Table 2):

Table 2. Fuzzy general rules for the forward strategy

Distance	Desired_Orientation
Far	R2
Close	R2
Touching	Straight
Less than touching	L1

That means that the car will try to move forward with an angle R2 (quite to the right) when it is far or close to the parking bay. When the car is touching the border of the lane, it will try to keep a straight angle. If the distance to the bay is less than zero, then the car will look for a left orientation angle, and so on.

Each of these rules of Table 2 is implemented by a set of fuzzy rules as in the previous example.

```
if(curbDist >= close & orientation <  R2)
    -> steering_fw_approach = R3;
if(curbDist >= close & orientation == R2)
    -> steering_fw_approach = straight;
if(curbDist >= close & orientation >  R2)
    -> steering_fw_approach = L3;

if(curbDist == touching & orientation < straight)
    -> steering_fw_approach = R3;
if(curbDist == touching & orientation > straight)
    -> steering_fw_approach = L3;

if(curbDist < touching & orientation >  L1)
    -> steering_fw_approach = L3;
if(curbDist < touching & orientation <= L1)
    -> steering_fw_approach = straight;
```

4.2.3 Forward Moving Away from the Rear Obstacle

When the car is far from the parking lane and the previous strategy was forward moving away, the next steps are defined in Table 3. There are no rules that define what happens when the car is close to or touching the border of the parking lane, because in that case this strategy is not going to be selected by the strategy selector (see next section).

Table 3. Fuzzy general rules (previous strategy = forward moving away)

Distance	Desired_Orientation
Far	Straight

The rules that implement this statement are:

```
if(orientation >  straight)
    -> steering_fw_moveaway = L3;
if(orientation == straight)
    -> steering_fw_moveaway = straight;
if(orientation <  straight)
    -> steering_fw_moveaway = R3;
```

4.2.4 Strategy Selector

Finally, the inference engine of the fourth fuzzy subsystem, strategy selector, decides which one of the three strategies must be applied. The reasoning is as follows,

- If the car is touching the border of the parking lane and its orientation is straight, then stop; it is successfully parked.
- If there is enough back space, then go backward.
- If the back obstacle is really near and the car is close to (or near) the parking bay, then go forward towards the bay.
- If the back obstacle is really near and the car is far from the bay, then go forward moving away from the rear obstacle.

These fuzzy sets are described in Figure 5, right. When the car is touching the back obstacle, it has to move forward. But then, it will not be touching the obstacle anymore, so the next movement is going backwards. That way, the car could be moving a little forward, then a little backward, then a little forward again, and so on. That is the reason why there is an input that is the "previous strategy".

Knowing the previous strategy, this logic can be added to the preceding rules:

If the previous strategy was forward (approaching the parking bay, or moving away), then continue moving forward applying the same strategy (approaching or moving away), unless the back obstacle is far, the front obstacle is very near, or the car is correctly parked.

The rules the describe the behaviour of the strategy selector are the following,

```
if(orientation == straight & curbDist == touching)
    -> newStrategy = stop;
if(backObstacleDist > touching & (!(orientation == straight &
curbDist == touching)))
    -> newStrategy = backward;
if(backObstacleDist <=  touching  &  curbDist  <=  close  &
(!(orientation == straight & curbDist == touching)))
    -> newStrategy = fwApproach;
if(backObstacleDist <=  touching  &  curbDist  >  close  &
(!(orientation == straight & curbDist == touching)))
    -> newStrategy = fwMoveAway;
```

```
if(prevStrategy  ==  fwApproach  &  backObstacleDist  <=  far  &
frontObstacleDist  >  touching  &  (!(orientation  ==  straight  &
curbDist == touching)))
        -> newStrategy = fwApproach;
    if(prevStrategy  ==  fwMoveAway  &  backObstacleDist  <=  far  &
frontObstacleDist  >  touching  &  (!(orientation  ==  straight  &
curbDist == touching)))
        -> newStrategy = fwMoveAway;
```

5   Results of the Car Parking Fuzzy Decision System with Different Fuzzy Operators

Different fuzzy logic operators have been applied to check which the best strategy for parking a four-wheeled vehicle with steering wheels is. The influence of the selection of one or another implementation of the fuzzy logic on the system response is shown. The results are compared in terms of number of iterations (that is, movements that the car has to carry out to park).

The following three logics are implemented to compute de conjunction (t-norm) operators in the premise of the rules [37]:

Table 4. Different implementation of the fuzzy operators

	and	or
Zadeh	$\min(x,y)$	$\max(x,y)$
Łukasiewicz	$\max(0, x+y-1)$	$\min(1, x+y)$
Product	$x*y$	$x + y - x*y$

Figure 10 shows the sequence of motions during an autonomous parallel parking for different fuzzy logics, when the car is far from the border of the parking lane and at different initial locations regarding the free bay (Behind, inFront, at_the_same_high, see Figure 3). The parking bay is between two objects. The rear wheel axle is plotted.

When the necessary "depth" is not reached, further iterative motion has to be carried out. The number of maneuvers is presented for each of the fuzzy logics (steps). The movements of the car are also printed in different colours (red for Zadeh, green for Product and blue for Lukasiewicz).

The interface that has been developed allows to place the car at any point of the scenario, i.e., at any length of the parking lane and at different starting locations. The simulation can be run step by step or in a go. It is possible to chose if the user wants to see the three paths (corresponding to the three implemented fuzzy logics) or just one of them.

In this example, it is possible to see how Lukasiewicz's logic gives the worst results (more number of iterations) in all the cases. The behaviour of the others is quite similar, although Zadeh's seems slightly better.

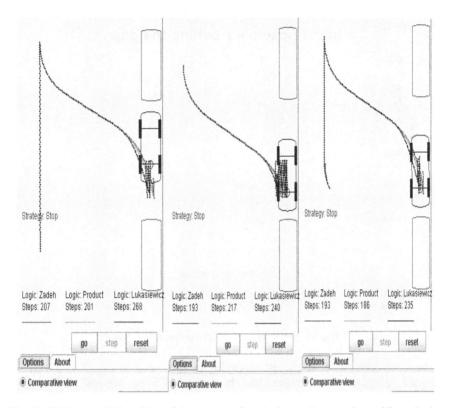

Fig. 10. Different initial position of the car regarding the bay and comparison of fuzzy logics

5.1   Discussion of the Decisions Based on the Fuzzy Logic Used for the Car Application

In the following figures (11 to 13), the results of applying different fuzzy logics for different distances of the car to the border of the parking lane and to the environmental objects are shown. The car is moving from an initial position near to the border of the parking lane (80) to a distance of 400 far from the bay, in the X-axis. The distance (80 to 400) does not correspond to any real unit. It stands for pixels; it is a convention to show the meaning of far or close regarding the bay. If a conversion to length units were to be given, 80 would be equivalent to around 4 m and 400 around 20 m. The size of the car in the scenario is 90x40 px. The Y-axis is fixed for each case (relative distance to the bay). When the number of movements is 1000 that means that the car did not achieve to park.

First, in Figure 11 the results when the initial position of the car is behind the free bay are shown. In this case, if the car is initially near the border of the parking lane (80), it is difficult to say which the best logic in terms of fewer motions for parking is; but while it is moving away from the border (from 230 to 400), Lukasiewicz performs worse and Zadeh's is the best one.

Some erratic lines appear in Figure 11, where the number of motions goes to zero at same distances. That means that the system was not able to park the vehicle properly. Near the parking bay, it strongly depends on where the car is initially placed.

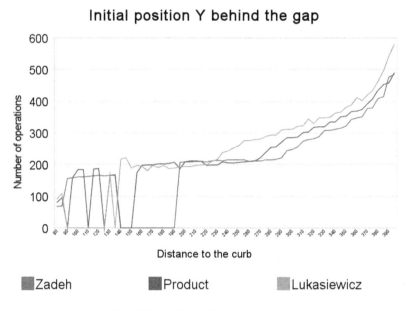

Fig. 11. Initial position of the car behind the bay

The results correspond to the left part of Figure 10.

Figure 12 shows the maneuvers that the car makes from an initial position at the height of the free bay, that is, between obstacles.

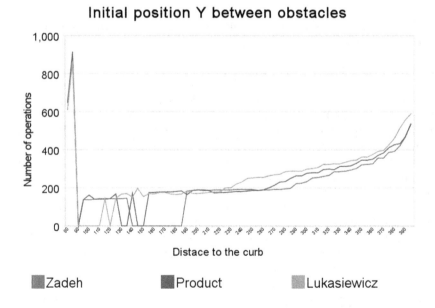

Fig. 12. Initial position of the car at the height of the bay

In this case, as it can be seen in Figure 12, if the car is near the parking lane, the best logic is again Zadeh's, but while it is moving away from the border of the parking lane (from 230 to 400), Product and Zadeh's are quite similar. In any case, the difference between the three implementations of the fuzzy logic when the car is far from the bay is very small.

Finally, in Figure 13 the results of applying different fuzzy logics when the car is in front of the bay at different distances to the border of the parking lane are presented.

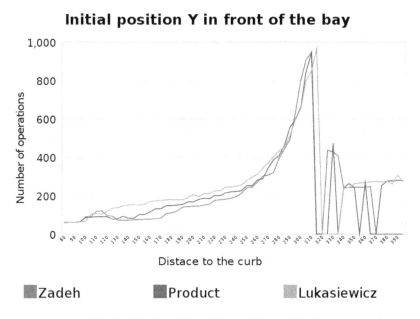

Fig. 13. Initial position of the car in front of the bay

In this case, the three fuzzy logics perform quite well until the distance to the border of the parking lane is around 240. In that case, the behavior gets worse for all of them, although it is slightly better for Zadeh's, especially for large distances (from 370 on).

The results of applying those different fuzzy logics allow us to draw some conclusions that are summarized in Table 5. The best logic is determined respect to the distance to the border of the parking lane.

Table 5. Results of applying different fuzzy logics (best logic)

Initial car position	Behind	inFront	at_the_same_height
Zadeh	far	near	far
Łukasiewicz	near	-	-
Product	medium	-	medium

The results of applying Zadeh implementation of the fuzzy operators gives good results when the car is far from the border of the parking lane, either if it is placed behind or at the same height of the bay. Nevertheless, Product gives the best results when the car is not near but not far from the border of the lane for any initial position regarding the bay. The Lukasiewicz's implementation does not perform well in this application.

6   Conclusions

Fuzzy logic has been proved very useful in order to developed decision systems when subjective implications are involved [38, 39].

The use of fuzzy logic has helped to develop a simple but robust and efficient decision system for car parking. The car in the simulation applet is able to park successfully from most of the initial situations.

The application of three different fuzzy logics allows us to show how the number of maneuvers that are needed to park the car depends on both the initial position and the chosen logic. For each initial position of the car, a specific fuzzy implementation of the operators gives the best results.

The implemented applet can by tested at http://www.fdi.ucm.es/profesor/lgarm end/SC/aparca/index.html.

References

1. Zadeh, L.A.: Fuzzy sets. Information and Control 8, 338–353 (1965)
2. Zadeh, L.A.: The concept of linguistic variable and its application to approximate reasoning, parts I, II, III. Inform. Sci. 8, 199–249 (1975)
3. Paromtchik, I.E., Laugier, C.: Autonomous Parallel Parking of a Nonholonomic vehicle. In: Proc. IEEE Intelligent Vehicles Symp., pp. 13–18 (1996)
4. Oentaryo, R.J., Pasquier, M.: Self-trained automated parking system. In: Proc. of the 8th IEEE International Conference on Control, Automation, Robotics and Vision, vol. 2, pp. 1005–1010 (2004)
5. Höglund, P.G.: Parking, energy consumption and air pollution. Science of The Total Environment 334-335, 39–45 (2004)
6. Gorinevsky, D., Kapitanovsky, A., Goldenberg, A.: Neural network architecture and controller design for automated car parking. IEEE Trans. Contr. Syst. Technol. 4, 50–56 (1996)
7. Cheng, C.W., Chang, S.J., Li, T.H.: Parallel-parking control of autonomous mobile robot. In: Proc. Int. Conf. Industrial Electronics, Control and Instrumentation, pp. 1299–1304 (1997)
8. Suruz, M.M., Wail, G.: Intelligent parallel parking of a car-like mobile robot using RFID technology. In: Proc. ROSE 2007 Int. Workshop on Robotic and Sensors, pp. 1–6 (2007)
9. Ono, S., et al.: Parking vehicle detection system by using laser range sensor mounted on a probe car. In: Proc. of Intelligent Vehicle Symposium (2002)
10. Xu, J., Chen, G., Xie, M.: Vision-guided automatic parking for smart car. In: Proc. of the IEEE Intelligent Vehicles Symposium, Dearbon, MI, pp. 725–730 (2000)

11. Zhu, C.H., Hirahara, K., Ikeuchi, K.: Street-Parking vehicle detection using line scan camera. IEEE (2003)

12. La Valle, S.M.: Planning algorithms. Cambridge University Press (2006)

13. Kanayama, Y., Hartman, B.I.: Smooth local path planning for autonomous vehicles. In: Proc. of the IEEE Int. Conf. on Robotics and Automation, Scottsdale, USA, pp. 1265–1270 (1989)

14. Laumond, J.-P., Jacobs, P.E., Taix, M., Murray, R.M.: A motion planner for nonholonomic mobile robots. IEEE Trans. on Robotics and Automation 10(5), 577–593 (1994)

15. Dolgov, D., Thrun, S., Montemerlo, M., Diebel, J.: Path planning for autonomous driving in unknown environments. In: Proc. of the 11th Int. Symp. On Experimental Robotics ISER, Athens, Greece (2008)

16. Likhachev, M., Ferguson, D.: Planning dynamically-feasible maneuvers for autonomous vehicles. In: Proc. of Robotics: Sciences and Systems IV, Zurich, Switzerland (June 2008)

17. Latombe, J.C.: Robot motion planning. Kluwer Academic Publishers (1991)

18. Valverde, L., Santos, M., López, V.: Fuzzy decision system for safety on roads. In: Intelligent Decision Making Systems, pp. 326–331. World Scientific (2009)

19. Driankov, D., Saffiotti, A.: Fuzzy Logic Techniques for Autonomous Vehicle Navigation. Physic-Verlag, Heidelberg (2001)

20. Pérez, J., Gajate, A., Milanés, V., Onieva, E., Santos, M.: Design and implementation of a neuro-fuzzy system for longitudinal control of autonomous vehicles. In: 2010 IEEE World Congress on Computational Intelligence WCCI FUZZ-IEEE, pp. 1112–1117 (2010)

21. Bentalba, S., El Hajjaji, A., Rachid, A.: Fuzzy Parking and Point Stabilization: Application Car Dynamics Model. In: Proc. 5th IEEE Mediterranean, Cyprus (1997)

22. Holve, R., Protzel, P.: Reverse Parking of a Model Car with Fuzzy Control. In: Proc. of the 4th European Cong. on Intelligent Techniques and Soft Computing - EUFIT 1996, Aachen, pp. 2171–2175 (1996)

23. Yasunobu, S., Murai, Y.: Parking control based on predictive fuzzy control. In: Proc. IEEE Int. Conf. Fuzzy Syst., pp. 1338–1341 (1994)

24. Chiu, C.S., Lian, K.Y.: Fuzzy gain scheduling for parallel parking a car-like robot. IEEE Trans. Control Syst., Technol. 13(6), 1084–1092 (2005)

25. Chang, S.J.: Autonomous fuzzy parking control of a car-like mobile robot. IEEE Trans. Syst. Man Cybern., Part A 33(4), 451–465 (2003)

26. Martín, C., Garmendia, L., Santos, M., González del Campo, R.: Influence of different strategies and operators on a fuzzy decision system for car parking. In: Intelligent Decision Making Systems, pp. 345–350. World Scientific (2009)

27. Montenegro, D., Santos, M., Garmendia, L.: An educational example for learning fuzzy systems. In: 7th IFAC Symposium on Advances in Control Education, Madrid, España (2006)

28. Baturone, I., Moreno-Velo, F.J., Sanchez-Solano, S., Ollero, A.: Automatic design of fuzzy controllers for car-like autonomous robots. IEEE Transaction on Fuzzy System 12(4), 447–465 (2004)

29. Zhao, Y., Collins, E.G.: Robust automatic parallel parking in tight spaces via fuzzy logic. Robotics and Autonomous Systems 51, 111–127 (2005)

30. Li, T.-H.S., Chang, S.-J., Chen, Y.-X.: Implementation of human-like driving skills by autonomous fuzzy behavior control on an FPGA-based car-like mobile robots. IEEE Transactions on Industrial Electronics 50(5), 867–880 (2003)

31. Miyata, H., Ohki, M., Yokouchi, Y., Ohkita, M.: Control of the autonomous mobile robot DREAM-1 for a parallel parking. Math. Comput. Simul. 41(1-2), 129–138 (1996)

32. Ohkita, M., Miyata, H., Miura, M., Kouno, H.: Travelling experiment of an autonomous mobile robot for a flush parking. In: Proceedings of the 2nd IEEE International Conference on Fuzzy Systems, San Francisco, CA, pp. 327–332 (1993)

33. Holve, R., Protzel, P.: Reverse parking of a mobile car with fuzzy control. In: Proceedings of the 4th European Congress on Intelligent Techniques and Soft Computing, pp. 2171–2175 (1996)

34. Demirli, K., Khoshnejad, M.: Autonomous parallel parking of a car-like mobile robot by a neuro-fuzzy sensor-based controller. Fuzzy Sets and Systems 160, 2876–2891 (2009)

35. Paromtchik, I.E., Laugier, C.: Motion generation and control for parking an autonomous vehicle. In: Proc. of the IEEE Int. Conf. on Robotics and Automation, Minneapolis, USA, pp. 3117–3122 (1996)

36. IMSE Centro Nacional de Microelectrónica. Herramientas de CAD para Lógica Difusa. Xfuzzy 3.0. Seville (2003), http://www.imse.cnm.es

37. Schweizer, B., Sklar, A.: Probabilistic Metric Spaces. North-Holland, NY (1960)

38. López, V., Santos, M., Montero, J.: Fuzzy Specification in Real Estate Market Decision Making. Int. J. of Computational Intelligence Systems 3(1), 8–20 (2010)

39. Farias, G., Santos, M., López, V.: Making decisions on brain tumour diagnosis by soft computing techniques. Soft Computing 14, 1287–1296 (2010)

Chapter 14
Risk-Based Decision Making Framework for Investment in the Real Estate Industry

Nur Atiqah Rochin Demong[1, 2] and Jie Lu[1]

[1] Decision Systems & e-Service Intelligence (DeSI) Lab
Quantum Computation & Intelligent Systems (QCIS) Centre
Faculty of Engineering and Information Technology
University of Technology, Sydney (UTS)
P.O. BOX 123, Broadway, NSW 2007, Australia
ndemong@it.uts.edu.au, jie.lu@uts.edu.au
[2] Faculty of Office Management and Technology
Universiti Teknologi MARA, Puncak Alam Campus
Puncak Alam, Selangor, 42300 Malaysia
rochin@salam.uitm.edu.my

Abstract. Investment in the real estate industry is subject to high risk, especially when there are a large number of uncertainty factors in a project. Risk analysis has been widely used to make decisions for real estate investment. Accordingly, risk-based decision making is a vital process that should be considered when a list of projects and constraints are being assessed. This chapter proposes a risk-based decision making (RBDM) framework for risk analysis of investment in the real estate industry, based on a review of the research. The framework comprises the basic concepts, process, sources and factors, techniques/approaches, and issues and challenges of RBDM. The framework can be applied to problem solving different issues involved in the decision making process when risk is a factor. Decision makers need to understand the terms and concepts of their problems and be familiar with the processes involved in decision making. They also need to know the source of their problems and the relevant factors involved before selecting the best and most suitable technique to apply to solve their problems. Furthermore, decision makers need to recognize the issues and challenges related to their problems to mitigate future risk by monitoring and controlling risk sources and factors. This framework provides a comprehensive analysis of risk-based decision making and supports decision makers to enable them to achieve optimal decisions.

Keywords: Risk-based decision making framework, risk analysis, investment, real estate industry, risk-based decision making technique.

1   Introduction

Decision making is a process of gathering input and processing the data collected for analysis to produce a list of outcomes based on given sources and limitations. All

J. Lu, L.C. Jain, and G. Zhang: Handbook on Decision Making, ISRL 33, pp. 259–283.
Springerlink.com   © Springer-Verlag Berlin Heidelberg 2012

decisions made will involve low, medium or high level risk based on the uncertainty factors affecting the analysis. The higher the uncertainty of factors related to the decision making, the higher the level of risk. If decision makers are familiar with the sources and factors that will affect the decision making, however, and know how to monitor and control the uncertainty factors, the risk will be lower. This chapter will propose a risk-based decision making framework for investment in the real estate industry which aims to reduce risk.

Risk-based decision making concepts and applications have been explored by many researchers who have applied different techniques and methods to support the decision making process in different fields. For example, a study into the practices of risk-based decision making for investment in the real estate industry has been conducted to investigate risk-related issues in which it was found that many decisions are made based on an investigation and analysis of factors, then weighting, calculating and selecting the best option based on a high performance index (Piyatrapoomi, Kumar & Setunge, 2004). Real estate projects are characterized by high risk, high returns and long cycles; thus real estate investors need to carefully research each project if they are to maximize the return and minimize the risk (Shiwang, Jianping & Na, 2009; Ren & Yang, 2009; Juhong & Zihan, 2009).

Risk analysis decision making is an important tool because such investments normally yield a high return but at the same time pose a high risk to success (Zhou, Li & Zhang, 2008; Zhi & Qing, 2009). There are many substantial studies related to risk analysis techniques and approaches for the real estate industry. In principle, risk-based decision making techniques involve risk analysis and support the decision making process. The literature shows that various risk-based decision making techniques have been integrated with decision support tools and intelligent agents to enhance the usefulness of the technique. Predicting and controlling risk has become the key to the success or failure of a project (Wanqing, Wenqing & Shipeng, 2009). Several techniques have been proposed and applied in e-service intelligence to evaluate, analyze, assess or predict risk, including the Analytic Hierarchy Process (AHP) and Monte Carlo Simulation, Markowitz Portfolio Optimization Theory, real options methodology (Rocha et al., 2007), and a Hidden Markov Model (Sun et al., 2008; Lander & Pinches 1998, cited in Rocha et al., 2007).

Risk analysis with uncertainty in the decision making process deals with the measurement of uncertainty and probability and the likely consequences for the choices made for the investment. The uncertainty of variables or factors that will affect the risk analysis process will impact the success of a project investment in the real estate industry. Techniques such as fuzzy set theory (Sun et al., 2008) and fuzzy-analytical hierarchy process (F-AHP) have been proposed to solve such problems.

This chapter proposes a risk-based decision making framework for investment pre-analysis in the real estate industry, as shown in Figure 1. The proposed risk-based decision making framework comprises five main elements namely: foundation concepts; process; sources and factors; techniques/approaches; and issues and challenges. These five elements will be explained in subsequent sections. This framework can be applied to different problems or issues in various industries and can support decision makers to render their decision making process more structured and manageable.

Fig. 1. The structure of the proposed risk-based decision making framework

The structure of this chapter is organized as follows. Following this introduction, risk-based decision making concepts are highlighted in Section 2. The risk-based decision making process is explained in Section 3. Section 4 elaborates on risk sources and risk factors that may affect the risk analysis for each project to be selected for investment. In Section 5, some examples of the risk-based decision making techniques currently applied in real estate project investment are described. Section 6 explains the issues and challenges of risk-based decision making, and the chapter is summarized in Section 7.

2   Risk-Based Decision Making Concepts

This section discusses the concepts of risk-based decision making including the definition of risk, types of risk, and a brief explanation of risk analysis. These concepts relate mainly to risk-based decision making for investment as applied in the real estate industry.

2.1   Definition of Risk and Risk-Based Decision Making

According to Aven (2007), risk is defined as the combination of possible consequences and associated uncertainties, as shown in Figure 2. Associated uncertainties refer to the uncertainties of the sources of risk. A source is a situation or event that carries the potential of a certain consequence, and 'vulnerability' is defined as the combination of possible consequences and associated uncertainties from a given source. There are three categories of sources: threats, hazards, and opportunities. Exposure of a system to certain threats or hazards can lead to various outcomes such as economic loss, the number of fatalities, the number of attacks and the proportion of attacks that are successful. Based on the identified vulnerabilities, risk can be described, using standard risk matrices showing the likelihood of threats and possible consequences (Aven, 2007). The vulnerability analysis literature has a focus on methods for identifying vulnerabilities and measures that can be implemented to mitigate these vulnerabilities. A common definition of vulnerability is a fault or weakness that reduces or limits a system's ability to withstand a threat or to resume a new stable condition. Vulnerabilities are related to various types of objects such as physical, cyber, human/social and infrastructure (Anton, et al., 2003 as cited in Aven, 2007).

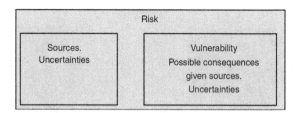

Fig. 2. Risk viewed as a combination of sources and vulnerability (Aven, 2007)

Risk-based decision making is a process based on the analysis of risk related issues. The result of the analysis of the choices on the list will vary depending on the level of uncertainty factors that will affect the decision making process. Risk arises because of possible consequences and associated uncertainties, and there are various risk sources that will affect the level of risk for given alternatives. The risk-based decision making for each investment project aims to minimize or eliminate unwanted outcomes to optimize the benefits of the investment.

2.2   Types of Risk

Two main types of risk affect the decision making process for investment in the real estate industry: *systematic risk* (beta) and *unsystematic risk*. According to Chauveau and Gatfaoui (2002), systematic risk is a measure of how the asset co-varies with the economy, and unsystematic risk also known as *idiosyncratic risk* which is independent of the economy. According to the Capital Asset Pricing Model (CAPM) (Lintner, 1965; Sharpe, 1963, 1964, cited in Lee & Jang, 2007), the total risks are calculated as follows:

$$\text{Total risk} = \text{Systematic risk} + \text{Unsystematic risk.} \tag{1}$$

2.2.1 Systematic Risk

Systematic risk refers to a type of risk that influences a large number of assets. It cannot be avoided despite stock portfolio diversification (Brealey & Myer, 2000, cited in Lee & Jang, 2007). According to Lee and Jang (2007), systematic risk can differ from period to period. Managerial decisions about operations, investments, and financing will influence the performance for the company, consequently affecting how its returns vary with market returns. The CAPM suggests that the expected rate of return on a risk asset can be obtained by adding the risk-premium to the risk-free rate; the expected risk premium varies in direct proportion to beta in a competitive market (Chen, 2003; Gencay, Selcuk & Whitcher, 2003; Lintner, 1965; Sharpe, 1963, 1964; Sheel, 1995, cited in Lee & Jang, 2007). Mathematically, the expected rate of return is described as

$$R_i = R_f + (R_m - R_f)\beta_i, \tag{2}$$

where R_i is the expected return on the ith security R_f the risk-free rate; R_m the return on the market portfolio; β_i the estimated beta of the ith security; $(R_m - R_f)\beta_i$ the risk premium. Based on CAPM, systematic risk refers to a type of unavoidable risk on the stock market. Systematic risk is presented by beta which is calculated by linear analysis between the daily prices of stocks and the security index of the stock market (Jian, Zhao & Xiu, 2006).

2.2.2 Unsystematic or Idiosyncratic Risk
Unsystematic risk, or idiosyncratic risk, is sometimes referred to as a specific risk which is sensitive to diversification, contrasting with systematic risk, which is undiversifiable. Idiosyncratic risk is significant for asset pricing because it inhibits the intergenerational sharing of aggregate risk (Storesletten, Telmer & Yaron, 2007).

CAPM and Arbitrage Pricing Theory (APT) assert that idiosyncratic risk should not be priced in the expected asset returns, and the recent surge of interest in the idiosyncratic risk of common stocks has generated substantial evidence on the role of idiosyncratic risk in equity pricing (Liow & Addae-Dapaah, 2010). The main reason for this interest is that most investors are under-diversified due to wealth constraints, transaction costs or specific investment objectives; as such, idiosyncratic risk may be important to less well-diversified real estate investors who wish to be compensated with additional risk premium. Such investors need to consider idiosyncratic risk (together with market risk) when estimating the required return and the cost of capital on assets or portfolios. Both systematic (market) and idiosyncratic volatility are relevant in stock asset pricing (Campbell et al., 2001, cited in Liow & Addae-Dapaah, 2010).

Various intelligent techniques including the Real Option method, Multi-State approach, variable precision rough set (VPRS), Condition Value-at-Risk (CVaR), AHP, Support Vector Machine (SVM), Radial Basis Function Neural Network, Fuzzy Comprehensive Valuation Method and Projection Pursuit Model based on Particle Swarm Optimization (PSO) have been applied to deal with and support unsystematic or idiosyncratic risk-based decision making.

2.3 Risk Analysis

Risk analysis is the process of identifying the security risks to a system and determining their probability of occurrence, their impact, and the safeguards that would mitigate that impact (Syalim, Hori & Sakurai, 2009). The risk analysis concept is present in business transactions, especially in the real estate industry which involves high cost and high capital (Wu, Guo & Wang, 2008). Regardless of the types of risk, the application of risk analysis has a positive effect in identifying events that could cause negative consequences for a project or organization and taking actions to avoid them (Olsson, 2007). Risk analysis is a vital process for project investment in the real estate industry which has low liquidity and high cost. It mainly consists of three stages: risk identification, risk estimation and risk assessment (Yu & Xuan, 2010).

Several risk analysis techniques, tools, and methodologies have been developed to analyze risk in different industries. Some of these techniques, such as the even swaps method, have been integrated with decision support tools called Smart-Swaps to support multi-criteria decision analysis and assist decision makers, in particular the project manager, to engage in optimal decision making (Mustajoki & Hamalainen, 2007).

2.3.1 Risk Identification

There are currently several risk identification techniques at present including the Delphi technique, brainstorming, Fault Tree Analysis, SWOT analysis and expert survey. Of these, Delphi is the most widely used.

2.3.2 Risk Estimation

Risk estimation quantifies the risks that exist in the process of investment in real estate projects. It uses risk identification, determines the possible degree of influence of such risks and objectively measures them to make evaluation decisions and subsequently choose the correct method to address the risks. The theoretical basis of risk estimation includes the Law of Large Numbers (LLN), the Analogy Principle, the Principle of Probabilistic Reasoning and the Principle of Inertia (Xiu & Guo, 2009).

2.3.3 Risk Assessment

The risk assessment or risk evaluation of investment in real estate projects refers to the overall consideration of the risk attributes, the target of risk analysis and the risk bearing capability of risk subjects on the basis of investment risk identification and estimation, thus determining the degree of influence of investment risks on the system (Xiu & Guo, 2009).

Risk occurs at different stages of the investment process. There are many techniques or approaches available for risk-based decision making, each of which has its own features or criteria and is used for quantitative or qualitative analysis, or both. Some researchers have combined or embedded these techniques to conduct both quantitative and qualitative analysis. For example, the real options method is strictly concerned with quantitative analysis. However, Information and Communication Technology (ICT) investments experience tangible and intangible factors, and the latter can be mainly treated by qualitative analysis. They have proposed a decision analysis technique which combines real options, game theory, and an analytic hierarchy process for analyzing ICT business alternatives under threat of competition (Angelou & Economides, 2009).

Risk analysis involves decision making in situations involving high risks and large uncertainties, and such decision making is difficult as it is hard to predict what the consequences of the decisions will be (Changrong & Yongkai, 2008; Ju, Meng & Zhang, 2009; Hui, Zhi & Ye, 2009; Chengda, Lingkui & Heping, 2001). There are two dimensions of risk analysis, namely, possible consequences and associated uncertainties (Aven, Vinnem & Wiencke, 2007).

An example of the risk analysis framework proposed by Li et al. (2007) including the risk factor system, data standards for risk factors, weights of risk factors, and integrated assessment methods was used to quantitatively analyze the outbreak and spread of the highly pathogenic avian influenza virus (HPAI) in mainland China. They used a Delphi method to determine the risk factors according to predetermined principles, and an AHP integrated with multi-criteria analysis was used to assess the HPAI risk.

Risk analysis is an important process that needs to be conducted to achieve optimal decision making. Real estate franchisors can achieve their goals and objectives if they fully understand and can identify the uncertain factors and variability that will affect the level of risk for each given alternative. The uncertain factors or variables will lead to probability and consequences and can be retained as a list of threats that will affect the risk level. It is therefore important to propose a framework of risk analysis as a guideline for investors in the real estate industry.

3 Risk-Based Decision Making Process

This section discusses the risk-based decision making process which includes the main risk-based decision making activities, the types of decision making process and the decision support technology for risk-based decision making. The decision support technology discussed briefly as this chapter focusing on the five main elements of risk-based decision making framework for investment in real estate industry.

3.1 Main Risk-Based Decision Making Activity

According to Busemeyer and Pleskac (2009), decision making processes become more complex, experience greater uncertainty, suffer increasing time pressure and more rapidly changing conditions, and have higher stakes. Thus, it is important to have a guidelines or step-by-step activity that will ensure all the requirements and elements for the risk-based decision making are clearly identified, defined and prioritized.

The main activity for the risk-based decision making need to be listed and perform accordingly as required to ensure the decision made is beneficial. Moreover, risk analysis needs to be performed carefully to ensure there are no undetected or potential problems on the horizon as the risk factors and its sources have the tendency to be uncertain. The uncertainty of risk factors will lead to probability and consequences to the outcome of the decision making process with risk. Risk can be distinguished from other events due to the unwanted effects associated with it, and its ability to change the outcome of the interaction in a negative way or towards an unwanted direction. The consequences are the outcome of an event expressed qualitatively or quantitatively, being a loss, injury, disadvantage or gain. There may be a range of possible outcomes associated with an event (Hussain et al., 2007). Figure 3 shows the main activities of risk-based decision making for investment in the real estate industry.

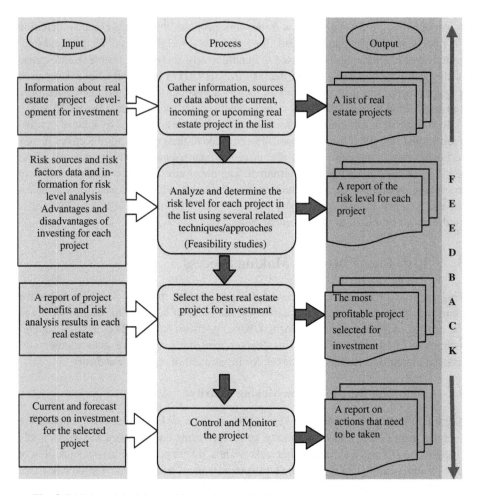

Fig. 3. Risk-based decision making main activity for investment in the real estate industry

The process starts from defining and gathering information sources or data about a current, incoming or upcoming real estate project in the list. For example, the details about the properties or property portfolio need to be collected from valid sources such as real estate agency. Information regarding the project available in the real estate industry should be gathered as much as possible for the risk-based decision making. A check list is used to determine all the elements of potential risk factors are identified and to ensure they are all covered. All the potential risk factors will be identified by asking questions such as what can happen, why it happens and how it happens. The risk identification processes include internal and external sources through brainstorming, inspections, professional consultations, case studies, audits and questionnaires.

The identified risk will be analyzed and the risk level for each project in the list will be determined using several related techniques. The analysis involves ranking and prioritizing the risks for each project based on the project's profile and the risk-consequences analysis. The risk analysis might use qualitative or quantitative techniques, or a combination of the two, to identify the risk level and the consequences of each project. Feasibility studies for all elements in the list will be checked and aim to uncover the strengths and weaknesses of the identified risk factors and determine whether or not the project will be successful. Once the analysis is done, the best project for investment will be selected and monitored. Feedbacks from the decision makers will be asked for each of the output of all of the risk analysis activities involved to ensure all elements or criteria have been met.

The output or decision point of the decision making process can be categorized into three choices or alternatives: hold, proceed (go) or discard (stop) (Strong et al., 2009). The hold state stipulates waiting for a better time to continue the process; the proceed (go) decision point is to proceed with the potential or actual line of business; and discard (stop) decision point terminates the process. Feasibility studies on all the sources and factors or variables that will affect the project investment need to be carried out to eliminate, hedge or mitigate the risk.

3.2 Types of Risk-Based Decision Making Process

Decision making process can be divided into two main types: *static decision making process* and *dynamic decision making process*. These two main types of decision making process are related to different investment strategies in the real estate industry: simultaneous strategy or sequential strategy. Simultaneous and sequential investments are common in the real estate market (Rocha et al., 2007). All decision making analysis will involve risk, and what matters is the level of risk involved in each solution chosen by decision makers. The risks that arise are an inherent implication of decision making processes which need to be analyzed carefully, and there is a need for a system that can cater for this problem, thus helping decision makers to lower the risk or to make wise decisions for every problem they face.

3.2.1 Static Risk-Based Decision Making Process

The static decision making process normally corresponds to the simultaneous investment strategy. It is a now-or-never decision where the two options are represented using the circle nodes and the triangle nodes represented the end of process in Figure 4. All irreversible resources are compromised at once, and are a process that is related to investment in the real estate industry. The simultaneous strategy is usually implemented during periods of increased demand and implies lower construction costs but, in turn, carries less certain returns. The industry has suffered bitter experiences with residential housing developments and mega-entertainment resorts that started simultaneously but have generated profits only after five or more years of construction. Figure 4 shows the static decision making process.

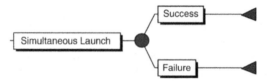

Fig. 4. Static (simultaneous) decision making process (Rocha et al., 2007)

3.2.2 Dynamic Risk-Based Decision Making Process

The dynamic decision making process is related to sequential investment strategy, in which risks are faced in sequence with relatively smaller increments at every phase of the project, but at the expense of relatively higher construction costs. In sequential investment strategy, the initial outflow is lower than in simultaneous investment strategy and expected inflows of a previous phase may finance subsequent ones. Dynamic decisions arise in many applications including military, medical, management, sports, and emergency situations. Dynamic risk prediction is an important process in achieving optimal decision making when dealing with investments with a limited budget plus time and other constraints. The dynamic decision making process for the prediction of risk level is generally related to three managerial flexibilities characteristics (information gathering, waiting option and abandon option) as shown by three different nodes (circle, square and triangle) of real option methods, as depicted in Figure 5. The circle nodes represent the options available for the next steps and the triangle nodes represent the end of the process for the particular decision made. The most important characteristic related to the dynamic decision making process is the waiting option represented in square nodes.

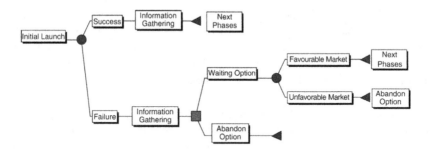

Fig. 5. Dynamic (sequential) decision making process (Rocha et al., 2007)

The challenge decision maker's face with is that the characteristics of each option are often unknown, especially when dealing with a higher uncertainty of risk factors or risk sources. Higher uncertainties of risk sources or risk factors will lead to higher probability of unsuccessful investment or consequences of the decision made for given options. Owing to the increased complexity of the decision, the uncertainty of evaluation also increases. In this situation, decision makers are unable to use precise numbers to express their evaluations, although they can still give approximate ranges of evaluations through their knowledge and cognition (Lin, Lee & Ting, 2008). To achieve optimal decision making, however, it is important for them to master good knowledge of risk analysis, and they need to analyze the risk sources and risk factors for the given options.

3.3   Decision Support Technology for Risk-Based Decision Making Process

The literature review indicates that there is much work to be done to develop an intelligent decision support system (IDSS) for handling risk-based decision making in business operations. Decision support tools help businesses to manage their business tasks efficiently and effectively, especially when managers deal with decision making processes in their daily routine (Lu, et al., 2007; Niu, Lu & Zhang, 2009). This is perhaps the most important concern for the future of risk analysis systems for managers, since it will promote vital and useful technology that helps decision makers to identify the risks involved in making decisions to meet their organization's goals and objectives.

A combination of tools in an IDSS will make the system more useful to organizations, especially when dealing with risk analysis. All decision making involves risk, and it is the level of risk involved in each chosen solution that matters to decision makers. The risks that arise need to be analyzed carefully using a system that helps decision makers to lower the risk and make wise decisions. There appears to be general agreement that the use of an intelligent agent as one of the artificial intelligence mechanisms integrated with decision support system tools will provide an IDSS that helps managers to make decisions effectively and efficiently. IDSS is an example of decision support technology which is flexible in use. IDSS for real estate investment can help real estate investors to make effective cost, fund and market analysis, and can assist investors to make scientific decisions. Furthermore, IDSS is reliable and can result in a profitable outcome because the factors in the decision support system are numerous and a large number of modules and methods are provided to help a policy maker analyze problems (Rui et. al., 1996).

Uncertainty and complexity are common conditions that have led to greater recognition of systemic and holistic approaches to problem solving (Cassaigne & Lorimier, 2007). Cassaigne and Lorimier have reported on the findings of exploratory research into the technical and organizational challenges facing IDSS for nonoperational decision making and have indicated directions for future research. They believe that due to the complexity of nonoperational decisions, it might be necessary for the decision maker to involve one or more domain experts to identify the possible characteristics of the problem, the decision technique, the possible solutions and their impacts. IDSS "would combine the knowledge-based reasoning method with formal methods of decision analysis" (Holtzman, 1999, cited in Cassaigne & Lorimier, 2007). Thus, the integration of different methods with decision support tools to solve uncertainties of risk factors should be applied to achieve the optimal decision. This is possible when dealing with semi-structured problems.

The literature also notes current developments in the field of decision support technology use for risk analysis. Most research efforts reported in popular journals or databases have tried to fulfill the need to develop IDSS for structured types of decisions to solve the daily routine activities of an organization. The systems related to

IDSS proposed by other researchers have focused more on how daily problems are solved in industry. For instance, Delen and Pratt (2006) developed an integrated IDSS for manufacturing systems that is capable of providing and independent model representation concept. They believe that managers need to integrate intelligent information systems that are capable of supporting them throughout the decision making life-cycle, which starts with structuring a problem from a given set of symptoms and ends with providing the information needed to make the decision. They also report on a collaborative research effort, the aim of which has been to fill this need by developing novel concepts and demonstrating the viability of these concepts within an advanced modeling environment.

According to Mora et al. (2007), research on how to design, build and implement IDSS from a more structured and software engineering/systems engineering perspective in absent in the entire research period from 1980–2004. They used an existing conceptual framework to assess the capabilities and limitations of the IDSS concept, and through a conceptual meta-analysis research of the Decision Support System (DSS) and artificial intelligence (AI) literature from 1980 to 2004, they developed a strategic assessment of the initial proposal called Capability Assessment Framework for decision making support system (DMSS). This framework identifies three dimensions as core structural components: the user-interface capability dimension (UICD); the data, information and knowledge-representation capability dimension (DIKCD); and the processing-capability dimension (PCD). They discovered that the DMSS community focused on the decision making process rather than on the development of a specific AI mechanism. Most of the decision support technologies developed have been integrated with intelligent agent to make the software system more useful and beneficial to users.

4   Risk Sources and Risk Factors in the Real Estate Industry

This section explains the five main categories of risk sources and risk factors for investment in the real estate industry, namely financial risk, economic risk, scheduled risk, policy risk, and technical risk and others. Each of these risk sources has its own risk factors as a sub element. The next part of this section discusses the risk factors based on the stages of real estate investment.

4.1   Financial Risk

Financial risk refers to the uncertainty of profits which originate from the process of financing, money allocation and transfer, and interest payments as financial aspects of a project. The financial risk consists of three sub-categories of risk factors: policy, engineering and market. The value of financial risk will normally be decreased because of the high rates of return in a short period of time that allow investors who have sufficient budget and capital to engage in investment in the real estate industry. Some investors acquire their capital for investment through mortgages, either from a bank or an organization; however, since this involves high cost and high capital, and properties are not easily sold, the risk level for the real estate investment will be very high. Over the past decade, real estate has become a hot investment area, but real estate projects are characterized by high risk, high return and long cycles, which require that real estate investors carefully research each project to maximize the return and minimize the risk. Risk analysis using scientific methods and tools to understand the risk situation thoroughly is therefore essential when making decisions on investment (Yu & Xuan, 2010).

According to Zhi and Qing (2009), financial risk includes own fund risk, bank loan risk, shares risk and financial structure risk. Financial risk analysis is the core of real estate investment risk analysis and will directly determine the decision making concerning the investment.

Real estate industry business processes include managing, buying and selling properties, rental services for properties, and advertising properties for sale or rent. Buying and selling properties involves high project costs or investment and careful consideration needs to be given to any transaction. Even though investment in the real estate industry incurs high cost and slow liquidity, it offers more value and a higher rate return on investment in a short period of time (Zhou, Wang & Li, 2010). The value of financial risk can be minimized with rises in the price and value of the property over time, especially when the property is located in an area which is still in a development phase. For example, if public transport or other amenities such as a children's playground or park, community centres, or schools are still in the development phase, the value of the property will increase once these facilities are completed. Property prices are also affected by many other factors, such as interest rate, land supply and inflation rate (Hui, Yu & Ip, 2010).

Real estate investment is subject to high risk because it is heavily dependent on bank loans, and the risks involved can be traced back to the asset security of bank loans (Xiaozhuang, 2008). Financial risk can also be mitigated by analysing the real estate portfolio, based on the financial requirements of the real estate investment (Wu et al., 2009).

According to Saleem and Vaihekoski (2008), currency risk can have very important implications for portfolio management, the cost of capital of a firm, and asset pricing, as well as currency hedging strategies, as any source of risk which is not compensated for in terms of expected returns should be hedged. Real estate investment is speculative and its return and risk are influenced by many factors, such as the natural environment, the socio-economic environment, the market, and enterprise purchasing capability (Liu, 2007).

4.2   Economic Risk

Economic risk includes regional development risk, market supply and demand risk, and inflation risk (Zhi and Qing, 2009). Li and Suo (2009) define economic risk factors as consisting of the sales cycle, industry competitiveness, economic operation, exchange rate and interest rate. Sun et al. (2008) propose a model based on general relationships among significant elements of dynamic risk prediction for real estate franchisors in the real estate industry using the real option method. One of the risk sources proposed in the model is economic risk, which includes finance, financing, market requirement and land price.

4.3   Scheduled Risk

Scheduled risk affects the degree of risk for the given alternatives (Sun et al., 2008). Real estate investment is normally related to capital risk and liquidity risk. Investors must have the knowledge to understand and manage these factors in the real estate industry. The list of projects for an investment need to be analyzed and priority is given to the most beneficial project according to the available budget and time. Liquidity risk is related to scheduled risk because both of these risk sources are dependent on time series.

Yu and Xuan (2010) define scheduled risk as the delay of part of the whole project process, or even the entire project, which is often accompanied by an increase in cost. The Critical Path method based on Work Breakdown Structure (WBS) is the most commonly-used methodology for scheduled risk.

4.4   Policy Risk

Policy risk refers to uncontrollable elements which can cause great harm as a source of risk in real estate development, even though there is only a small likelihood of their occurrence (Yu & Xuan, 2010). Jin (2010) highlights policy environment risk as the lifecycle risk impact factors of a real estate project. Organization policy and industrial policy are also examples of the variables or factors that might affect the result of risk analysis (Sun et al., 2008). Zhi and Qing (2009) define policy risk as including monetary policy risk, industrial policy risk, land policy risk, housing policy risk, tax policy risk, and town planning risk. Li and Suo (2009) point out that policy factors consist of environmental policy, tax policy, financial policy, and industrial policy.

4.5   Technical Risk and Others

Technical risk refers to the harm and danger caused by technical deficiencies or defects (Yu and Xuan, 2010), and tendering management, design change and project construction are the source of risk factors for technical risk (Sun et al., 2008). Leifer et al. (2000, cited in Strong et al., 2009), define three major dimensions of uncertainty that are relevant for all innovation development projects targeting new lines of business: technological, market organizational, and resource uncertainties.

Other risk factors include political risk, construction risk, location factors, and settlement risk. Sun et al. (2008) propose political risk as a major source of risk. Risk factors for political risk include industrial policy, housing and regulation reform, and social risk: city planning, zone development and public interference.

Zhi and Qing (2009) describe construction risk as the first-level index, other than financial, policy and economic risk that should be selected for risk evaluation after investigating residential real estate markets. Construction risk includes nature condition risk, the risk of project delays, project quality risk, development cost risk and construction claim risk.

Li and Suo (2009) highlight two other main risk factors for real estate investment: the location factor, and settlement risks: sales return sum, settlement ability, settlement period. Moreover, Yu and Xuan (2010) suggest that another source of risk is management risk; that is, risks that originate from errors or changes in management, or that are linked to how a project is organized, managed and implemented. Xiaobing and Haitao (2009) state that the risk indicators for the early stage of real estate projects includes purchasing land risks, removing and resettlement risks, survey risks, design risks, financing risks, bidding risks, contract risks and approval risks. Based on the review of literature, Table 1 depicts the risk factors that will affect risk analysis for investment in the real estate industry.

Table 1. A summary of risk factors that will affect the risk analysis for investment in the real estate industry

Risk Analysis Factors	
Sociological risk	Technical risk
Organizational policy risk	Economic risk
Contractual risk	Behavioral risk
Types of user (decision makers)	Organizational risk
Goals and objectives of decision maker	Scheduled risk
Political risk	Currency risk
Social risk	Technological risk
Financial risk	Policy risk
Market organizational risk	Resource uncertainties risk
Construction risk	Psychological risk
Regulatory risk	Natural environment risk
Socio economic environment risk	Market risk
Enterprise purchasing capability risk	Nature condition risk
Location factor risk	Settlement risk
Management risk	Purchasing land risk
Removing and resettlement risk	Survey risk
Design risk	Bidding risk
Contract risk	Approval risk

4.6   Risk Factors Based on Stages of Real Estate Investment

Li et al. (2009) divided risk factors into four stages of real estate project investment. 1) Risk factors during the investment decision process: development opportunity risk, risk of regional economic environment, risk of regional social environment and risk of project positioning. 2) Risk factors during the process of obtaining land: risk of market supply and demand, risk of development cost, risk of financing and risk of levy land and remove. 3) Risk factors during the construction process: risk of project quality, risk of project duration, risk of development cost, risk of contracting, risk of project technology, risk of construction claim, risk of natural conditions and risk of contract mode. 4) Risk factors during the rent and sale management process: risk of marketing opportunity, risk of sales planning, risk of operating contract and risk of natural disasters or other contingencies.

Jin (2010) illustrates the lifecycle risk impact factors according to the different stages of a real estate project as follows. 1) The risk in the investment decision stage, including: policy environment risk, investment opportunity choice risk and investment property type choice risk. 2) The risk in the land acquisition stage, including: land price change risk, land idle risk, levy land and dismantle risk and raise funds risk. 3) The risk in the construction stage, including: inviting public bidding mode risk, contract way risk, contract bargain risk, quality risk, schedule delay risk, development cost risk, construction safe risk and construction claim risk. 4) The risk in the lease and sale stage, including: lease and sale opportunity risk, lease and sale contract risk. 5) The risk in the property operation stage, including: natural disaster risk and contingency risk.

5   Risk-Based Decision Making Techniques for Real Estate Project Investment

Investment in the real estate industry in emerging economies demonstrates tight working capital, low liquidity, slow payback, high sunk cost, capital intensive outflows that are not immediately recovered, enduring uncertainties about demand, price/m^2, land costs, and short to medium construction times. It is very important for investors to have an approach or technique to analyze the real estate project investment to minimize the uncertainty or risks that will affect their profits and margins (Rocha et al., 2007; Sun et al., 2008). This section will explain some of the risk-based decision making techniques for investment in the real estate industry, which are divided into three categories, namely, quantitative, qualitative and hybrid technique.

5.1   Quantitative RBDM Technique

Quantitative technique refers to the analysis of variables or elements that can be measured using either discrete or continuous numerical data involving statistical data analysis. Some examples of quantitative RBDM techniques for investment in the real estate industry include Beta, Projection Pursuit model based on Particle Swarm Optimization (PSO), condition value-at-risk (CVaR), Maximal Overlap Discreet Wavelet Transform (MODWT), Markowitz's Portfolio Analysis, Regression Analysis, Statistical Stepwise Regression Analysis and Neural Network Sensitivity Analysis.

5.1.1   Beta

Beta is a risk measurement for systematic risk (Li & Huang, 2008). Beta measures the degree of co-movement between the asset's return and the return on the market portfolio. In other words, beta quantifies the systematic risk of an asset (Xiong et al., 2005).The systematic risk, as denoted by β_i, is a measure of the slope of a regression line between the expected return on the ith security (R_i) and the return on the market portfolio (R_m) such as Standard and Poor's 500 (S&P 500 cited in Lee & Jang, 2007) and Stock Index and New York Stock Exchange (NYSE) Index. Mathematically, the beta (β_i) is expressed as

$$R_i = \beta_0 + \beta_i R_m + e_i. \tag{3}$$

Based on the formula given above, an asset with a higher beta should have a higher risk than an asset with a lower beta (Tang & Shum, 2003).

5.1.2   Projection Pursuit Model Based on Particle Swarm Optimization (PSO)

The Projection Pursuit model based on Particle Swarm Optimization (PSO) is used to process and analyze high dimensional data, especially non-linear and non-state high-dimensional data. PSO can be used to solve a large number of non-linear, non-differentiable and complex multi-peak optimization problems and has been widely used in science and engineering. The Projection Pursuit model can make exploratory analysis and is also referred to as the deterministic analysis method (Shujing & Shan, 2010). Modeling steps for the Projection Pursuit Model are as follows: Investment risk assessment program of the normalized values; Projection index structure function; Optimized projection target function; Scheme selection.

5.1.3   Condition Value-at-Risk (CVaR)

A dynamic condition value-at-risk (CVaR) technique is one of the new tools of risk measurement for studying investment in real estate. This technique was proposed by Rockafellar and Uryasev as cited in Meng et al. (2007) and has many good properties, such as being computable, convex and more efficient than the Markowitz value-at-risk technique for portfolio investment.

5.1.4   Maximal Overlap Discreet Wavelet Transform (MODWT)

Maximal Overlap Discreet Wavelet Transform (MODWT) provides a natural platform for investigating the beta or systematic risk behavior at different time horizons without losing any information (Xiong et al., 2005). They proposed this method to decompose a time series of any length into different timescales and listed the advantages of MODWT over the Discreet Wavelet Transform (DWT) as follows: 1) The MODWT can handle any sample size, while the Jth order DWT restricts the sample size to a multiple of 2^J; 2) The detail and smooth coefficients of a MODWT multi-resolution analysis are associated with zero-phase filters; 3) The MODWT is invariant to circularly shifting the original time series; 4) The MODWT wavelet variance estimator is asymptotically more efficient than the same estimator based on the DWT.

5.1.5   Markowitz's Portfolio Analysis and Regression Analysis

Lin and Chen (2008) carried out a study on the identification of default risk as a systematic risk based on Chinese stock markets using two analyses, namely portfolio analysis and regression analysis to check whether default risk is systematic and to discover the relationship between the expected return and the default risk. Regression analysis was used to examine whether default risk is systematic in the Chinese stock market. They determined that default risk is not a systematic risk factor of the Chinese stock market; however, these two analyses can be used to analyze and identify the systematic risk for investment in an industry.

5.1.6   Statistical Stepwise Regression Analysis and Neural Network Sensitivity Analysis

Based on the research study by Wang, Hsiao and Fu (2000) exploring the relationship between a firm's systematic risk and its long-term investment activities, the results of these techniques show that systematic risk is reduced for investment activities in the fibre industry. For the electronic industry, however, the systematic risk is higher, as firms increase their long-term investment ratio. Companies with a higher portion of long term investment in assets will show a more significant difference. They use step-wise regression analysis to explore the impact of independent variables on systematic risk and neural network sensitivity analysis to analyze the non-linear relationship between a firm's systematic risk and its long-term investment activities. Their findings indicate that if an industry is somewhat mature and does not have many investment opportunities, the long-term investments in the industry are more diversified.

5.2   Qualitative RBDM Technique

Qualitative technique is a method for analyzing variables or elements that cannot be measured using numerical data; these variables or elements will instead be given a category such as low, medium or high. Some examples of qualitative RBDM techniques for investment in the real estate industry include the fuzzy comprehensive valuation method and variable precision rough set (VPRS) technique.

5.2.1   Fuzzy Comprehensive Valuation Method

The fuzzy comprehensive valuation method is used to evaluate the risk degree of a real estate project. Jin (2010) applied this method for comprehensive risk evaluation which would be beneficial and practical for the real estate projects lifecycle. Fuzzy comprehensive valuation is also used to estimate the lifecycle of the project's identified risk factors to confirm the risk level (highest risk, higher risk, general risk, lower risk, low risk), the first class index evaluation and index weight of all classes. The results of the paper shows that the total risk of each stage of a real estate project reduces gradually with the development of real estate projects, and the risk of same stage reduces gradually with the development of real estate projects. This would provide a foundation data to dynamic deal scheme decision for risk for real estate projects. This technique has also been used to obtain the value of the risk of real estate investment and has significance in theory and practice for investment risk analysis (Li & Suo, 2009).

5.2.2   Variable Precision Rough Set (VPRS)

Xie et al. (2010) designed an adaptive algorithm for dynamic risk analysis in a petroleum project investment based on a variable precision rough set (VPRS) technique. Their intention was to develop a risk ranking technique to measure the degree of risk for individual projects in a portfolio for which experts are invited to identify risk indices and support decision makers in evaluating the risk exposure (RE) of individual projects. Their investigation includes the definition of multiple risks involved in any petroleum project investment using multi-objective programming to obtain the optimal selection of projects with minimum risk exposure. The significance of risk indices is then assigned to each of the corresponding multi-objective functions as a weight.

5.3   Hybrid RBDM Technique

In order to provide a comprehensive evaluation of risk analysis, the combinations of qualitative and quantitative techniques have been useful for generating better decisions and takes into account all possible uncertainty factors.

5.3.1 Radial Basis Function Neural Network

Radial Basis Function (RBF) Neural Network is an example of an evaluation model for development risk in the real estate industry. RBFs are embedded in a two-layer feed-forward neural network; the input into an RBF neural network is nonlinear while the output is linear. The RBF neural network consists of one hidden layer of basic functions, or neurons and the chosen RBF is usually a Gaussian. At the input of each neuron, the distance between the neuron's centre and the input vector is calculated. The output of the neuron is then formed by applying the basis function to this distance. The RBF network output is formed by a weighted sum of the neuron outputs and the unity bias shown. It empirical analysis shows that the evaluation model is characterized as good data approximation, with high stability and normalization. RBF neural network has the advantage of automatically defining the initial weights and reducing the influence of overlay depending on the experience and knowledge of experts (Zhi & Qing, 2009).

5.3.2 Support Vector Machine

The SVM modelling approach has been proposed by Li et. al. (2009) to predict risk for real estate investment. Firstly, the merits of the structural risk minimization principle and the small study sample and non-linear case are used to analyze the risk factors during the investment stage in real estate projects. A model based on SVM in real estate investment risk is then built up. SVM learning training samples are usually based on a project proposal, project feasibility study report, project evaluation reports and other information.

According to Tao and Yajuan (2010), the main idea of SVM is to transform the input space into a higher dimension space with the nonlinear transformation of inner product function definition, then seeking out the nonlinear relation of the input variables and output variables in the higher dimension. They agreed that SVM can solve such problems as small samples, nonlinear case and higher dimensions, and that it provides a global optimal solution since SVM is a convex quadratic programming problem.

5.3.3 Analytic Hierarchy Process

AHP is a multi-criteria decision analysis technique that is commonly used for risk analysis. It aims to choose from a number of alternatives based on how well these alternatives rate against a chosen set of qualitative as well as quantitative criteria (Saaty & Vargas, 1994; Schniederjans, Hamaker & Schniederjans, 2005, cited in Angelou & Economides 2009). AHP has also been employed to determine the weight of every index to deal with the uncertainty of the risk analysis of real estate investment (Li & Suo, 2009).

5.3.4 Real Option Method

The application of a real option method seeks to examine the changes in uncertainty that will affect the optimal timing for investment. There are three managerial flexibility criteria of the real option method that influence optimal timing: information gathering, waiting option and abandon option (Rocha et al., 2007).

Wong (2007) examines the effect of uncertainty on investment timing in a canonical real options model. His study shows that the critical value of a project that triggers the exercise of the investment options exhibits a U-shaped pattern against the volatility of the project. It is found that there is a positive relationship between the risk factor and return factor when the volatility of the project increases.

According to Xie et al. (2010), a related factor that makes the timing of a project crucial is the irreversibility of the investments because, for example, the sunk cost cannot be recovered even if market conditions change adversely. One way to avoid regret for irreversible investments under uncertainty is to 'wait and see what happens'.

There are other RBDM techniques, not discussed here, which can be applied in the real estate industry, but this chapter focuses on the RBDM framework. Each technique has its own limitations and benefits because decision makers must have the knowledge of how to apply the particular technique when making decisions. Decision makers need to choose the best technique to suit their problem solving situation.

6   Issues and Challenges of Risk-Based Decision Making

The issues and challenges of RBDM need more consideration and would be a relevant focus for future research. The risk analysis of investment in real estate projects refers to the overall consideration of the risk attributes, the target of risk analysis and the risk bearing capability of risk subjects on the basis of investment risk identification and estimation, which determine the degree of influence of investment risks on the system (Xiu & Guo, 2009). There are several methodologies or techniques proposed by other researchers to evaluate, analyze, assess or predict the risk. Some of these are the Monte Carlo method, fuzzy set theory (Sun et al., 2008), Markowitz, fuzzy-analytical hierarchy process (F-AHP), a real option method (Rocha et al., 2007), and a hidden Markov model. There are a number of issues related to these methods or models. The first is that they have different characteristics, advantages and limitations when applied in different fields (Sun et al., 2008; Lander & Pinches 1998, cited in Rocha et al., 2007). For example, real option methodology has problems in the practical implementation of risk analysis, such as lack of mathematical skills, restrictive modelling assumptions, increasing complexity and limited power to predict investment in competitive markets (Lander & Pinches, 1998 cited in Rocha et al., 2007).

Availability of high quality data is the second issue for RBDM. Zeng, An and Smith (2007) believe that high quality data are a prerequisite for the effective application of sophisticated quantitative techniques. They therefore suggest that it is essential to develop new risk analysis methods to identify major factors, and to assess the associated risks in an acceptable way in various environments in which such mature tools cannot be effectively and efficiently applied.

The third issue is that real estate investment risk evaluation is a complex decision making problem with multiple factors and multiple targets (Zhou, Zhang and Li, 2008). The majority of existing real estate investment risk evaluations give priority to single-goal decision making, use single indices such as the maximum expectation, the largest variance, the minimum standard deviation rate to evaluate the real estate investment. These evaluating methods are easy to understand, but they cannot comprehensively evaluate the quality of an overall program. There are also those who use Multi-element Analysis Model (MAM) for real estate investment risk evaluation. The traditional MAM is based on the assumption that the whole of the distribution is subject to the normal distribution, yet the whole distribution of a real estate investment program is uncertain; thus, it is imprecise to use MAM for real estate investment risk analysis. Furthermore, many evaluation programs or models involve many evaluation indices, such that the dimensions are different and the weights are difficult to determine, and there are therefore difficulties in the practical application.

The fourth issue is related to incomplete risk data availability. In decision making, the correct methodology is important to ensure that the right decision is made, and that it will be beneficial to investors, users or agents. More formal methodology is thus necessary in decision making processes (Hussain et al., 2007; Zeng, An & Smith, 2007). Formal methodologies are needed to make sure that any decision can be assessed effectively and efficiently. Many risk analysis techniques currently used in the UK construction industry are comparatively mature, such as Fault Tree Analysis, Event Tree Analysis, Monte Carlo Analysis, Scenario Planning, Sensitivity Analysis, Failure Mode and Effects Analysis, Programme Evaluation and Review Technique (Zeng, An & Smith, 2007). In many circumstances, however, the application of these tools may not give satisfactory results due to the incompleteness of risk data.

The fifth issue is the need for an effective and efficient technique. New risk analysis methods to identify major factors and to assess the associated risks in an acceptable way in various environments are needed, as older tools cannot be effectively and efficiently applied.

The sixth issue is the non scientific method proposed. The methods of risk analysis which have been used by domestic real estate developers so far, such as risk survey, break-even analysis and sensitivity analysis, are based on the discounted cash flow and net present value (NPV). These methods are far from scientific and easily lead to faults, and furthermore, to the severe consequences of failure (Yu & Xuan, 2010).

7   Summary

This chapter suggests a risk-based decision making framework that is applicable for investment in the real estate industry. Risk-based decision making is an important area of focus in real estate investment, which involves high risk and high cost. Risk with high uncertainties will lead to the occurrence of a higher percentage of probabilities and consequences. The uncertainties of a number of risk factors and risk sources contribute to the level of dynamic risk prediction, which is dependent on what takes place from the initial investment to the later stages of the real estate development.

The decision support technology review for the framework indicates that there is much work to be done to develop an intelligent decision support system that can be used for handling risk-based decision making in business operations, or as a tool for businesses managing their business tasks, particularly when they deal with decision making processes in their daily routine as a manager. This is perhaps the most important concern for the future of an information system related to risk analysis or risk aggregation for managers who deal with decision making processes, because it will promote vital and useful technology to help decision makers identify the risk involved in making certain decisions to meet an organization's goals and objectives. Thus, it is vital to explore a new technique for risk analysis using decision support technology such as Intelligent Decision Support System for investment in the real estate industry.

References

1. Angelou, G.N., Economides, A.A.: A multi-criteria game theory and real-options model for ir-reversible ICT investment decisions. Telecommunications Policy 33(10-11), 686–705 (2009)
2. Aven, T.: A unified framework for risk and vulnerability analysis covering both safety and security. Reliability Engineering and System Safety 92(6), 745–754 (2007)
3. Aven, T., Vinnem, J.E., Wiencke, H.S.: A decision framework for risk management, with application to the offshore oil and gas industry. Reliability Engineering & System Safety 92(4), 433–448 (2007)
4. Busemeyer, J.R., Pleskac, T.J.: Theoretical tools for understanding and aiding dynamic decision making. Journal of Mathematical Psychology 53(3), 126–138 (2009)
5. Casaigne, N., Lorimier, L.: A challenging future for i-DMSS. In: Intelligent Decision-making Support Systems Foundations, Applications and Challenges, Part III, pp. 401–422 (2007)
6. Changrong, D., Yongkai, M.: Study of Chinese real estate stock daily return under different market condition. In: 4th International Conference on Wireless Communications, Networking and Mobile Computing, WiCOM 2008, pp. 1–4 (2008)
7. Chauveau, T., Gatfaoui, H.: Systematic risk and idiosyncratic risk: a useful distinction for valuing European options. Journal of Multinational Financial Management 12(4-5), 305–321 (2002)
8. Chengda, L., Lingkui, M., Heping, P.: New applications and research of GIS in the real estate. In: International Conference on Info-tech and Info-net, ICII 2001, Beijing, vol. 1, pp. 261–266 (2001)
9. Delen, D., Pratt, D.B.: An integrated and intelligent DSS for manufacturing systems. Expert Systems with Applications 30(2), 325–336 (2006)
10. Hui, E.C.M., Yu, C.K.W., Ip, W.C.: Jump point detection for real estate investment success. Physica A: Statistical Mechanics and its Applications 389(5), 1055–1064 (2010)
11. Hui, S., Zhi, Q.F., Ye, S.: Study of impact of real estate development and management risk on economic benefit. In: 16th International Conference on Industrial Engineering and Engineering Management, IE&EM 2009, pp. 1299–1303 (2009)
12. Hussain, O.K., Chang, E., Hussain, F.K., Dillon, T.S.: Quantifying the possible financial consequences of failure for making a risk based decision. In: 2nd International Conference on Internet Monitoring and Protection, pp. 40–46 (2007)
13. Hussain, O.K., Chang, E., Hussain, F.K., Dillon, T.S.: Quantifying failure of risk based decision making in digital business ecosystem interactions. In: 2nd International Conference on Internet and Web Applications and Services, Mauritius, pp. 49–56 (2007)
14. Jian, G.Z., Zhao, M.W., Xiu, X.: Recognizing the pattern of systematic risk based on financial ratios and rough set-neural network system. In: 2006 International Conference on Machine Learning and Cybernetics, pp. 2408–2412 (2006)
15. Jin, C.H.: The lifecycle comprehensive risk evaluation for real estate projects – Application. In: IEEE International Conference on Advanced Management Science (ICAMS), vol. 2, pp. 438–440 (2010)
16. Ju, Y.J., Meng, Q., Zhang, Q.: A study on risk evaluation of real estate project based on BP neural networks. In: International Conference on E-Business and Information System Security, EBISS 2009, pp. 1–4 (2009)
17. Juhong, G., Zihan, W.: Research on real estate supply chain risk identification and precaution using Scenario analysis method. In: 16th International Conference on Industrial Engineering and Engineering Management, IE&EM 2009, pp. 1279–1284 (2009)
18. Lee, J.S., Jang, S.: The systematic-risk determinants of the US airline industry. Tourism Management 28(2), 434–442 (2007)

19. Li, J., Wang, J.F., Wu, C.Y., Yang, Y.T., Ji, Z.T., Wang, H.B.: Establishment of a risk assessment framework for analysis of the spread of highly pathogenic avian influenza. Agricultural Sciences in China 6(7), 877–881 (2007)
20. Li, W., Zhao, Y., Meng, W., Xu, S.: Study on the risk prediction of real estate investment whole process based on support vector machine. In: 2nd International Workshop on Knowledge Discovery and Data Mining, WKDD 2009, pp. 167–170 (2009)
21. Li, Y., Huang, L.: Financial constraints and systematic risk: theory and evidence from China. In: 15th Annual International Conference on Management Science and Engineering, ICMSE 2008, pp. 1113–1119 (2008)
22. Li, Y., Suo, J.: Model on Risk Evaluation of Real Estate Investment. In: 6th International Conference on Fuzzy Systems and Knowledge Discovery, FSKD 2009, vol. 3, pp. 138–140 (2009)
23. Lin, H., Chen, X.P.: Is default risk a systematic risk of Chinese stock markets? In: 15th Annual International Conference on Management Science and Engineering, ICMSE 2008, pp. 1169–1174 (2008)
24. Lin, Y.H., Lee, P.C., Ting, H.I.: Dynamic multi-attribute decision making model with grey number evaluations. Expert Systems with Applications 35(4), 1638–1644 (2008)
25. Liow, K.H., Addae-Dapaah, K.: Idiosyncratic risk, market risk and correlation dynamics in the US real estate investment trusts. Journal of Housing Economics 19(3), 205–218 (2010)
26. Liu, L., Zhao, E., Liu, Y.: Research into the risk analysis and decision-making of real estate projects. In: International Conference on Wireless Communications, Networking and Mobile Computing, WiCom 2007, pp. 4610–4613 (2007)
27. Lu, J., Zhang, G., Ruan, D., Wu, F.: Multi-objective group decision making: methods, software and applications, vol. 1. Imperial College Press, London (2007)
28. Meng, Z.Q., Yu, X.F., Jiang, M., Gao, H.: Risk measure and control strategy of investment portfolio of real estate based on dynamic CVaR model. Systems Engineering - Theory & Practice 27(9), 69–76 (2007)
29. Mora, M., Forgionne, G., Gupta, J.N.D., Garrido, L., Cervantes, F., Gelman, O.: A strategic descriptive review of the intelligent decision-making support systems research: the 1980–2004 period. In: Gupta, J.N.D., Forgionne, G.A., Mora, M.T. (eds.) Intelligent Decision-making Support Systems Foundations, Applications and Challenges, Part III, pp. 441–462 (2007)
30. Mustajoki, J., Hamalainen, R.P.: Smart-Swaps – a decision support system for multicriteria decision analysis with the even swaps method. Decision Support Systems 44(1), 313–325 (2007)
31. Niu, L., Lu, J., Zhang, G.: Cognition-Driven Decision Support for Business Intelligence - Models, Techniques, Systems and Applications, 1st edn. Springer, Heidelberg (2009)
32. Olsson, R.: In search of opportunity management: is the risk management process enough? International Journal of Project Management 25(8), 745–752 (2007)
33. Piyatrapoomi, N., Kumar, A., Setunge, S.: Framework for investment decision-making under risk and uncertainty for infrastructure asset management. Research in Transportation Economics 8, 199–214 (2004)
34. Ren, H., Yang, X.: Risk measurement of real estate portfolio investment based on CVaR model. In: International Conference on Management and Service Science, MASS 2009, pp. 1–4 (2009)
35. Rocha, K., Salles, L., Garcia, F.A.A., Sardinha, J.A., Teixeira, J.P.: Real estate and real options – a case study. Emerging Markets Review 8(1), 67–79 (2007)
36. Rui, S.L., Qi, Y.W., Jian, H.H., Li, G., Liang, L.: An intelligent decision support system applied to the investment of real estate. In: IEEE International Conference on Industrial Technology, ICIT 1996, pp. 801–805 (1996)

37. Saleem, K., Vaihekoski, M.: Pricing of global and local sources of risk in Russian stock market. Emerging Markets Review 9(1), 40–56 (2008)
38. Shiwang, Y., Jianping, W., Na, G.: Application of project portfolio management in the real estate corporations. In: 16th International Conference on Industrial Engineering and Engineering Management, IE&EM 2009, pp. 1225–1228 (2009)
39. Shujing, Z., Shan, L.: Projection pursuit model based on PSO in the real estate risk evaluation. In: 2010 International Conference on Intelligent Computing and Cognitive Informatics, ICICCI, pp. 235–238 (2010)
40. Storesletten, K., Telmer, C.I., Yaron, A.: Asset pricing with idiosyncratic risk and overlapping generations. Review of Economic Dynamics 10(4), 519–548 (2007)
41. Strong, R., Paasi, J., Ruoyi, Z., Luoma, T.: Systematic risk management for the innovative enterprise. In: 42nd Hawaii International Conference on System Sciences, HICSS 2009, pp. 1–9 (2009)
42. Sun, Y., Huang, R., Chen, D., Li, H.: Fuzzy set-based risk evaluation model for real estate projects. Tsinghua Science & Technology 13(S1), 158–164 (2008)
43. Syalim, A., Hori, Y., Sakurai, K.: Comparison of risk analysis methods: Mehari, Magerit, NIST800-30 and Microsoft's security management guide. In: International Conference on Availability, Reliability and Security, ARES 2009, vol. 5, pp. 726–731 (2009)
44. Tang, G.Y.N., Shum, W.C.: The relationships between unsystematic risk, skewness and stock returns during up and down markets. International Business Review 12(5), 523–541 (2003)
45. Tao, L., Yajuan, L.: A risk early warning model in real estate market based on support vector machine. In: International Conference on Intelligent Computing and Cognitive Informatics, ICICCI 2010, pp. 50–53 (2010)
46. Wang, K., Hsiao, T.P., Fu, H.C.: Using statistical and neural network methods to explore the relationship between systematic risk and firm's long term investment activities. In: 2000 IEEE Signal Processing Society Workshop on Neural Networks for Signal Processing X, vol. 2, pp. 851–858 (2000)
47. Wanqing, L., Yong, Z., Wenqing, M., Shipeng, X.: Study on the risk prediction of real estate investment whole process based on support vector machine. In: 2nd International Workshop on Knowledge Discovery and Data Mining, WKDD 2009, pp. 167–170 (2009)
48. Williams, D.J., Noyes, J.M.: How does our perception of risk influence decision-making? Implications for the design of risk information. Theoretical Issues in Ergonomics Science 8(1), 1–35 (2007)
49. Wong, K.P.: The effect of uncertainty on investment timing in a real options model. Journal of Economic Dynamics and Control 31(7), 2152–2167 (2007)
50. Wu, C., Guo, Y., Wang, D.: Study on capital risk assessment model of real estate enterprises based on support vector machines and fuzzy integral. In: Control and Decision Conference, CCDC 2008, pp. 2317–2320 (2008)
51. Wu, X.Y., Li, H.M., Niu, J.G., Liu, Z.Q.: Memetic algorithm for real estate portfo-lio based on risk preference coefficient. In: 16th International Conference on Industrial Engineering and Engineering Management, IE&EM 2009, pp. 1245–1249 (2009)
52. Xiaobing, Z., Haitao, L.: Extension evaluation on risks in earlier stage of real estate project. In: 16th International Conference on Industrial Engineering and Engineering Management, IE&EM 2009, pp. 1229–1233 (2009)
53. Xiaozhuang, Y.: Real estate financial risk prevention and countermeasure. In: 4th International Conference on Wireless Communications, Networking and Mobile Computing, Wi-COM 2008, pp. 1–4 (2008)

54. Xie, G., Yue, W., Wang, S., Lai, K.K.: Dynamic risk management in petroleum project investment based on a variable precision rough set model. In: Technological Forecasting and Social Change (2010) (in press, corrected proof)
55. Xiong, X., Xiao, T.Z., Wei, Z., Cui, Y.L.: Wavelet-based beta estimation of China stock market. In: International Conference on Machine Learning and Cybernetics, vol. 6, pp. 3501–3505 (2005)
56. Xiu, L.T., Guo, D.L.: A tentative study on risk estimation and evaluation of investment in real estate projects. In: 2009 International Conference on Machine Learning and Cybernetics, vol. 1, pp. 561–566 (2009)
57. Yu, J., Xuan, H.: Study of a practical method for real estate investment risk decision making. In: 2010 International Conference on Management and Service Science, pp. 1–4 (2010)
58. Zeng, J., An, M., Smith, N.J.: Application of a fuzzy based decision making methodology to construction project risk assessment. International Journal of Project Management 25(6), 589–600 (2007)
59. Zhi, Q.M., Qing, B.M.: The research on risk evaluation of real estate development project based on RBF neural network. In: 2nd International Conference on Information and Computing Science, ICIC 2009, vol. 2, pp. 273–276 (2009)
60. Zhou, S., Wang, F., Li, Y.: Risk assessment of real estate investment. In: 2nd International Asia Conference on Informatics in Control, Automation and Robotics, vol. 1, pp. 412–414 (2010)
61. Zhou, S.J., Li, Y.C., Zhang, Z.D.: Self-adaptive ant algorithm for solving real estate portfolio optimization. In: International Symposium on Computational Intelligence and Design, ISCID 2008, vol. 1, pp. 241–244 (2008)
62. Zhou, S.J., Zhang, Z.D., Li, Y.C.: Research of real estate investment risk evaluation based on fuzzy data envelopment analysis method. In: International Conference on Risk Management & Engineering Management, ICRMEM 2008, pp. 444–448 (2008)

Chapter 15
Risk Management in Logistics

Hui Ming Wee[1], Mauricio F. Blos[2], and Wen-Hsiung Yang[3]

[1] Industrial and Systems Engineering Department,
Chung Yuan Christian University, Chungli, Taiwan
weehm@cycu.edu.tw
[2] SONY do Brasil Ltda, Supply Chain Solutions, Manaus – AM, Brasil
blos31@yahoo.com.br
[3] Management School, University of Liverpool
Liverpool L69 3BX, UK
jhs.why@gmail.com

Abstract. Due to the complex market and business environment, undesirable disruptions in logistics can affect enterprises and weaken its business strength. Risk Management has become the key in avoiding business losses. Logistics interruption can come from an unforeseen exogenous event such as an earthquake or from an endogenous event, like the Toyota Quality recalls in 2010 that interrupted enterprise logistic operations and degraded its performance (Trkman & McCormack, 2009). In this chapter, the risk management in logistics is studied from the process flow perspective. The topics discussed consist of logistics processes, risk management strategy, risk management process in logistic, and enterprise performance evaluation. Several risk management theories and framework from the literature are presented in the chapter. The aim is to provide valuable insights for enterprises through understanding the essential risk management concepts in logistics.

Keywords: Risk management, Logistics, Risk drivers, Vulnerability, Risk preference, Mitigation, Consequence/probability matrix.

1   Introduction

Due to the complex market and business environment, the increasing globalization resulted in a complex logistics network. To adapt this complexity, logistics has provided more services, such as vendor managed inventory (VMI) and cross-docking warehouse distribution as a business strategy. At the same time, undesirable disruptions in logistics affect enterprises and weaken its business strength. For example, the 2010 Iceland volcano eruption disrupted flight schedules; other factor such as wage increase may force manufacturing plants to relocate.

Logistics is a process linking the activities of product manufacturing from suppliers to customers. The Journal of Logistics Managements (1998) defined logistics as the "process of planning, implementing, and controlling the efficient and effective

J. Lu, L.C. Jain, and G. Zhang: Handbook on Decision Making, ISRL 33, pp. 285–305.
Springerlink.com

flow of storage goods, services, and related information from the point of origin to the point of consumption for the purpose of conforming to customer requirements". From the perspective of the logistics function, logistics is managing the physical delivery and storage in an efficient way to fulfil enterprise business requirements.

According to the Council of Supply Chain Management, logistics management should aim to achieve seven goals: deliver the right product to the right place at the right time with the right quantity and at the right quality and at the right price to the right customer. It is not easy to coordinate all the seven goals for all internal units and external partners. Therefore, to comply with all these goals, a global business must rely upon the collaboration of both the upstream and downstream partners.

Managing logistics in an effective and efficient manner has become a business strategy to sustain enterprise development. Loutenço (2005) stated that "the key to successful logistics management requires heavy emphasis on integration of activities, cooperation, coordination and information sharing throughout the entire supply chain, from suppliers to customers". In terms of information sharing, incessant enhancement of information technology has supported the logistics management on data exchange and communication across the entire process flow, such as the application of bar code and radio-frequency identification (RFID), point of sale (POS); electronic data interchange (EDI), virtual private network (VPN), and the enterprise resource planning (ERP) system. This not only reduces the complexity of physical process flow, but also lowers the uncertainties in the supply chain.

Even so, no logistic system can avoid being affected by the uncertainty of a risk attack. To compete in the low profit margin market, the company must sustain operation performance in all circumstances. The following sections discuss the logistics processes, the risk management strategy, the risk management process in logistics, and the enterprise performance evaluation.

2   Scope

Logistics play an important role in sustaining business competency through process efficiency improvement. Logistics link the internal functions and collaborate with the upstream and downstream partners to achieve synergistic result. Logistics management is adaptive to global networking (Nilsson & Waidringer, 2005) in order to meet market demand, provide prompt services and manage logistics effectively.

Due to the influence of risks in logistics performance, implementing risk management has been a critical issue recently. Risk management strategy can be viewed from the mitigation strategy approach and the contingency strategy. Mitigation strategy means taking pre-actions to avoid or lower the likelihood of a risk, whereas contingency strategy pertains to post-actions taken to handle damage in a quick and short time with minimal expense. This chapter focuses on mitigation strategy to set out risk management processes.

2.1   Logistics Processes

Logistics systems and management are diverse from one industry to another. No single type of logistics management can fit into all types of industries. There are two kinds of process flows in the logistic chains: the physical flow and the information flow. The physical flow carries the raw materials from the suppliers, transforms it into finished products, and then delivers the products to the customers. Nowadays, information technology helps in the collection of data from each node of logistics in a speedy and timely manner in order to support a complex global delivery network. The logistics network of physical and information flow is illustrated in Figure 1.

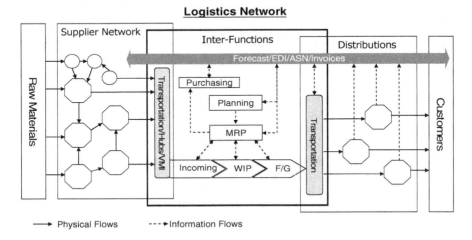

Fig. 1. Physical and information flows in logistics network. (Bowersox et al. 2005).

2.1.1   Physical Flow

Physical flow carries the materials from suppliers and delivers the final products to the customers. Mentzer (2008) classified the physical flow of logistics management functions into seven elements as listed in Table 1. Each function collaborates with other functions in an efficient manner to move products to the customers in accordance with the seven goals.

Table 1. A summary of logistics management functions (Mentzer, et al. 2008)

Transportation management	Network design (inbound and outbound), 3PL/4PL management, commercial tariff/regulation control, vehicle routing
Warehousing management	Location selection, design, management, shelf system and bounded warehouse
Material handling management	Product packing, labeling, picking system, delivery scheduling, dispatching and cross-docking management
Inventory management	FIFO, strategic inventory control, JIT, VMI, consignment and total inventory cost control
Order management and fulfillment	Ordering system, data mining and documents exchange
Procurement	Supplier management, cost negotiation and sourcing
Customer Service	Customer satisfaction, delivery scheduling and data mining

The logistics chains include many activities and involve different partners in the chain. Autry et al. (2008) defined the activities in the pipeline of logistics, including the material moving processes and management control network. Currently, the global business environment and the diversity of market demand on product design encourage enterprises to provide more services at local markets that result in a complex logistics network as well as adding more processes in the chains. There are three complexities- product, network, and process - which result in the difficulty of controlling logistic activities (Hofer & Knemeyer, 2009). In addition, the information flow consists of all the activities from the business plan to the front-end data collection. Many advanced information technologies have been developed to reduce the complexities in logistics flows.

2.1.2   Information Flow

Information is important in every business and organization, and it can be generated from any form of data such as market survey, production figures and shipment details. However, the contribution of information is in its value to the enterprise, and not in the volume of information. Each enterprise uses information technology at different levels, such as the EDI system which is implemented to reduce order cycle time and inventory (Willersdorf, 2007) or collaborative planning, forecasting and replenishment (CPFR) is used to improve forecast accuracy and to enhance collaboration with partners.

In logistics flow, besides involving many functions and activities, it also exchanges numerous data that need to be collected systematically and compiled in order to obtain valuable information for management decision-making. From the logistics process flow perspective, data are exchanged in three levels: daily operational processes, managerial control, and business planning.

- **Daily Operation:** the data collection starts from receiving the customer PO up to the delivery of products to the customers.
- **Managerial Control:** includes inventory stock monitoring, material In/Out, transportation scheduling and document exchange.
- **Planning:** integrates data from operation and processes based on business strategy and market status for planning.

In addition to information consolidation, information sharing among each node in the chain has been identified as a support to inventory control and to reduce the bullwhip effect. Nevertheless, both physical flow and information flow should move from one node to another efficiently. Any disruption may break the process flow or lower the performance of the logistics system.

2.2   Risk Management

Recently, risk management has been widely discussed. Many disturbances have been found in daily operations. Examples of disturbances are earthquakes, Iceland's volcano eruption, labour wage increase in China, BP oil leakage, and so on. Each disturbance may affect business operations and result in the adjustment of the logistics system.

The impact of accidental events can be controlled if the enterprise has a plan already in place. In general, risk management can be classified into two approaches. One is mitigation strategy and another is contingency strategy. Husdal (2008) illustrated the relationship of mitigation and contingency strategy in Figure 2. Mitigation action manages the risk sources while contingency action handles the consequences of a risk.

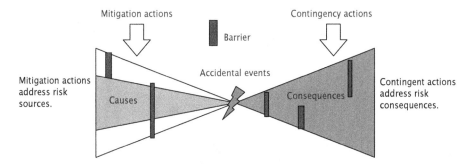

Fig. 2. Risk Mitigation and Contingency strategies. (Husdal, 2008).

In order to minimize the impact of uncertain events, enterprises should be prepared in responding to potential risks, and to take action to decrease the likelihood of disruption or reduce the level of impact when risk is unavoidable. When an enterprise is being attacked by a risk, how well the enterprise can resume its normal operation quickly and minimize its losses relies on a well designed contingency strategy.

2.2.1   Mitigation Strategy

Mitigation strategy is a set of plans to act against uncertain events prior to its occurrence. The purpose is either to prevent disturbance or to reduce the impact when it occurs. A mitigation strategy should tailor specific disturbances or threats, and it is expected that certain costs would be incurred in order to avoid disruption or lower losses. In addition, the costs and benefits of a mitigation strategy should be assessed before any decision is made. Moreover, the trade-off between strategies is another factor that needs to be assessed during the decision processes.

In enterprise risk management (ERM), there are some principle approaches that have been adopted in mitigation strategy which can also be applied in logistics management processes. The Committee of Sponsoring Organizations (COSC) provides guidance in response to a risk attack; it encompasses four approaches as shown below:

- **Accept** => monitor
- **Avoid** => eliminate (get out of situation)
- **Reduce** => institute controls
- **Share** => partner with someone

The above four approaches are set in a general direction, but the right mitigation strategy should be tailored to a specific risk. Chopra and Sodhi (2004) identified the source of threats in the supply chain, assessed the vulnerabilities by a stress test, and then defined the seven mitigation approaches and the tailored strategies. Other mitigation strategies have been identified by Tang (2006) and Faisal (2006) and so on.

A clear and comprehensive risk identification plan should be considered in developing an appropriate mitigation strategy that minimizes the impact of a risk. Another factor that could affect the mitigation strategy is the enterprise business strategy decision. Companies set forth their business competition strategy to win the market by selecting the best mitigation strategy.

Identifying and mitigating risks is a pre-requisite for an enterprise to reduce the impacts of risk. Pettit et al. (2010) demonstrated a zone of resilience composed of supply chain capability and vulnerability (see Figure 3). A company should aim to balance its portfolio of capabilities to match the pattern of vulnerabilities. The less vulnerable the system is, the lower the likelihood of being attacked by risk.

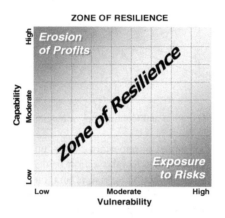

Fig. 3. Zone of Resilience (Pettit et al., 2010)

Table 2. Definition of Contingency plan (Department of Health and Human Services-US)

The National Institute of Standards and Technology (NIST)	As management policies and procedures designed to maintain or restore business operations, including computer operations, possibly at an alternate location, in the event of emergency, system failure, or disaster.	-Business Continuity Plan, -Business Recovery Plan, -Continuity of Operations Plan, -Continuity of Support Plan, -Crisis Communications Plan, -Cyber Incident Response Plan, -Disaster Recovery Plan, -Occupant Emergency Plan,
The Information Technology Infrastructure Library (ITIL)	As a series of processes that focus only upon the recovery processes, principally in response to physical disaster, that are contained within business continuity management (BCM).	
The Department of Health and Human Services (HHS) Enterprise Performance Life Cycle (EPLC)	As the strategy and organized course of action that is to be taken if things don't go as planned or if there is a loss in the established business product or system due to disasters such as a flood, fire, computer virus, or major failure.	

2.2.2   Contingency Strategy

Contingency strategy aims to respond to disruption rapidly and also to conduct recovery quickly with minimum expenses. Setting up a contingency plan is an essential policy to manage natural disasters in government institutions. Many organizations have defined its contingency plan for multiple purposes, and use different terms for the recovery processes (see Table 2).

The contingency strategy can be an extension of the mitigation strategy plan. A good mitigation strategy plan should be able to either reduce the likelihood of an event or to reduce the costs of contingency actions. This chapter focuses on mitigation strategy, and the processes of risk mitigation management are discussed in the following sections.

3   Risk Management Process in Logistic

Risk management should be led by enterprise high level management and be set as a strategic process across all hierarchy levels. The risk assessment framework of each functional department should be in accordance with the enterprise risk management strategy and in line with the customer and stakeholder value. Some risk management standards have been established such as the Committee of Sponsoring Organization (COSO) ERM, International organization for standardization ISO-31000:2009, or Australian standard AS/NZS-4630.

In Information Technology (IT) system, there are more standard procedures that have been applied to manage the risks in daily system operations. Such as Basel II +IT control objective, OCTAVE, ISO-2700n, CRAMM, CERT SNAP, ISF, and ISO/IEC TR13335. The distribution of the risk management standard is shown in Figure 4.

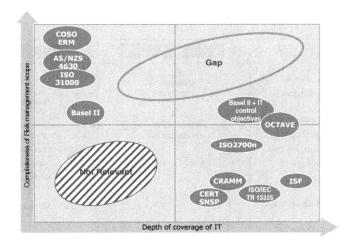

Fig. 4. Risk management system distribution ISACA/ITGI (2009)

The risk management of the logistics department should be adopted from the enterprise risk management strategy. In accordance with the organization functions, Nadler & Slywotzky (2008) classified risks in five categories (Strategic, Financial, Operational, Human Capital, and Hazard). The sources of risks are diverse in different business environments as well as in the company's internal systems. From a broad application, COSO (2004) provided a three dimension cubic to manage risks and built management processes across the entire enterprise (see Figure 5).

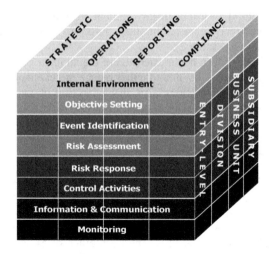

Fig. 5. A cubic of enterprise risk management (COSO, 2004)

The COSO model defined four risk categories in a firm (Strategic, Operations, Reporting, and Compliance), and applied it to every hierarchy level of an enterprise. Logistics department is a risk owner who assesses potential risks in these four categories in accordance with the enterprise's risk management strategy.

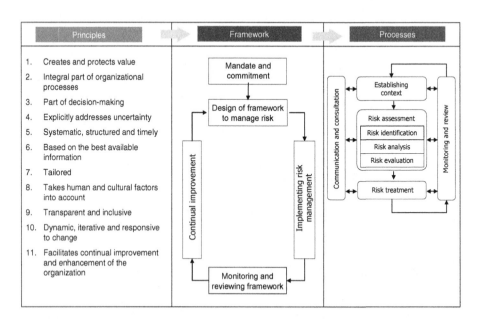

Fig. 6. ISO-31000 risk management standard (ISO)

Another international risk management standard (ISO-31000:2009) is issued by the International Organization for standardization (ISO). This is a continuous improvement process based on the PDCA (Plan-Do-Check-Act) continuous improvement cycle (see Figure 6). The COSO's framework organizes all the different units into one direction. The ISO's risk management standard is more practical and explicitly defines each process step.

An information system links all the departments of a firm. As mentioned above, there are many risk management procedures for a developed IT system. According to the Information Systems Audit and Control Association (ISACA), an IT risk management framework includes three aspects that correspond to business objectives.

- **Risk governance** => Integrated with ERM, risk awareness.
- **Risk evaluation** => Risk analysis, collect data, and maintain risk profile.
- **Risk response** => Manage risks; react to events, and articulate risk.

In this chapter, the COSO-ERM, ISO-31000, and ISACA risk IT framework are used to discuss risk management in logistics. An assessment framework is shown in Figure 7. For managing the risks in the logistics chains, this chapter assesses the risks from the process flow perspective, both the physical and the information flow. In accordance with the process steps of ISO31000, the following section discusses risk identification, evaluation, mitigation strategies, implementation, and monitoring.

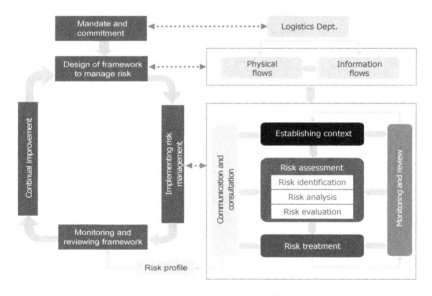

Fig. 7. Risk management framework of logistics (COSO, ISO, and ISACA)

3.1   Risks Identification

The first step of risk management is to identify the sources or drivers of risks. The disturbances can happen in any way at any time. Enterprises need to collect all possible disruptions or threats systematically. The risks can be found in different aspects, either from external environment or internal operations. In logistics chains, the chance of exposure to risk is higher than other departments. The disruption not only happens in the process flows but also at the suppliers and customers site.

The losses resulted from risk occurrence are determined not only by its source (driver) but also by the vulnerability of a system. The vulnerability is the weakness of an enterprise operation. It can be a process, function, or a certain condition, and it may not harm the system until it is attacked by risk. For example, having a lean production can reduce the inventory holding cost, but may increase the risk of material shortage due to volcano eruption or earthquake. In addition, the results of every company facing risk may vary due to their different vulnerabilities.

3.1.1 Sources or Drivers of Risk

The sources or drivers of risks are defined differently. Besides the OCSC definition of risk, many other studies assess risks from different points of view. Blos, et al. (2009) studied the supply chain of the automotive industries in Brazil, and classified risks into four categories such as strategic, financial, operational, and hazard. Cavinato (2004) identified risk from the physical flow, financial flow, information flow, relational, and innovation perspectives. In order to collect all possible sources of risks in logistics chains, the process flow and its functions for identifying the risk source in each node of logistics flows are used. A framework for risk identification is illustrated in Figure 8.

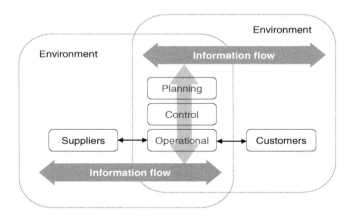

Fig. 8. Risk framework- process flow perspective

Following this framework to assess the potential risk might affect the performances of each logistics management function. As each enterprise has different risk awareness and vulnerability, a matrix (Table 3) can help to describe most of the sources of risk and assess its vulnerability against each identified risk. Departments may focus on the risks that may have a direct impact on their department's operations.

Table 3. A matrix of risks in each of logistics management function

Risk categories \ Functions	Transportation management	Warehouse management	Material Handling	Inventory Management	Order management	Purchasing management	Customer Services	Information management
Environment	✓	✓	✓	✓				✓
Suppliers				✓	✓	✓		✓
Operational	✓		✓			✓	✓	✓
Control		✓		✓	✓			✓
Planning				✓	✓	✓	✓	✓
Customers					✓		✓	✓
Information	✓	✓	✓	✓	✓	✓	✓	✓

3.1.2   Vulnerabilities

Depending on their vulnerability to risks, enterprises may not be able to avoid disruptions, but they can control the effects of risk. For example, when an earthquake occurs, an enterprise located in that area would be affected. Though another enterprise is not located in the disaster zone, yet its suppliers in that area could not deliver the materials. Therefore, the vulnerability of the two enterprises is not the same. The way they manage the disaster will be different as well.

As a consequence, the way of handling risks is varied. Since there is no standard vulnerability among companies, finding the vulnerability of each risk is a challenge. Chopra and Sodhi (2004) have taken a stress test to assess the vulnerability of each given risk (see Table 4) where each department can evaluate its own vulnerabilities.

Table 4. Stress test on production disruption (Chopra & Sodhi, 2004)

Risk source		Stress testing
Production disruption	Supplier	A supplier capacity dropped by 20% overnight
	Internal- operation	Key plant shut down unexpectedly for one month
	Customer	Demand goes up by 20% for a key product; Demand goes down by 20% for a key product.

As mentioned above, each department might have different risk awareness. It would be necessary to collect and review all the vulnerabilities of each department. This can be used to assess the overall risks and vulnerabilities of a firm; and evaluate the level of risk so as to set the strategy priority as well as resource allocation.

3.2   Risk Analysis and Evaluation

After identifying the risk's source and vulnerability, the next step is to identify the loss severity and the likelihood of risk. However, learning the impact of a risk depends on the nature of the risks such as earthquake, or foreign exchange rate fluctuation. The diversity of risks requires different techniques for risk evaluation, a general calculation of risk can be illustrated by the likelihood of risk and its severity level.

3.2.1   Assessment Techniques

The type of risks varies and requires different techniques for decision-making. The ISO-31010:2009 has summarized a set of risk assessment techniques in Table 5. The application can be classified and applied to each step of the risk assessment process. Moreover, the technique should be applied in accordance with the nature of risks. It not only selects the right technique but also identifies the risk level for decision-making. As many methods can be used for risk assessment, the discussion in this chapter is based on a common application across three stages, the consequence/probability matrix, to evaluate the likelihood and severity of risks.

3.2.2   Risk Evaluation

In general, risk naturally exists in a daily business operation. The risk information is collected from historical data and the assumption is that the disruption would happen in the future in the same way. The enterprise should be able to reduce or avoid either the possibility of risk or its impact. In addition, some risks may not be repeating threats to an enterprise, for example, the transfer of a production plant to new locations away from a danger zone can avoid the same danger happening. Hence, the risks should be evaluated for a period of time and be profiled for further actions.

Table 5. Assessment techniques in each stage (ISO-31010/FDIS IEC)

Stage	Assessment technique (Strong applicable)
Identification	Brainstorming, Structure or Semi-Structure Interview, Delphi, Check-list, Primary hazard analysis, HAZOP, HACCP, *Environment risk assessment, Structure-What if-SWIFT*, Scenario analysis, *FMEA*, Cause-and-effect analysis, *Human reliability analysis, Reliability centered maintenance, Consequence/probability matrix.*
Analysis	HAZOP, HACCP, *Environment risk assessment, Structure-What if-SWIFT*, Scenario analysis, Business impact analysis, Root cause analysis, *FMEA*, Fault tree analysis, Event tree analysis, Cause and consequence analysis, Cause-and-effect analysis, LOPA, Decision tree, *Human reliability analysis*, Bow tie analysis, *Reliability centered maintenance*, Markov analysis, Bayesian statistics and Bayes Nets, FN curves, Risk indices, *Consequence/probability matrix*, Cost/Benefit analysis, Multi-criteria decision analysis.
Evaluation	HACCP, *Environment risk assessment, Structure-What if-SWIFT*, Root cause analysis, *FMEA, Reliability centered maintenance*, Mento Carlo simulation, Bayesian statistics and Bayes Nets, FN curves, Risk indices.

Risks could be assessed according to the possibility of risk occurrence and its severity level. As discussed in the previous sections, the enterprise vulnerability is a key factor that leads to its losses. For example, a computer hacker may succeed particularly when the system is vulnerable to such attacks. Therefore, the likely solution should include the frequency of hacker attacks and the possibility of the system's vulnerability. In addition, the severity should not only solve the physical losses but also the concealed expenses such as the production downtime and the expenditure for recovery. The consequence/probability matrix method collects the information in accordance with each identified risk source/driver, and assesses the likelihood as well as the level of severity. An example of data collection is shown in Table 6. In arriving at the score of each risk, the management team can then set the priority for each risk and develop a tailored mitigation strategy accordingly. (The higher the score, the higher the risk level)

Table 6. Risk level calculation table (Asbjornslett & Rausand, 1999)

Scenario		Likelihood (5-1)		Consequences of Scenario (5-1)								Resources to mitigate, rebuild, restore, etc.		
No.	Dscription	Source (frequency)	Vulnerability (possibility)	Transportation management	Warehouse management	Material Handling	Inventory Management	Order management	Purchasing management	Customer services	Information management	Internal (5-1)	External (5-1)	Total Score
1	Economic Recession	Demand dropped 20% (2)	Product % over 50% (5)				5		3	3		5	2	180
2	Production line down	Supplier failed delivery (5)	Single source (1)				3	3	2	5		2	3	90

Afterwards, by using a grid which is composed of the possibility and the severity level of each risk, the grid classifies risk into nine levels. The priority of each risk could be identified based on the nine levels in the evaluation results. For example, the risk score in C2, C3, and B3 are the first priority that needs actions. If the risk score in A1 is of less priority and enterprise can either ignore it or handle it evenly.

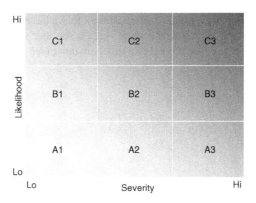

Fig. 9. Risk level grid

3.3   Risk Treatment-Mitigation Strategy

Prevention is better than the cure. The mitigation strategy is designed to manage the risks prior to its occurrence and to minimize the losses. Meanwhile, there are two factors, risk preference and trade-offs, that should be considered when selecting a mitigation strategy. Also, there are two kinds of tradeoffs in selecting strategies: the cost-benefit and the tradeoff strategy.

3.3.1   Preference

The preference on risks means that companies would choose the risk which they are willing to deal with. This is highly associated with the enterprise's capability and its business strategy. When an economic recession happens, some enterprises would try to cut expenses; others would invest more on new technologies or new markets in order to boost when economy upturn. There are no standard criteria to follow in selecting the preference suitable for an enterprise. It would be a challenge for the enterprise to set a proper risk management policy and to achieve logistic performance requirements. In order to lower the impact of risk, a tailored mitigation strategy is needed, but it may not be able to comply with all the needs of each department or worst against others' benefit because the preference of each department is not the same. Therefore, it should be a collective risk mitigation strategy against both long-term and short-term performance requirements.

3.3.2   Trade-Off

It is not possible to use a single strategy for all the risks. In other words, it will have more strategies or actions to deal a specific risk. While deciding a tailored risk mitigation strategy for an identified risk, two factors should be assessed and answered. One is the cost and benefit trade-off of the strategy. Another is the trade-off between strategy options.

3.3.2.1   Costs/Rewards. A certain investment for mitigation strategy is needed in order to lower the likelihood of risk or reduce losses. In general, the value of investment should surpass the potential losses when no actions are taken to reduce the risks.

According to Chopra and Sodhi (2004), managers should "move to a higher level of efficiency by reducing risk while increasing rewards". For example, in order to reduce the material shortage risk, whether having more suppliers (from location x to B), or investing in suppliers as a joint venture (from location x to A), the investment can lower the risk of the material shortage, and at the same time, a supplier's early involvement can help to launch the products to the market in a timely manner (see Figure 10).

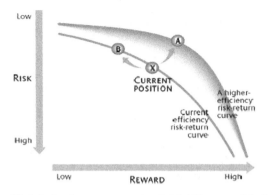

Fig. 10. Relation between reward and risk (Chopra and Sodhi, 2004)

3.3.2.2   Selecting Strategies. There could be more than one mitigation strategy needed to manage the risks. Selecting only one strategy may lead to other risks or may benefit one department but not other departments. Hence, the costs/rewards analysis is used to evaluate the impact of all decisions to each function of the logistics flow so as to decide which strategy is most beneficial to the entire enterprise.

3.3.3   Tailored Mitigation Strategies

To select a correct mitigation strategy is not easy especially when the risk is characterized by uncertainty. The risk mitigation strategy should be tailored to a specific risk, and most of the time, no single action can solve all the risks. Therefore, collective strategies to mitigate the risks are needed. The relationship between mitigation strategies is not in a linear cause-and-effect chain but in a complex interaction among the strategies.

Causal loop diagrams are used to explain the tailored mitigation strategy development and the relationships between strategies. For example, consider the risk of demand uncertainty, there could be a number of uncertainties (internal as well as external factors) that affect demand. Moreover, it is difficult to predict if the demand would change. Besides the uncertain characteristic of risk, the interaction of risk strategies also increases the risk uncertainty and affects the result of risk management. For example, in order to meet customer demand on time, keeping extra inventory could be a strategy but this may lead to the risk of material devaluation or obsolescence, not to mention the cost of holding inventory. Therefore, a collective risk mitigation strategy would be needed in order to achieve a comprehensive synergic result. A causal loop diagram example is illustrated in Figure 11.

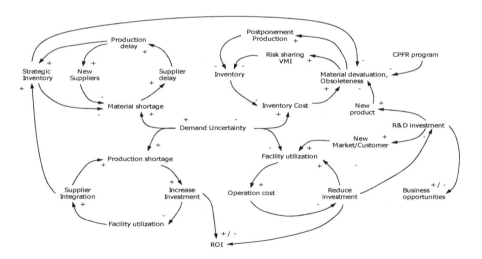

Fig. 11. Risk mitigation strategies interaction (Yang, 2010)

3.4   Implementation

Risk management can be successful if the policy protects the enterprise from potential losses effectively. An efficient risk management relies on the management involvement to control as well as to monitor all the processes. Therefore, the implementation process could become another vulnerability to the success of risk management.

3.4.1   Governance

Governance is built on an effective management process. The primary objective of logistics management is to serve the customer effectively and efficiently. At the same time, logistics management should aim to reduce the cost in each node of the chain. Therefore, balancing the cost and service levels are a synergy performance result of the logistics management. In addition, the processes in the logistics chain are linked with one another, and any failure in the chain flow can lead to a serious impact on the entire logistics chain performance.

With this, a robust governance mechanism is necessary for an enterprise to link up all factors. Besides integrating the processes from suppliers to customers, it also ensures that risk management sustains the logistics performance targets. The risk management in logistics is a part of the enterprise risk management system. From the functional perspective, the governance of risk management should be taken from the enterprise's policy and strategy to set and evaluate logistics performance in accordance to its objectives.

From a process perspective, the governance is "a process by which the board sets the objectives for an organization and oversees progress toward those objectives" (Zoelick & Frank 2005), and the governance should provide "a base foundation for facilitating the effective communication and collaboration needed to achieve success on complex process projects" (Richardson, 2006).

There is no standard governance procedure for all enterprises or logistics firm. Zoellick & Frank (2005) provided a guideline of Governance, Risk management, and Compliance (GRC) that can be used as a reference for risk management in logistics function. The procedure is shown in Figure 12.

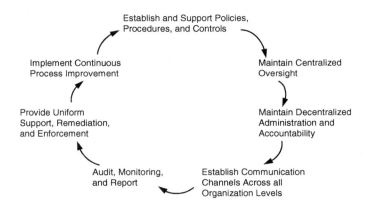

Fig. 12. Governance procedure (Zoellick & Frank, 2005)

The governance cannot be successfully executed without a solid control, monitoring mechanism and continuous improvement process. The control, monitoring, and continuous improvement processes are aim to sustain the logistics management performance so as to satisfy customer needs.

3.4.2   Control and Monitoring

Once the risk management governance has been set and implemented down to the front-end employees, the next action is to control the risks and to monitor the results. In general, a mitigation strategy can be categorized into five approaches such as accept, reduce, alliance, insure, and avoid. The control is based on mitigation strategies tailored to the identified risk, and the actions should be put in place prior to the risks occurrence. The results of mitigation strategies are seen after the risk occurs. There is a need to monitor as well as to periodically collect the data from the logistics activities in order to ensure that risk management is performed. The result should be recorded and also profiled if new risks are identified during the process.

3.4.3   Continuous Improvement

The risk status is subject to environment, business competition, and/or a company's decision. When there is an emerging market that needs local deliveries, the enterprise needs to set up a warehouse near this market. In this condition, the risk of shipment delay would be lower, but the inventory holding cost and risk of inventory obsolesce would increase. Business competition is changing all the time. Having a continuous improvement mechanism can help companies to adjust the strategies in a timely manner so as to become an adaptive organization which controls the changes or enhances the enterprise's resilience. The continuous improvement in risk management also maintains the risk profile and reviews the mitigation strategies in line with the enterprise's strategies.

Managing a change or strategy is not easy, especially in collaborating with different functional departments. Since each department has its own interest, strong and authorized governance mechanisms are needed to conduct and align all of the activities in the same direction. Therefore, an enterprise performance level index would be the basis for all departments in setting risk management strategy.

4   Performance Evaluation

Up to this point, the processes of risk management and its mitigation strategies have been discussed. However, the benefit of these actions to the business operations should be evaluated effectively. Fugate et al. (2008) provided the role of logistics in three aspects, including efficiency, effectiveness, and differentiation to evaluate the performance of logistics management. Other specific performance measurement indexes can be different for each company. Moreover, the key logistics performance index would be affected by the enterprise's strategy. Some performance indexes have been identified in the studies listed in Table 7. The enterprise should choose the key performance index in accordance with their business strategy. For example, to have a lean system, the inventory turnover rate should be higher and the stock-keeping unit (SKU) should be less. However, this may not fulfill the increasing customer demand in a timely manner.

Table 7. Logistics performance indexes

World Bank LPI	Griffis, S.E., et al.,(2007)	Fawcett, & Copper, (1998)
• Efficiency of the clearance process	• On-time delivery %	• Total logistics cost
• Quality of trade and transport related infrastructure	• Logistics costs / sales	• On-time delivery
	• Days order late	• Cost trend analysis
• Ease of arranging competitively priced shipments	• Inventory turnover ratio	• Customer satisfaction
	• Complete order fill rate	• Actual versus budget
• Competence and quality of logistics services	• Average order cycle time	• Stock out
	• Order cycle time variability	• Customer complaints
• Ability to track and trace consignments	• Items picked per person per hrs	• Inventory levels
	• Average line item fill rate	• Inventory turns
• Timeliness of shipments in reaching destination within the scheduled or expected delivery time	• Weeks of supply	• Cost per unit
	• Average backorder fill time	• Delivery consistency
	• Sales lost due to stockout	
	• Percent error pick rate	
	• Logistics cost per unit	

Regarding the risk management performance setting, the first priority should be to sustain the logistics performance. Similar to risk preference, companies need to choose the risk that they are willing to accept. For example, in comparing company #1 and #2 as shown in Figure 13, both company's risk preference are not the same, when the risk #4 moves to the first quadrant, it is accepted by company #1 but is not good enough for company #2.

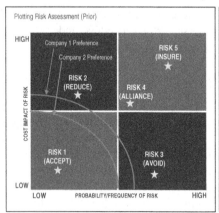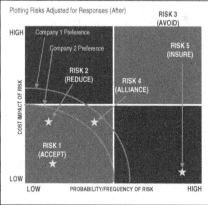

Fig. 13. Plotting risk mitigation strategies (Ballou & Heitger, 2005)

In addition, individual risk should be assessed and reported. Based on the five approaches, observe if the likelihood or losses are reduced or controlled (see Table 8). Finally, identify whether there are any risks that have not been found.

Table 8. Risk management assessment

Strategies	Likelihood	Losses
Accept	--	--
Reduce	↓	↓
Alliance	↓	↓
Insure	--	↓
Avoid	✖	✖

↓ Decreasing; ✖ Will not happen; -- Unchanged

5   Conclusions

Risks resulting in logistic disruptions are inevitable. Therefore, it is important to understand the applications of risk management in an enterprise. Besides risk awareness, the vulnerability of the system and preference on managing risks affect the results of logistic risk management. The company business strategy plays a key role in deciding logistic management operations as well as the risk management implementation. Nevertheless, how to decide the right mitigation varied between companies. This chapter studies the risk management in logistics by discussing the logistics processes, the risk management strategy, the risk management process in logistic, and the enterprise performance evaluation. This gives an understanding on the important risk management principles in dealing with logistics disruption.

References

1. Asbjornslett, B.E., Rausand, M.: Assess the vulnerability of your production system. Production Planning & Control 10(3), 219–229 (1999)
2. ASSE. TECH 482/535 Class Notes: Industrial Safety Engineering Analysis, http://www.ceet.niu.edu/depts/tech/sse/tech482/index.html (retrieved June 20, 2010)
3. Autry, C.W., Zacharia, Z.G., Lamb, C.W.: A logistics strategy taxonomy. Journal of Business Logistics 29(2), 27–51 (2008)
4. Ballou, B., Heitger, D.L.: A Building-Block Approach for Implementing COSO's: Enterprise Risk Management-Integrated Framework. Management Accounting Quarterly 6(2), 1–10 (2005)
5. Blos, M.F., et al.: Supply chain risk management (SCRM): a case study on the automotive and electronic industries in Brazil. Supply Chain Management: An International Journal 144(4), 247–252 (2009)
6. Bowersox, D.J., Closs, D.J., Drayer, R.W.: The digital transforma-tion: Technology and beyond. Supply Chain Management Review 9(1), 22–29 (2005)
7. Cavinato, J.L.: Supply chain Logistics Risks: From the back room to the board room. International Journal of Physical Distribution & Logistic Management 34(5), 383–387 (2004)
8. Chopra, S., Sodhi, M.S.: Managing Risk to Avoid Supply-Chain Breakdown. MIT Sloan Management Review 46(1), 53–62 (2004)

9. Department of Health and Human Services-US. Enterprise performance life cycle framework: Practice guide- Contingency Plan,
    http://www.hhs.gov/ocio/eplc/EPLC%20Archive%20Documents/
    36-Contingency-Disaster%20Recovery%20Plan/
    eplc_contingency_plan_practices_guide.pdf (retrieved June 21, 2010)
10. Faisal, M.N., Banwet, D.K., Shankar, R.: Supply chain risk mitiga-tion: modeling the enablers. Business Process Management Journal 12(4), 535–552 (2006)
11. Faisal, M.N., Banwet, D.K., Shankar, R.: Information risks man-agement in supply chains: an assessment and mitigation framework. Journal of Enterprise Information Management 20(6), 677–699 (2007)
12. Fugate, B.S., Mentzer, J.T., Flint, D.J.: The role of logistics in ket orientation. Journal of Business Logistics 29(2), 1–26 (2008)
13. Griffis, S.E., et al.: Aligning logistics performance measures to the in-formation needs of the firm. Journal of Business Logistics 28(2), 35–56 (2007)
14. Hofer, A.R., Knemeyer, A.M.: Controlling for logistics complexity: scale development and validation. The International Journal of Logistics Management 20(2), 187–200 (2009)
15. Husdal, J.: The dark side of supply chain,
    http://www.husdal.com/2008/11/12/supply-chain-risk/
    (retrieved January 5, 2008)
16. ISACA/ITGI. Risk-IT-Overview, http://www.isaca.org/Knowledge-Center/
    Risk-IT-IT-Risk-Management/Pages/Risk-IT1.aspx (retrieved June 22, 2010)
17. Logistics world. What is Logistics?,
    http://www.logisticsworld.com/logistics.htm (retrieved June 1, 2010)
18. Lourenço, H.R.: Logistics Management: an opportunity for Metaheuristics. In: Rego, C., Alidaee, B. (eds.) Metaheuristics Optimization via Memory and Evolution, pp. 329–356. Kluwer Academic Publishers (2005)
19. Mentzer, J.T., Stank, T.P., Esper, T.L.: Supply chain management and its relationship to logistics, marketing, production, and operations management. Journal of Business Logistics 29(1), 31–46 (2008)
20. Nadler, Slywotzky: Risk and the Enterprise. MMC Journal Viewpoint 1, 1–11 (2008)
21. Nilsson, F., Waidringer, J.: Toward adaptive logistics management. In: Proceedings of the 38th Hawaii International Conference on System Science. IEEE (2005),
    http://www.computer.org/portal/web/csdl/doi/10.1109/
    HICSS.2005.629 (retrieved June 20, 2010)
22. Pettit, T.J., et al.: Ensuring supply chain resilience: development of a conceptual framework. Journal of Business Logistics 31(1), 1–21 (2010)
23. Richardson, C.: Project Performance Corporation. Process governance best practices: building a BPM center of excellence (2006),
    http://www.ppc.com/Documents/ProcessGovernanceBestPractices.
    pdf (retrieved June 20, 2010)
24. Spekman, R.E., Davis, E.W.: Risky business: expanding the discus-sion on risk and the extended enterprise. International Journal of Physical Distribution & Logistics Management 34(5), 414–433 (2004)
25. Tang, C.S.: Robust Strategies for Mitigating Supply Chain Disruptions. International Journal of Logistics: Research and Applications 9(1), 33–45 (2009)
26. The institute of internal auditors, COSO. Applying COSO's Enterprise Risk Management — Integrated Framework (2004),
    http://www.coso.org/documents/COSO_ERM.ppt (retrieved June 23, 2010)

27. Trkman, P., McCormack, K.: Supply chain risk in turbulent environ-ments—A conceptual model for managing supply chain network risk. International Journal of Production Economics 119(2), 247–258 (2009)
28. Willersdorf, R.G.: Adding Value through Logistics Management. Logistics Information Management 3(4), 6–8 (1993)
29. World Bank. Logistics Performance Index,
    http://info.worldbank.org/etools/tradesurvey/Mode2a.asp
    (retrieved June 25, 2010)
30. Yang, W.H.: Identification of risk mitigation strategies in supply chains of electronics manufacturing services companies, Master Thesis: University of Liverpool (2010)
31. Zoellick, B., Frank, T.: Compliance Consortium Whitepaper. Governance, Risk Management, and Compliance: An Operational Approach (2005),
    http://www.securitymanagement.com/archive/library/
    compliance_consortium0805.pdf (retrieved June 20, 2010)

Part III

Risk Assessment and Response Systems

Chapter 16
Natural Disaster Risk Assessment Using Information Diffusion and Geographical Information System

Zhang Jiquan, Liu Xingpeng, and Tong Zhijun

Institute of Natural Disaster Research,
College of Urban and Environmental Sciences,
Northeast Normal University, Changchun, Jilin, 130024, P.R. China
{zhangjq022,liuxp912,gis}@nenu.edu.cn

Abstract. With the social and economic development, the losses caused by nat-ural disasters were more and more serious. Natural disaster assessment, man-agement and research are the important field, developing direction and hotspot issues on disaster science and geo-science in resent year. However, because most of the natural disasters are a small sample of events, and uncertainty of natural disasters, so natural disaster risk assessment is particularly difficult based on historical data. Information diffusion theory is useful method for natu-ral disaster risk assessment based on small sample even; it is a fuzzy approach to quantitative analysis of the probability of natural disaster risk. Therefore, the information diffusion theory has unique advantages in natural disaster risk as-sessment and management. This chapter presents a Geographical Information Systems (GIS) and information diffusion theory-based methodology for spatio-temporal risk assessment of natural disasters, taking grassland fire disasters in the Northern China as the case study. Firstly, we discuss connotation and form-ing mechanism of natural disaster risk, basic theory and framework of natural disaster risk assessment and management. Secondly, we introduce information diffusion theories and Geographical Information Systems (GIS) in the form of definitions, theorems and applications comprehensively and systemically. Final-ly, we give the case study on application of information diffusion theory and Geographical Information Systems (GIS) on grassland fire disasters in the grassland area of the Northern China. We employed information matrix to ana-lyze and to quantify fuzzy relationship between the number of annual severe grassland fire disasters and annual burned area. We also evaluated the conse-quences of grassland fire disaster between 1991 and 2006 based on historical data from 12 Northern China provinces. The results show that the probabilities of annual grassland fire disasters and annual damage rates on different levels increase gradually from southwest to northeast across the Northern China. The annual burned area can be predicted effectively using the number of annual se-vere grassland fire disasters. The result shows reliability as tested by two-tailed Pearson correlation coefficient. This study contributes as a reference in decision making for prevention of grassland fire disaster and for stockbreeding sustaina-ble development planning. The fuzzy relationship could provide information to make compensation plan for the disaster affected area.

J. Lu, L.C. Jain, and G. Zhang: Handbook on Decision Making, ISRL 33, pp. 309–330.
springerlink.com   © Springer-Verlag Berlin Heidelberg 2012

1   Introduction

A natural disaster is the effect of a natural hazard (e.g., flood, grassland fire disaster, hurricane, volcanic eruption, earthquake, or landslide) that affects the environment, and leads to financial, environmental and/or human losses. The resulting loss depends on the capacity of the population to support or resist the disaster, and their resilience [1]. This understanding is concentrated in the formulation: "disasters occur when hazards meet vulnerability"[2]. A natural hazard will hence never result in a natural disaster in areas without vulnerability, e.g. strong earthquakes in uninhabited areas.

Risk is the future safety. In quantitative terms, risk is often defined as the probability of an undesired outcome. Natural disaster risk is the appearance probability of some kind if disaster during a certain period. It is a function of disaster intensity and the appearance (probability). Risk assessment of the natural disaster is an important part of disaster reduction. Many international object of disaster reduction have involved this research. In many cases, however, it is practically impossible to precisely get the probability distribution we need. Sometimes the estimated values of the probabilities may be so imprecise as to be practically useless if we still regard them as crisp values.

For risk assessment of natural disaster, Completeness of the information is very important. However, in many cases, data are only a part of the facts, and information carried by them is incomplete. For example, a small sample which contains a few observations is incomplete data when we use them to study the natural disaster. Before there was fuzzy set theory, statisticians used to consider incomplete data in random uncertain viewpoint. For incomplete information, statistical methods are often ineffective. Fuzzy set theory provides a unifying framework for fuzzy information processing includes studying incomplete data. Information diffusion theory is an important method to process small sample using fuzzy set theory, and it was used in many areas such as risk analysis and disaster assessment.

Geographic Information Systems (GIS) are computer-based systems that store and process (e.g. manipulation, analysis, modeling, display, etc.) spatially referenced data at different points in time [3]. GIS may be used in archaeology, geography, cartography, remote sensing, land surveying, public utility management, natural resource management, precision agriculture, photogrammetry, urban planning, emergency management, military, navigation, aerial video, and localized search engines. The fuzzy processing and GIS were independent of each other in the past. At present, through the integration of information diffusion theory and GIS, fuzzy classification and identification in GIS, it can be widely used for many different problem domains including evaluating risk, risk zone, soil classification, crop-land suitability analysis, identifying and ranking wild fire risk.

Take grassland fire disaster as an example, this chapter presents the application of information diffusion theory in natural disaster risk analysis of small sample. Computations based on this analytical grassland fire disaster risk model can yield an estimated probability and burned areas value. This study indicates that the aforementioned model exhibits fairly stable analytical results, even when using a small set of sample data. The results also indicate that information diffusion theory and technology is highly capable of extracting useful information and therefore improves system recognition accuracy. This method can be easily applied and the analytical results produced are easy to understand. Results are accurate enough to act as a guide in disaster situations.

2   Basic Theory and Method of Information Diffusion

2.1   Information Diffusion

Information distribution [4], a method by which incomplete data is analyzed better, is a new way which allows us to divide an observation into two parts to belong to two subsets. The fundamental view of information diffusion is to cancel the restriction that an observation just belongs to two subsets in the domain. In information diffusion view, an observation can belong to all the subsets in the domain by different membership.

Let x_i ($i = 1,2,3,\cdots,n$) be observations of natural disaster. Let $X = \{x_1, x_2, \ldots, x_n\}$, be a given sample and the universe of discourse be U. A mapping from X $\times$U to [0, 1], We call X a sample, and n its size.

$$\mu : X \times U \rightarrow [0,1]$$

$$(x,v) \mapsto \mu(x,v), \forall (x,v) \in X \times V$$

is called information diffusion of X on U, if it satisfies

(1) $\forall x_i \in$ X, if $u_o = x_i$, then $\mu(x_i, u_o) = \sup_{u \in U} \{\mu(x_i, u)\}$

(2) $\forall x \in$ X, $\mu(x,u)$ is a convex function about u ;

$\mu(x,u)$ is called an information diffusion function of X on U. When U is discrete, the function also can be written as $\mu(x_i, u_j)$. μ is called a diffusion function and V is called a monitoring space. When $V = U$, we say that $\mu(x,v)$ is sufficient. The set $D(X) = \{\mu(x,u) | x \in X, u \in U\}$ is called the sample of fuzzy sets derived from X on U by diffusion μ .

Given an origin x_0 and a bin width h , the histogram is a frequency distribution on the intervals $A_j = [x_0 + (j-1)h, x_0 + jh], j = 1,2,\ldots,m$. When X is incomplete, the corresponding histogram must be too rough. Information distribution method can improve it by using an allocation function instead of the associated characteristic function.

Let $\mu_j (j = 1,2,3,\ldots,m)$ be the centers of the intervals $[x_0 + (j-1)h, x_0 + jh]$ of the histogram, the allocation function of X can be defined by

$$\mu_h(x_i, u_j) = \begin{cases} 1 - |x_i - u_j| / h & if |x_i - u_j| \le h \\ 0 & if |x_i - u_j| > h \end{cases} \tag{1}$$

In fact, $\mu_h(x_i, u_j)$ is a membership function which indicates that x_i can belong to more than one interval.

A more reasonable estimation can be obtained by using Eq.(2).

$$\bar{f}(A_j) = \frac{1}{n}\sum_{i=1}^{n}\mu_h(x_i,u_j) \tag{2}$$

For $D(X)$, we have the principle of information diffusion to deal with small sample. This principle holds, at least, in the case of estimating a probability density function. For estimating a probability density function, in fact, the diffusion estimate is just a Parzen kernel estimate, but, there are many ways to do diffusion estimate. On the other hand, the whole of the kernel theory focuses on what properties the estimate possesses. Normal distribution, Poisson distribution and exponential distribution are most popular principle of information diffusion.

Take normal diffusion as example, we researched the similarities of information and molecules in diffusion action, and obtain a partial differential equation to represent the information diffusion. Solving the equation, we obtain the normal diffusion function

$$\mu(x,u) = \frac{1}{h\sqrt{2\pi}}\exp\left[-\frac{(x-u)^2}{2h^2}\right] \tag{3}$$

According to the relation between the membership and possibility, $\mu(x,u)$ would be regarded as some possibility that u will occur if x has occurred. The nearer the u is to x; the larger the possibility that u will occur. The largest possibility would be 1. For x, normalizing $\mu(x,u)$ along u, we have a membership function with respective to u,

$$\mu'(x,u) = \exp\left[-\frac{(x-u)^2}{2h^2}\right], u \in R. \tag{4}$$

Based on the two-point criterion and average distance assumption, we have Eq. (5) to calculate coefficient h.

$$h = \begin{cases} 0.6841(b-a) & for \quad n=5; \\ 0.5404(b-a) & for \quad n=6; \\ 0.4482(b-a) & for \quad n=7; \\ 0.3839(b-a) & for \quad n=8; \\ 2.6851(b-a)/(n-1) & for \quad n\geq 9, \end{cases} \tag{5}$$

where

$$b = \max_{1\leq i\leq n}\{x_i\}, a = \min_{1\leq i\leq n}\{x_i\}.$$

Strictly speaking, the normalized distribution in Eq. (4) with the h calculated by Eq. (5) would be called simple normalized normal diffusion function. In short, we still call it normal diffusion function and write the corresponding fuzzy set A with x being centroid as

$$\mu_A(u)= \exp\left[-\frac{(x-u)^2}{2h^2} \right], u \in R. \tag{6}$$

The concept sketch of normal information diffusion can be shown as Fig.1.

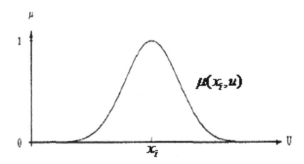

Fig. 1. Information diffusion function of x_i on a continuous universe of discourse U

2.2   Information Matrix

In the process of risk analysis of natural disaster, sometimes we should to determine the relationship between two variables, and in most cases, this relationship is not accurate, but a kind of fuzzy relations. Information matrix [5] is a useful tool to estimate the fuzzy relationships between natural disaster indicators.

Let $(x_i, y_i), i=1,2,......,n$, be observations of a given sample X with domain U of input and range V of output. Let $A_j, j=1,2,......,t$ and $B_k, k=1,2,......,l$ be fuzzy sets with membership functions $\mu_{A_j}(u), \mu_{B_k}(v)$, respectively. Let

$$\begin{cases} U = \{A_j | j=1,2,......,t\}; \\ V = \{B_k | k=1,2,......,t\}; \end{cases} \tag{7}$$

Their Cartesian product $U \times V$ is called an illustrating space. (A_j, B_k) is called an illustrating point.

Let $\circ$ be an operator defined on R (the set of real numbers).

$$q_{jk}(x_i, y_i) = \mu_{A_j}(x_i) \circ \mu_{B_k}(y_i) \tag{8}$$

is called information gain of (x_i, y_i) at (A_j, B_k) with respect to the operator.

Let

$$Q_{jk} = \sum_{i=1}^{n} q_{jk}(x_i, y_i). \tag{9}$$

Then

$$Q = \begin{array}{c} \\ A_1 \\ A_2 \\ \vdots \\ A_t \end{array} \begin{pmatrix} B_1 & B_2 & \cdots & B_l \\ Q_{11} & Q_{12} & \cdots & Q_{1l} \\ Q_{21} & Q_{22} & \cdots & Q_{2l} \\ \vdots & \vdots & \vdots & \vdots \\ Q_{t1} & Q_{t2} & \cdots & Q_{tl} \end{pmatrix} \tag{10}$$

is called an information matrix of X on $U \times V$.

3   The Application of GIS in Risk Assessment Natural Disaster

GIS (Geographic Information System) is an important tools, techniques and disciplines to acquisition, storage, analysis and management of geospatial data, in recent years has been widespread concerned and rapid development. Its powerful spatial analysis functions of geographic information are playing an increasingly important role in the GPS and route optimization. It is based on geospatial database, with the support of the computer's hardware and software, using the theory of systems engineering and information science, scientific management and comprehensive analysis of geographic data with spatial content, to provide information for management and decision-making. Simply, GIS is a technical system of comprehensive treatment and analyze geospatial data. Natural disaster management is a process that involves time, space and need to analysis of large amounts of data, so GIS has a wide range of application in natural disasters researches.

3.1   Components and Functions of GIS

GIS consists of hardware, software, data, personnel and methods. Hardware and software provide environment for GIS; Data is an important aspect to GIS; Methods provide solutions for the GIS; Personnel are the key and activity factors, it directly impact and coordinate other components.

Just for GIS itself, which mostly function is normally have five kinds of basic function, they are 1) The function of data collection and editing, 2) Attribute data editing and analysis, 3) The Function of Mapping, 4) The Function of Spatial database Management, and 5) The Function of Spatial Analysis.

3.2   Applications of GIS

Within the last 30 years, GIS has made remarkable development, it widely used in resources, environmental assessment, disaster prediction, land management, urban planning, telecommunications, transportation, military police, water electricity, utilities management, forestry, animal husbandry, statistics, business finance and almost all areas.

1) GIS and natural disasters risk information management

Because the factors influenced natural disasters is complexity, the natural disasters risk management data include real-time data monitoring data, historical disaster data, environmental data and regional socio-economic background data. These data have significant spatial geographical features. These regional and spatial characteristics for natural disasters management is a very important, such as natural background of different regions, population and pasture resources, the hut, and the spatial information distribution of other elements. GIS have the advantages to describe, manage, analyze and operate the spatial data, such as display the spatial data, inquire the correlative attribute data and orient goals. Particularly GIS powerful spatial analysis functions offer the effective tool for natural disasters risk management. Natural disasters risk assessment applying GIS mainly contains the following aspects:

(1) Natural disaster data maintenance

The natural disasters data input, and the output, updating and maintenance based on the electronic map. Maintain contents include regional basis geographical data (for example, administrative divisions, river and lake areas, settlements, population and contour lines), regional data on the natural environment (such as the average number of dry thunderstorms, drought index, wind speed, temperature, sunny days), natural disasters thematic data, natural disasters historical data (for example, when and where occurred fire, the yield loss, loss of livestock, the affected population and the impacted socio-economic, etc.). The data maintenance more intuitive and convenient based on the electronic map.

(2) Natural disasters data display and query

Using the function of positioning and layer management, the system could realize the display the all kinds of data from two-dimensional vector graphics applying electronic map, and even 3-D display if we have all kinds of needed data. Users could take zooming, roaming, and other operations, and could operate attribute data, such as integration operation.

(3) Natural disasters spatial analyze

Using specific spatial analysis of GIS, we could make a correlative analysis, such as buffer analysis for natural disasters influence area, Superposition analysis, and so on.

(4) Establish the risk management policy and emergency response plan

Based on upon operations, we could get adequate information on the natural disasters risk. It can offer provide for advancing measure and developing emergency response plans.

2) Natural disasters risk assessment is an important means and an important part of natural disasters risk management and a new perspective of the natural disasters risk Research. Natural disasters risk is a quantitative characteristic of natural disasters hazard and the possibility of the consequences become reality. According as the formation mechanism of the natural disaster risk, summarizes the formation mechanism of the natural disasters risk for the result of the interaction of hazard(H),exposure (E),vulnerability (V)and emergency response and recovery capability(R). Based on

the hazard (H), exposure (E), vulnerability (V) and emergency response and recovery capability(R) of the natural disasters risk, Choose the main influencing factors of the natural disasters that in including the natural factor and the social factor. Adopt to various methods such as linearity, Weighted Mathematical Model, the Analytic Hierarchy Process(AHP),the Weighted Comprehensive Analysis(WCA) , Information Diffusion Theory and Information Matrix, and so on , obtains the natural disasters risk index, and then to natural disasters risk degree in the study area for quantitative risk assessment. On this basis, supported by GIS, use its space analysis, study area is divided into a number of natural disasters risk areas, and conveys it in the form of digital maps, to verify the result.

3) GIS-based system of natural disasters risk management

The goal of this system constructs is the use advanced satellite remote sensing technology, the geographic information system, the database technology, the multimedia technology, the network technology , to geography space data comes from the different data pool, the different form, the different type carries on the unification processing. Integrating disaster risk assessment model, a powerful GIS spatial analysis functions in support, realize natural disasters hazards analysis, impact assessment and evaluation and prediction of loss, natural disasters vulnerability analysis, mitigation analysis, risk assessment and evaluation, for the fire risk integrated management and emergency response for the country and sub-regional providing the information and decision support.

The overall objective of system design is that friendly interface, easy operation and stable, reliable performance, a reasonable system structure and function, data safe and intelligent decision support for the natural disasters departments provide a practical, efficient, stable tool.

4   Case Analysis

Take grassland fire disaster as an example, this study presents a Geographical Information Systems (GIS) and information diffusion-based methodology for spatiotemporal risk analysis of grassland fire disaster to livestock production in the grassland area of the Northern China. Information matrix was employed to analyze and to quantify fuzzy relationship between the number of annual severe grassland fire disasters and annual burned area. We also evaluated the consequences of grassland fire disaster between 1991 and 2006 based on historical data from 12 Northern China provinces.

4.1   The Study Area and Statistical Analysis of Grassland Fire Disasters

Grassland fire disaster is a critical problem in China due to global warming and human activity. The northwestern and northeastern China face more challenges for mitigation of grassland fire disasters than other regions due to broad territory combined with the effects of complex physiognomy. According to statistical analysis of historical data of grassland fire disaster from 12 northern China provinces between 1991 and 2006, grassland fire disasters have been increasing gradually with economic development and population growth. The increased grassland fire disasters had significant impacts on the national stockbreeding economy. One of the main challenges is to establish the grassland fire disaster risk system so that the distribution of limited resources for disaster reduction and economic assistance can be made. Risk assessment is one of important means of natural disaster management. Risk assessment of natural disasters is defined as the assessment on both the probability of natural disaster occurrence and the degree of damage caused by natural disasters. In recent years, an increasing number of studies focus on natural disaster risk analysis and assessment of floodings, earthquakes and droughts among others. However less attention has been focused on grassland fire disasters. The occurrence of grassland fire disasters is due to both natural and human factors and their interactions. Traditional studies of grassland fire disaster are often limited to models in fire behavior, fire hazards and fire forecast. Disaster risk assessment has been used to manage wild fires. For example, Finney [6] calculated the wild fire risk by combining behavior probabilities and effects. Castro et al. [7] simulated the moisture content in shrubs to predict wild fire risk in Catalonia. Jaiswal et al. [8] applied a color composite image from the Indian Remote Sensing Satellite (IRS) and GIS for forest fire risk zoning in Madhya Pradesh, India. The grassland fire potentials depend on factors such as fuel type and density, topography, humidity, proximity to settlements and distances from roads. Thus grassland fire disasters occur randomly with uncertainties. In order to manage grassland fire disasters and compensate losses effectively, it is important to obtain probability and losses of grassland fire disasters on different risk levels.

In general, a disaster risk is defined as the outcome of probability multiplying potential losses. The main issues of risk assessment are implemented through estimating the probability distributions based on the historical data, which are usually substituted by frequencies. However, as grassland fire disasters are considered as the small sample event, it may not be unreasonable of using a frequency as a substitute instead of using an actual probability. Information diffusion was first put forward and proved with number theory. It is an effective method that can transform a traditional sample-point into a fuzzy set to partly fill the gap caused by incomplete data. Thus, information diffusion is suitable to deal with the small sample data. Yi et al. [9] estimated a frequency analysis method of flood disaster losses based on fuzzy mathematics theory of information diffusion with short time series of flood disaster samples. Huang et al. [10] discussed the benefit of the soft risk map calculated by the interior-outer-set model and proved that the soft risk map is better than the traditional risk map. Liu and Huang [11] established a fuzzy relationship between fire and surroundings based on the winter data of Shanghai. The purposes of the this part are to: (i) analyze the situation of grassland fire disasters, (ii) obtain the probabilities of annual grassland fire disasters and annual damage rates in different levels, and (iii) establish the fuzzy relationship between the number of annual severe grassland fire disasters and annual burned areas. The methodology in this study can be applied to study other natural disasters as well. The information from this study is potentially useful reference in decision making of grassland fire disaster prevention and stockbreeding sustainable development planning. The fuzzy relationship could provide information to make compensation plan for the disaster area.

Study Area

The main grassland areas in China lie in the arid and semi-arid northern 12 provinces of Inner Mongolia, Jilin, Heilongjiang, Liaoning, Xinjiang, Ningxia, Qinghai, Sichuan, Shaanxi, Shanxi, Hebei, and Gansu. Grassland fire disasters are one of the most concerned natural disasters there. According to historical data, 6,801 grassland fire disasters were reported over a period of 16 years between 1991 and 2006. About 6.22×10^8 ha of grasslands had been burned out in the northeastern and northwestern China, which represented an annual average of 3.89×10^7 ha.

Statistical Analysis of Grassland Fire Disasters

The grassland fire disaster is a natural disaster and a threat related to environment, people's livelihood, and socio-economic development, which is caused by the combination of both a natural hazard and a societal vulnerability. Spatial and temporal distributions of grassland fire disasters are determined by the interactions between natural condition and human activity. Historical data of grassland fire disaster in northern China from 1991 to 2006 illustrate the fluctuant trend of the number of grassland fire disasters and severe grassland fire disasters (Fig. 2). The trend indicates that the number of grassland fire disasters increased gradually, while the severe grassland fire disasters decreased (Fig.2). In recent years, although the provinces in northern China have strengthened management in fire prevention, grassland fire disaster risks are in increasing trend with economic growth and increased trend of human activities.

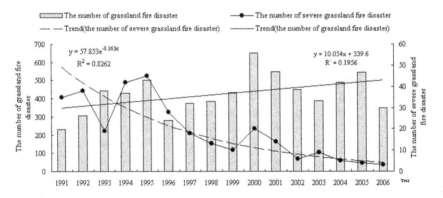

Fig. 2. Variation of grassland fire disasters and severe grassland fire disasters from 1991 to 2006 in northern China

Statistical data of fire seasons (from March to May and September to November) in northern China show a strong relationship between the number of grassland fire disasters and the burned area (r = 0.98, p = 0.01). The highest value emerged in April, followed by October, March, and May (Fig.3). The occurrence frequencies of April, October, March, and May account for about 41%, 16%, 13%, and 12%, respectively, while the burned area of April, October, March, and May account for about 47%, 14%, 8%, and 17%, respectively (Fig. 3).

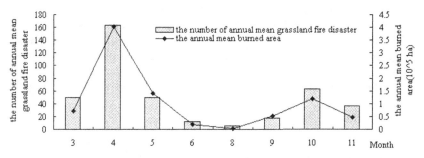

Fig. 3. Status of grassland fire disasters in fire seasons in northern China

For spatial distribution, grassland fire disasters in the provinces of Inner Mongolia, Hebei, Heilongjiang, and Xinjiang occurred more frequently than other provinces in northern China. The Inner Mongolia was the province with the most annual burned area (Fig.4). According to annual damage rates ($r = D_i/S_i$, r is annual damage rates, D_i is annual damage areas of province i, and S_i is grassland areas of province i), Heilongjiang and Inner Mongolia are the regions that are most likely affected by grassland fire disasters (Fig.5).

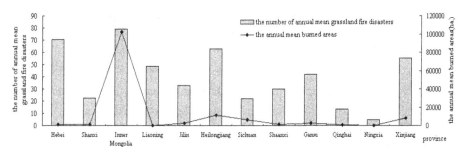

Fig. 4. Spatial distribution of grassland fire disasters in northern China provinces

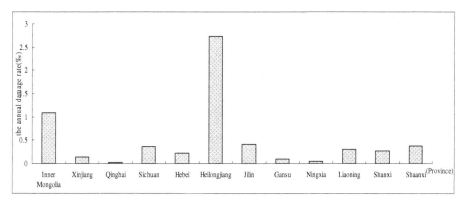

Fig. 5. Spatial distribution of annual damage rates of the grassland fire disaster in northern China provinces

In China, grassland fire disasters are classified into four grades with respect to its harmful level, i.e., grassland fire alarm, general grassland fire disaster, serious grassland fire disaster, and extraordinarily serious grassland fire disaster. Because serious grassland fire disaster and extraordinarily serious grassland fire disaster always have severe impacts on the environment, livelihood and socio-economic development we integrated those two grades into one category as the severe grassland fire disaster.

Data and Methods

Actual data sets are often incomplete due to various reasons in the study of grassland fire disaster [12]. Compared to accurate quantitative model, fuzzy mathematical model is a useful, objective technique to describe non-linear relationship in the grassland fire disaster. Information diffusion is an effective tool to deal with small sample event and establish the fuzzy nonlinear relationship. It has been widely used in disaster research such as earthquake and flooding among others. As a fuzzy mathematics method, information diffusion can offset the information deficiency in small sample event through transforming an observation sample into a fuzzy set.

Information Diffusion

Information diffusion is a fuzzy mathematics method that offset the lack of information in small sample event. The basic principle of information diffusion is to transform the crisp observations into fuzzy sets for partly filling up gaps caused by a scarcity of data so that the recognition of relationship between input and output could be improved. The detailed computational process is illustrated as follows.

Let $Y = \{y_1, y_2, y_3, \cdots y_n\}$ be a set of observations, called a given sample, and $U = \{u_1, u_2, u_3, \cdots u_m\}$ be the chosen framework space. If the observations cannot provide sufficient information to identify the precise relationship that is needed, then Y is called an incomplete data set. For any $y \in Y$, $u \in U$, the Eq.11 is called normal information diffusion.

$$\tilde{f}_j(u_i) = \frac{1}{h\sqrt{2\pi}} \exp[-\frac{(y_j - u_i)^2}{2h^2}]$$ (11)

Where y_j is a given sample, the universe U is the monitoring space, u_i is the controlling points, h is the diffuse coefficient, and h is calculated using Table 1 [4].

Table 1. Relationship of the number of sample and diffuse coefficient

n	h	n	h
5	0.8146(b-a)	9	0.3362(b-a)
6	0.5690(b-a)	10	0.2986(b-a)
7	0.4560(b-a)	≥ 11	2.6851(b-a)/(n-1)
8	0.3860(b-a)		

h is diffuse coefficient, n is the number of sample, and b and a represents maximum and minimum value of sample, respectively.

It is assumes that

$$C = \sum_{i=1}^{n} \tilde{f}(u_i) \qquad (12)$$

and the membership function of fuzzy subset is obtained as follows.

$$\mu_{y_j}(u_i) = \tilde{f}_j(u_i)\Big/C \qquad (13)$$

Where $\mu_{y_j}(u_i)$ is the normalizing information distribution of sample y_j.

Assumption

$$q(u_i) = \sum_{j=1}^{m} \mu_{y_j}(u_i) \qquad (14)$$

The probability of u_i is calculated as follows.

$$p(u_i) = q(u_i)\Big/ \sum_{i=1}^{m} q(u_i) \qquad (15)$$

In theory, $\sum_{i=1}^{m} q(u_i) = m$. The exceed probability ($p(u \geq u_i)$) of u_i is calculated as follows.

$$p(u \geq u_i) = \sum_{k=i}^{n} p(u_k) \qquad (16)$$

Information Matrix

Let $H = \{(x_1, y_1), (x_2, y_2), \cdots (x_n, y_n)\}$ be a given sample, each sample has two elements of x as the input and y as the output. The input domain is U, $U = \{u_1, u_2, u_3, \cdots u_j\}$ j=1, 2, 3..., m, and output domain is V, $V = \{v_1, v_2, v_3, \cdots v_k\}$ k=1, 2, 3..., n. Eq.17 and Eq.18 are used to deal with information distribution.

$$q_{ijk} = \begin{cases} (1 - \dfrac{|u_j - x_i|}{\Delta x})(1 - \dfrac{|v_k - y_i|}{\Delta y}), & |u_j - x_i| < \Delta x \quad and \quad |v_k - y_i| < \Delta y \\ 0, & otherwise \end{cases} \qquad (17)$$

$$Q_{kj} = \sum_{i}^{n} q_{ijk} \qquad (18)$$

With Eq.17 and Eq.18, the simple information matrix ($Q = \{Q_{jk}\}_{m \times n}$) can be obtained on a $U \times V$ space. Based on the simple information matrix and applying Eq.19 and Eq.20, the fuzzy information matrix ($R = \{r_{jk}\}_{m \times n}$) can be obtained.

$$Q = \begin{matrix} u_1 \\ u_2 \\ \vdots \\ u_m \end{matrix} \begin{bmatrix} \overset{v_1}{Q_{11}} & \overset{v_2}{Q_{12}} & \cdots & \overset{v_n}{Q_{1n}} \\ Q_{21} & Q_{22} & \cdots & Q_{2n} \\ \vdots & \vdots & \vdots & \vdots \\ Q_{m1} & Q_{m2} & \cdots & Q_{mn} \end{bmatrix} \qquad (19)$$

$$\begin{cases} S_k = \max_{1 \le j \le m} \{Q_{jk}\} \\ r_{jk} = Q_{jk} / S_k \\ R = \{r_{jk}\}_{m \times n} \end{cases} \qquad (20)$$

Fuzzy information matrix ($R = \{r_{jk}\}_{m \times n}$) establishes the fuzzy relationship of variables. It expresses the approximate relation of input-output. The basic element of fuzzy information is the fuzzy set described by a membership function. By max-min algorithm, the membership function of output variable could be described as Eq.21.

$$\tilde{I}_{x_i} = \bigvee_{x} [\mu(x_i, u_j) \wedge R] \qquad (21)$$

Where, $\vee$ is symbol of max-product algorithm, $\wedge$ is symbol of min-product algorithm and the operator "$\wedge$" is defined as expression, $\mu(e_i) = \min\{1, \sum \mu(x_i) r_{ij}\}$, R is fuzzy information matrix, and $\tilde{I}_{x_i}$ is the fuzzy output, and $\mu(x_i, u_j)$ is 1-dimension linear information distribution function. Let $X = \{x_1, x_2, x_3, \cdots x_m\}$ be a given sample, and $U = \{u_1, u_2, u_3, \cdots u_m\}$ be the chosen framework space with $\Delta = u_j - u_{j-1}$, $j = 2, 3 \ldots$, m. For any $x \in X$, and any $u \in U$, the 1-dimension linear information distribution could be described as Eq.22.

$$\mu(x_i, u_j) = \begin{cases} (1 - \dfrac{|u_j - x_i|}{\Delta}), & |u_j - x_i| < \Delta \\ 0, & otherwise \end{cases} \qquad (22)$$

Eq.21 is a fuzzy set about degree of membership calculated by fuzzy approximate reasoning. Because the information is too dispersive to apply in decision making, Eq.23 could calculate the fuzzy inference values and was employed to concentrate the information.

$$A_{x_i} = \sum_{k=1}^{n} \tilde{I}_{x_i}^{\alpha} \cdot v_k \Bigg/ \sum_{k=1}^{n} \tilde{I}_{x_i}^{\alpha} \tag{23}$$

Where, $\tilde{I}_{x_i}$ is fuzzy information distribution, v_k is the controlling points, α is constant, generally $\alpha = 2$, and A_{x_i} is the fuzzy inference value.

According to information diffusion theory, we calculated the probabilities of annual grassland fire disasters and annual damage rates in 12 northern China provinces based on the historical data from 1991 to 2006 collected by the Department of Grassland Fire Prevention, Supervision Center and Ministry of Agriculture of the People's Republic of China. Due to the significant importance of the annual burned area for policy makers and the strong correlations between the number of annual severe grassland fire disasters and annual burned area (r=0.728, p=0.01), we used the number of annual severe grassland fire disasters to predict the annual burned area in this study. The information matrix was employed during this process considering the efficiency of analyzing the fuzzy and nonlinear relationship between the annual several grassland fire disasters and annual burned area.

Results and Discussions

Probability Analysis of the Grassland Fire Disaster Risk

Traditionally, the probability of an event was calculated by frequency histogram. The frequency histogram is reasonable when the samples are abundant. However as a grassland fire disaster belongs to small sample event, we employed information diffusion to deal with the data of grassland fire disasters. By putting observation values of annual severe grassland fire disasters into the information diffusion function (Eq.11-Eq.16), Fig. 6 illustrates the exceed probabilities on different levels exceed probabilities.

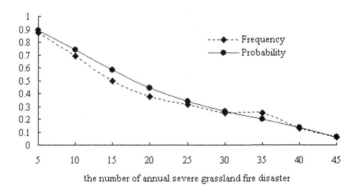

Fig. 6. Probabilities of annual severe grassland fire disasters based on information diffusion and frequencies

Fig.6 shows the scatter plot of measured (frequencies calculated by frequency histogram) and predicted (probabilities calculated by information diffusion) values of test samples in northern China. The predicted values agreed well with the measured values.

Fig.7 shows the correlation between measured and predicted annual probabilities of severe grassland fire disasters. The two-tailed Pearson correlation coefficient (r value) reached to 0.992 (p = 0.01). This shows that information diffusion is able to accurately predict the probabilities of grassland fire disasters on different levels.

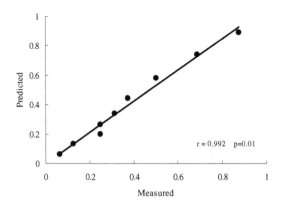

Fig. 7. Correlation between frequencies and probabilities of annual severe grassland fire disasters

The results in Fig.6 and Fig.7 indicate that probabilities obtained by normal information diffusion and frequency histogram are consistent. It means that normal information diffusion is useful to analyze probabilities of grassland fire disaster. If in the condition of having a complete dataset, the two methods should perform equally well. As the grassland fire disaster belong to the fuzzy events with incomplete data, therefore, the normal information diffusion is better than frequency histogram to analyze probabilities of the grassland fire disaster.

According to above method and the historical data of northern China, we calculated the probabilities of the number of annual grassland fire disasters and annual damage rates in 12 provinces using GIS. The results of spatial distributions are illustrated (Figs. 8, 9) show that the probabilities of annual grassland fire disasters and annual damage rates on different levels increased gradually from southwest to northeast across northern China. The Inner Mongolia and Heilongjiang are more easily affected by grassland fire disasters with high frequencies. The Qinghai and Shanxi provinces have less grassland fire disasters than other provinces in northern China.

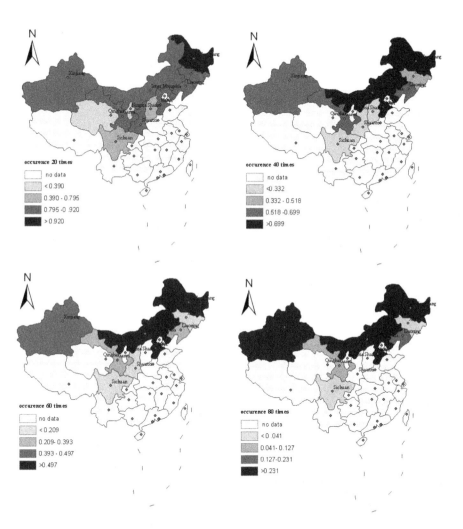

Fig. 8. Probabilities of annual grassland fire disasters on different levels in northern China

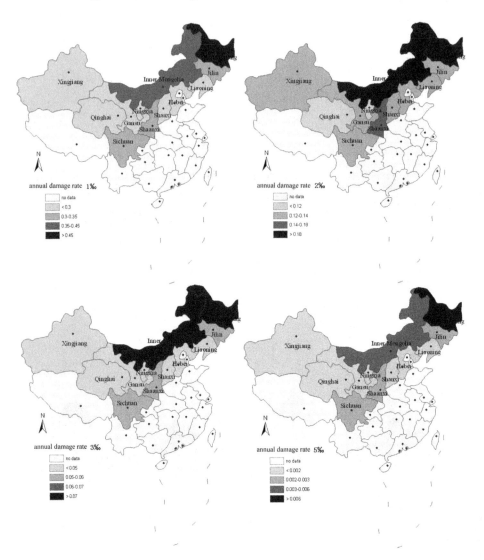

Fig. 9. Probabilities of annual damage rates on different levels in northern China

Prediction of Annual Burned Area

According to the method of information matrix, let X be the sample of annual severe grassland fire disasters, and Y be the sample of annual burned area.

$$X = \{35,38,19,42,45,28,18,13,10,20,14,6,9,5,4,3\}$$
$$Y = \{32.4,68.9,41.9,129.0,52.0,138.4,43.0,27.4,\cdots,4.74\}$$
$$H = \{X,Y\}$$

Correspondingly, we suppose that the chosen framework space of annual severe grassland fire disasters in input space is U, and the chosen framework space of annual burned area in output space is V. Annual severe grassland fire disasters range from 3 to 45, and nine controlling points are U= {5, 10, 15... 45}, step length Δ= 5. Chosen annual burned area V= {10, 20, 30... 140}, step length Δ=10.

Applying the observation values of two factors and information diffusion function (Eq.17-Eq.20), the simple information matrix and fuzzy matrix between two factors can be calculated as follows.

$$
Q = \begin{array}{c} \\ 5 \\ 10 \\ 15 \\ \vdots \\ 45 \end{array}
\begin{array}{cccc} 10.0 \quad 20.0 \quad 30.0 \quad \cdots \quad 140.0 \\ \left[\begin{array}{ccccc} 1.63 & 0.00 & 0.30 & \cdots & 0.00 \\ 1.06 & 1.08 & 0.44 & \cdots & 0.00 \\ 0.54 & 0.41 & 0.89 & \cdots & 0.00 \\ \vdots & \vdots & \vdots & \vdots & \vdots \\ 0.00 & 0.00 & 0.00 & \cdots & 0.00 \end{array} \right] \end{array}
\qquad
R = \begin{array}{c} \\ 5 \\ 10 \\ 15 \\ \vdots \\ 45 \end{array}
\begin{array}{cccc} 10.0 \quad 20.0 \quad 30.0 \quad \cdots \quad 140.0 \\ \left[\begin{array}{ccccc} 1.00 & 0.00 & 0.00 & \cdots & 0.00 \\ 0.65 & 1.00 & 0.33 & \cdots & 0.00 \\ 0.33 & 0.38 & 0.50 & \cdots & 0.00 \\ \vdots & \vdots & \vdots & \vdots & \vdots \\ 0.00 & 0.00 & 0.00 & \cdots & 0.00 \end{array} \right] \end{array}
$$

With the fuzzy relation matrix and Eq.21, we could obtain the corresponding annual burned area by a given observation of annual severe grassland fire disasters. For example, if we have an observation $x_i =18$, according to function Eq.22, the 1-dimension linear information distribution of observation $x_i =18$ can be obtained as follows (In all the formula, the signs "+" represent "and", the numerator represents subject degree, and the denominator represents discrete controlling point).

$$
\mu(18, u_j) = \frac{0}{5} + \frac{0}{10} + \frac{0.4}{15} + \frac{0.6}{20} + \frac{0}{25} + \frac{0}{30} + \frac{0}{35} + \frac{0}{40} + \frac{0}{45}
$$

According to the min -product algorithm, we obtain:

$$
e = \frac{0.13}{10} + \frac{0.21}{20} + \frac{0.80}{30} + \frac{0.77}{40} + \frac{0.20}{50} + \frac{0}{60} + \frac{0}{70} + \frac{0}{80} + \frac{0}{90} + \frac{0}{100} + \frac{0}{110} + \frac{0}{120} + \frac{0}{130} + \frac{0}{140}
$$

In order to concentrate the information, the max-product algorithm defined as follows.

$$
\overline{e} = S_k \cdot e = (1.63 \times 0.13, 1.08 \times 0.21, 0.89 \times 0.80, \cdots, 0.5 \times 0) = (0.21, 0.23, 0.71, 0.82, \cdots, 0)
$$

Normalize the $\overline{e}$, the final information distribution is obtained as follows.

$$
\tilde{I}_{18} = \frac{0.27}{10} + \frac{0.28}{20} + \frac{0.87}{30} + \frac{1}{40} + \frac{0}{50} + \frac{0.19}{60} + \frac{0}{70} + \frac{0}{80} + \frac{0}{90} + \frac{0}{100} + \frac{0}{110} + \frac{0}{120} + \frac{0}{130} + \frac{0}{140}
$$

In order to find the core of the information, the values of annual burned area can be calculated based on fuzzy inference (Eq.23).

$$
A_{18} = \frac{10*0.27^2 + 20*0.28^2 + 30*0.87^2 + 40*1^2 + 50*0 + \cdots + 120*0 + 130*0 + 140*0}{0.27^2 + 0.28^2 + 0.87^2 + 1^2 + 0.19^2} = 34.40
$$

According to above analysis, for one observation, $x_i = 18$, the annual occurrence probability is about 0.5 (from Fig.6); the maximum probable of annual burned area is 40; the secondary probable of annual burned area is 30; and the core of the information of annual burned area is 34.40. Compare with traditional probabilistic method, multi-valued risk result can provide more characteristics of risk system when we analyze the risk of system. By the methods described above, the results of annual burned area were calculated for the year of 1991 through 2006 (Fig.10).

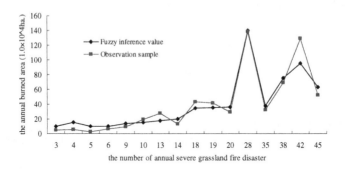

Fig. 10. A comparison between sample and fuzzy inference values of information diffusion matrix

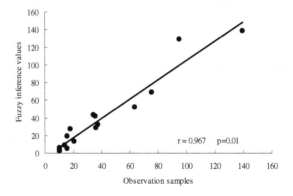

Fig. 11. Correlation between sample values and fuzzy inference values of annual burned areas

Fig.10 shows that the annual burned area simulated by information matrix matched well with the actual samples. The trends of annual burned areas between observation samples and fuzzy inference values are also consistent. From Fig. 10, it can be seen that annual burned area changes randomly, and it is impossible to be fitted by a linear function. It confirms the possibility of using the information diffusion theory for dealing with the fuzzy relationship of the grassland fire disaster. Fig.11 shows the correlation between observation samples and fuzzy inference values using informa-tion matrix. The two-tailed Pearson correlation coefficient (r value) reached to 0.967 (p = 0.01). This shows that information matrix is viable to accurately predict the annual burned area of grassland fire disasters on different levels. Because the relationship of between the number of annual severe grassland fire disasters and annual burned area is fuzzy, nonlinear, and could not be expressed by a linear relationship, the information matrix was employed and the annual burned area was predicted effectively using the number of annual severe grassland fire disasters with significant P values. The result could help in strategic decision makings to manage grassland fire disasters.

5   Conclusions

Grassland fires occur frequently in northern China provinces and cause significant property losses. In order to implement a compensation and disaster reduction plan, the spatio-temporal risk distribution of grassland fire disaster and the losses caused by grassland fires are among critically important information to grassland fire disaster managers. This study applied approximate reasoning of information diffusion to estimate probabilities and fuzzy relationship with scanty, incomplete data of the grassland fire disaster. Using information diffusion and information matrix, given observations, we can assess the spatio-temporal risk of grassland fire disaster and give an improved result to support risk management than traditional probability method.

Grassland fire disaster risk is an uncertain and complicated system. Considering the imprecise or incomplete information of grassland fire disaster, and the limits by the existing technique and experiment method, the analysis result could be inefficient and imprecise by employing only the traditional accurate model. If the risk values were estimated by classical statistics tool, we would lose information included in the given sample and could obtain wrong results. This study shows that information diffusion theory is effective to implement the impact and risk assessment associated with grassland fire hazards. The results could be used by the different level of authorities to develop grassland fire disaster risk protection plans. While the methods have certain advantages the accuracy in predicting could be improved by involving the impact factors of grassland fire disasters.

References

1. Bankoff, G., Frerks, G., Hilhorst, D.: Mapping Vulnerability: Disasters, Development and People (2003)
2. Wisner, B., Blaikie, P., Cannon, T., Davis, I.: At Risk - Natural hazards, people's vulnerability and disasters. Routledge, Wiltshire (2004)
3. Aronoff, S.: Geographic Information Systems: a Management Perspective. WDL Publications, Ottawa (1989)
4. Huang, C.F.: Principle of information diffusion. Fuzzy Sets and Systems 91, 69–90 (1997)
5. Huang, C.F., Moraga, C.: Extracting fuzzy if–rules by using the information matrix technique. Journal of Computer and System Sciences 70(1), 26–52 (2005)
6. Finney, M.A.: The challenge of quantitative risk analysis for wildland fire. Forest Ecology and Management 211, 97–108 (2005)
7. Castro, F.X., Tudela, A., Sebastià, M.T.: Modeling moisture content in shrubs to predict fire risk in Catalonia (Spain). Agricultural and Forest Meteorology 116, 49–59 (2003)

8. Jaiswal, R.K., Mukherjee, S., Raju, K.D., Saxena, R.: Forest fire risk zone mapping from satellite imagery and GIS. International Journal of Applied Earth Observation and Geoinformation 4, 1–10 (2002)
9. Yi, C., Huang, C.F., Pan, Y.Z.: Flood disaster risk analysis for Songhua river basin based on theory of information diffusion. In: 7th International Conference (ICCS 2007), part 3 (2007)
10. Huang, C.F., Inoue, H.: Benefit of soft risk map made by using information diffusion technique. In: International FLINS (Fuzzy Logic and Intelligent Technologies in Nuclear Science) Conference (2004)
11. Liu, J., Huang, C.F.: An information diffusion technique for fire risk analysis. Journal of Donghua University (Eng. Ed.) 21(3), 54–57 (2004)
12. Chen, J.N., Huang, H.K., Tian, F.Z., et al.: A selective Bayes classifier for classifying incomplete data based on gain ratio. Knowledge-Based Systems 21, 530–534 (2008)

Chapter 17
Applications of Social Systems Modeling to Political Risk Management

Gnana K. Bharathy and Barry Silverman

Ackoff Collaboratory for Advancement of Systems Approaches (ACASA)
University of Pennsylvania
120 Hayden Hall, 220 South 33rd Street, Philadelphia, PA 19104-6315
{bharathy,basil}@seas.upenn.edu

Abstract. Risk is inexplicably linked to complex inter-related structural and behavioral factors. Of these, human factors tend to be overarching and predominant. Any model that would assist with exploring the inter-relationships among structural and human factors would be immensely valuable to risk management. A social system model constructed with a valid set of software agent framework and complete with factions, institutions and other organizational descriptors, all based on best-of-breed social science theories, can act as the desired testbed to evaluate effects that may arise from alternative courses of action. Through our past case studies, we describe in this article, how social systems modeling and associated intelligent system tools could be applied to assess and manage political risk (and by extension other social systems based risks). We also enumerate the challenges of such a testbed and describes best-of-breed theories drawn from across the social sciences and synthesized and implemented in an agent-based framework. These predictions are examined in a real world cases (Bangladesh) where the agent models are subjected to a validity check and the political risks are estimated.

Keywords: risk management, socio-cultural simulation, agent-based models, social systems, political risk assessment, white-box models for risk.

1 Introduction: Issues with Risk Assessment and Motivation

Social systems are complex and risk is a complex and abstract concept. The combination of social system and risk, found in such applications as political risk management[1], is rather bewildering. With numerically and relationally complex set of actors, groups, causes, and emergent effects, it is difficult to trace through them all in a top down fashion. A common practice with such complex systems is attempting to capture the complexity in the statistical relationships, but these do not give any insight into plausible workings of such a model.

---

[1] Social system based perspective is relevant to all risks, but is particularly central to political risk.

J. Lu, L.C. Jain, and G. Zhang: Handbook on Decision Making, ISRL 33, pp. 331–371.
Springerlink.com   © Springer-Verlag Berlin Heidelberg 2012

The purpose of this paper is to illustrate how a social system model could potentially be employed to identify and elicit risk in a country, making the risks transparent to policy makers and leaders. This is the novel claim of the paper. This paper is neither about cognitively deep agent based modeling frameworks nor modeling methodology (including validation) for such social systems models, although we have briefly described the framework and methodology just enough to make the article stand on its own.

The paper has been organized into three major sections. In section one (this one), we explain the motivations for employing social system framework; in section two, we explain our social system modeling framework, model construction methodology and introduce the case study that will be used illustrate risk assessment through social system model; in section three, we describe the risk assessment process of risk conception, hazard identification, risk assessment and risk management.

Our thesis in this paper is that there are number of issues pertaining to risk assessment that could benefit by a social system modeling approach based primarily on cognitively detailed agents.

In order to understand this premise, it is useful to look at the origins of risk and understand some of the key assumptions in risk methodology. In the last 50 years, probabilistic risk assessment (PRA) and risk management have come to be applied to virtually every field, but they were originally developed for managing nuclear plant safety. For this, various related paradigms have evolved (NRC-NAS, 1983).

One of the unstated assumptions of these traditional risk assessment models such as fault-tree analysis, event tree analysis etc. is that one can conceptualize and causally link chains of events in risk identification, and perhaps, later in assessment. Identification of risk is considered the most important stage of risk assessment. Assessment of risk including quantitative treatment of risk is important primarily for providing systematic immersion in the system posing risk through a transparent methodology. The numbers that come out of the assumption are "nice-to-haves"; however, real emphasis is placed on the process of assessment. Making decisions based on the understanding one gains, losses and the mechanism by which they occur by going through a systematic process is deemed more important than the probabilistic estimates.

Of course, identifying risks in the physical realm involving safety hazards, is often easier and could be carried out systematically. A host of tools are available for traversing through the physical system and identifying and assessing risk. Most importantly, a process or nuclear plant is a concrete entity that exists in front of one's eyes!

On the other hand, identifying and assessing political, business or financial risks are clouded in mystery, and hence are far trickier. And as the financial meltdown indicates, over-reliance on abstract models has very serious consequences and has missed various key signals in the environment (Freedman, 2011). The financial, market, political and business risk assessment/ management is different to say the least. These fields have adopted risk assessment and risk management, but have limited means of actually identifying and assessing risks. That is not their fault. These risks are treated rather abstractly at a high level. The purpose of the assessment is primarily to optimize portfolio rather than continuously improving the existing system. Is this difference in approaches to risk assessment because of the availability of tools and techniques? If relatively simpler domains such as engineering are concerned about understanding the hazards rather than the numbers that come out of a PRA, shouldn't more complex domains pay more attention to the same?

Various critics have now argued that "mis-application" of PRA was one of the contributors to financial market meltdown (Salmon, 2009). We have worked on several cases of industrial and safety risk studies. One of the striking features of these risk studies is that they attempts to model failures through a mechanistic model[2]. Later, when we gained some understanding of financial risks, we were uncomfortable with the sole reliance on statistics without any regard for any underlying mechanisms.

This is well articulated in two popular books today. In *The Failure of Risk Management: Why It's Broken and How to Fix It,* Hubbard (2009) explains: *"PRA is what economists would call a "structural model" The components of a system and their relationships are modeled for example, in Monte Carlo simulations. If valve X fails, it causes a loss of backpressure on pump Y, causing a drop in flow to vessel Z and so on. Financial models, such as option theory (OT) and modern portfolio theory (MPT) have at least one important conceptual difference from PRA done for nuclear power. For MPT and OT, there is no attempt to explain an underlying structure to price change. Various outcomes are simply given probabilities. And if there is no history of a particular system-level event like a liquidity crisis, there is no way to compute the odds of it... In Finance, there is an attempt [to] find correlations among various factors, but without the attempt to understand much about the underlying* mechanism."Similarly, in another book entitled "How Markets Fail", another author Cassidy (2009) echoes similar sentiments that blind reliance on, and misuse of statistical models, as among the causes leading to failure of financial market of 2008.

This is also true of political risk, which is the key focus of this paper. The understanding of political risk is important to businesses and political leaders as well as ordinary citizens. Numerous studies commissioned by the World Bank and surveys of CEOs are examples that underscore political risk as the primary concern among investors (MIGA, 2009). Despite its importance, a consensus on its definition is slow to come (Fitzpatrick, 1983; Jakobsen, 2010).

Almost all of the extant literatures on political risk are investor or business centric, and simply treat the hazardous events contributing to risk as exogenous to the model. With this premise, different studies often provide different or conflicting opinions on the definition and scope of political risk. Accordingly, the term 'political risk' refers to the possibility that investors will lose expected returns due to political decisions, conditions, or events occurring in the country or emerging market in which they are investing.

Robock (1971) proposed an operational definition, "in which political risk in business exists when discontinuities, which are difficult to anticipate, occur in the business environment as a result of political change..."

---

[2] In quantum physics, mechanistic models have been considered as amateurish representations. We neither imply this context nor scale, but a relatively macro scale dynamic representations where not only correlations, but potential causations and emergence can be tested through the application of scientific process (Boudon, 1998; Mayntz 2002). More details are forthcoming.

Typically (Green 1972; Zink 1973; Nehrt 1970; Shapiro 1978; Rodriguez and Carter 1976), political risk is defined with respect to multi-nationals operating in a country or region, and is frequently ascribe to two broad classes of hazards, namely:

- environmental factors such as direct political violence and instability and,
- impediments to business operations including currency inconvertibility, discriminatory taxation, public sector competition and expropriation or nationalization.

In this paper, we will focus on the environmental factors such as political violence and instability, and take a broader, or at least a more encompassing, meaning of political risk as applicable to not only multi-nationals, but also political, military and civil leadership and the population as a whole.

World Bank's Multilateral Investment Guarantee Agency (MIGA, 2009), refers to political risk as: "Probability of disruption of operations of MNEs (Multi-National Enterprises) by political forces or events when they occur in host countries, home countries, or as a result from changes in the international environment. In host countries, political risk is largely determined by uncertainty over the actions of governments and political institutions, but also minority groups, such as separatist movements. In home countries, political risk may stem from political actions directly aimed at investment destinations, such as sanctions, or from policies that restrict outward investment." Details of definitions aside, they all focus on government interference with business operations or undesirable consequences of political instability, but do not deal with the mechanisms behind the events.

There is a price to be paid for a high level treatment of political risk, or for that matter any risk. A number of cases in the literature (Fatehi et.al. 1989; Kobrin, 1979; Loree and Guisinger 1995; Olibe et al 1997; Woodward and Rolfe, 1993) which treat the impact of political risks at an aggregate level (e.g Foreign Direct Investment) report contradictory, and henceforth, inconclusive results. We are not surprised that these studies are contradictory and inconclusive. For us, predicting social system behavior based on such data is the concern. Instead, we argue that one models social system, in a descriptive fashion along with mechanisms at least one or two levels deep, so that one could explore the social system testbed, instead of making predictions.

This reliance on high level, blackbox statistical models at the expense of studying (and attempting to gain an understanding of) the phenomenon is characteristic of financial and related risk management techniques. The objective and scope of statistical models are significantly different from those of other techniques, in that statistical models do not address the mechanisms behind the model. Instead, they directly represent the data correlation. As Lindley (2000) succinctly summarized: "... it is only the manipulation of uncertainty that interests us. We are not concerned with the matter that is uncertain. Thus we do not study the mechanism of rain; only whether it will rain."

When trying to predict rare events in social systems (such as the instability of a government), a forecast of constancy (tomorrow will be like today) may or may not be quite accurate. Even when the model does predict accurately, it will fail at offering any insights into causality. Likewise, the available lightweight statistical models may perform well at forecasting certain tasks, yet they will fail to support the criteria/considerations we have in mind here in Section 2.1.

Adopting an approach that had originally evolved for forecasting to risk management can be misleading, giving to a sense of misplaced security in sheer numbers.

Greatest benefit in risk modeling comes from the transparency and understanding it brings (Keey 2000). Abstracted statistical models, however sophisticated, do not provide this. On the other hand, white-box models with provision for drilling down into mechanisms will have a better chance of providing explanation (Hedström and Swedberg 1998).

In above context, we would like to describe a potential modeling methodology that would assist with understanding social system risk. We demonstrate it though an application in political risk due to uncertainty brought about by instability.

As with traditional risk assessment in process safety, prediction and forecasting are not the prime objective of a well conceived social systems model; instead, understanding, description, transparency, storytelling, analysis, prescription, policy making and decision support become more important goals of the modeling effort than making predictions.

Since our framework and social system model are central to our thesis, we describe the framework before we delve into risk assessment. Later, using these framework and model constructed in this framework, we will outline a risk assessment case study.

2   Social Systems Modeling Framework

We have developed a model that can represent purposeful actors with bounded rationality, rich cognitive processes, personality and value system, as well as heterogeneity among actors and social networks. Such a model can help in developing insight into the behavior of social systems, particularly for conflict laden environments.

Some of the long term, broader aims of building the framework include: providing a generic game simulator to social scientists and policymakers; creating plausible AI models of human behavior; and improving the science by synthesizing best-of-breed social science models with subject matter expert knowledge, exposing their limitations and showing how they may be improved.

2.1   Considerations for a Social Systems Model

In the recent years, in a much needed and welcome break from the dominant tradition of "correlational school" (aka neo-Humean positivist school) that had once monopolized the quantitative research landscape in social sciences, the mechanism-centered explanation and analysis is emerging as a contender and has found a legitimate place in the mainstream discourse (Hedström and Swedberg 1998; Mahoney 2001; Bunge 1997). Unlike probabilistic association between variables hypothesized by positivist, correlational school, mechanism-centered school (and its variants such as "causal reconstruction") advocate(s) white-box models and drilling down to explanation using mechanism (Boudon, 1998; Mayntz 2002). Causal reconstruction, which, at least in theory, would deal with processes (not correlations), intends to produce historical narratives using causal mechanisms (In this chapter, we simply refer to them as mechanistic models). In time, such approaches would result in robust models. Depending on their scope and coverage, one or more of these models can be synthesized and be employed in social system models. Along these lines, in the discussion below, we find it useful to introduce some key considerations in the design of our model. These have also been discussed earlier in Silverman (2010) and Silverman et.al. (2010).

Causality: In order to facilitate risk identification and assessment, the types of models we wish to study must be rich in causal factors that can be examined to see what leads to particular outcomes. We wish to understand mechanisms.

Scientific (Esoterics): The models must be based on best currently available scientific theories of social systems and the other types of systems involved. This is because our goal is to see if models based on theory will advance our understanding of a social system, and if not, what is missing.

Descriptive (Exoterics) : Parsimony has lead to the failure of game theory to help in asymmetric warfare, to the inability of statistical forecasting models to explain causality and mechanism, and to the kinds of overly simple quantitative risk models that helped cause the financial collapse (Salmon, 2009). Parsimonious thinking has lead to numerous diplomatic mis-adventures. We wish to take the alternative route and rely on data-rich, descriptive approaches. We seek approaches that drag in as much exoteric detail as possible about the actual stakeholders and personalities in the scenarios being studied – their issues, dilemmas, conflicts, beliefs, mis-perceptions, etc.

Model of Models: The modeling frameworks we wish to examine generally must make use of the model of models approach. The value of a model of models approach is that one can include the models one wants in a given forecasting exercise. Scientific (or engineering) models can be added, deleted, altered and perturbed in an effort to learn more about the social system in question and how alternative policies might impact and influence it. We think that study of these types of models is inherently different from single model systems and deserves its own methodology, though this is yet to be proven. It may turn out that the methodology we assemble will work equally well on single model systems.

Design Inquiry: The modeling framework must support explorations into the effects of what emerges when different policies and courses of action are attempted. Here we wish to explore the question whether esoteric social science models (with the aid of exoteric knowledge) can adequately capture and predict the phenomena underlying social dilemmas. In military parlance, these models also must support the study of effects-based operations and shaping actions.

Scientific Inquiry: Ideally, models created with scientific backing should also be usable to experiment and explore the dynamics of what emerges in a social system. Modeling and instrumentation in physics and chemistry enabled the discovery of new theories in those fields. The model of models approach is thought to serve a similar purpose for social science. Here we wish to ask the question whether social science itself can be advanced with the use of the model of models approach. Thus we wish to use a given model-of-model system to examine how alternative policies affect the social system, and what new theories can be derived from the emergence of alternative macro-behaviors. We wish to understand the performance of the system and mine the model base for mechanisms.

In addition, we also address the prevailing concerns in the field. For example, The National Research Council's review of agent-based modeling points out three major limitations on this way of conducting research. The three limitations are found in the following three realms of concern: (1) Degree of Realism; (2) Model Trade-Offs; and (3) Modeling of Actions (NRC, 2008).

Degree of Realism: With regard to the first concern, namely the degree of realism, we believe that our particular approach to agent-based modeling, highlighted by our recent effort in building CountrySim, has achieved a level of realism that has been an improvement on any previous models.

We use realistic cognitive-affective agents built by combining best available principles and conjectures from relevant sciences. The rules and equations that govern the interactions of these agents are also derived from the best available principles and practices, with particular attention to their realism. When populating our virtual countries with agents and institutions, we triangulated three sources: (1) existing country databases from the social science community and various government agencies and non-governmental organizations; (2) information collected via automated data extraction technology from the web and various newsfeeds; and (3) surveys of subject matter experts. This approach has yielded the best possible approximations for almost all parameter values of our model (See Silverman et al, 2009 for details).

Model Trade-Offs: Model trade off pertains to selection of modeling techniques and the tradeoffs one has to live with as a consequence. Two such trade offs are worth investigating. One being the use of cognitively sophisticated agents versus simple agents. If one were to model at a detailed level, we miss the bigger picture and if one were to model at the bigger picture, then one misses the details and nuances. The second, but related, trade off is the level of details in the models. For cognitive agents versus simple agent trade off, we combine both types of agents judiciously. The key actors are based on cognitively detailed agents who deliberate over decisions. Large populations who only contribute votes are based on simple agents. As for the level of details, we construct multiple models (for example, country level, district level and village levels). We again believe that the above approaches, highlighted by our recent effort in building CountrySim, has good theoretical and practical justification. The verdict is still out.

Large Feature Space and Curse of Dimensionality: For descriptive modeling intended to provide learning, exploration and immersive training through social systems, "curse of dimensionality" is a given. These models have very large feature spaces (on the order of 1000), giving rise to "curse of dimensionality" (De Marchi, 2005). Frequently, an argument is made in favor of reducing dimensionality by building simple, yet elegant toy models adhering to Keep It Simple, Stupid (KISS) principles.

Social system models, by definition, are complex, with imprecise, incomplete and inconsistent theories (Silverman et.al. 2009a). However, given the complexity of social systems, many researchers argue otherwise, arguing that simplicity, brought about for mathematical tractability and elegance, is not sufficient to explore social systems. They point out the need for complex, descriptive models in social system. This has resulted in an equally important and convincing emerging paradigm named KIDS ("Keep it Descriptive Stupid") (see Edmonds and Moss, 2004 for example).

De Marchi (2005) points out that with simpler models, frequently the assumptions are doing most of the "heavy lifting", resulting in the "curse of simplicity". This is Simple or stylized models are unable to encode domain information, particularly the depth of the social system. For example, human value systems are almost always assumed, hidden, or at the best, shrunk for the purpose of mathematical elegance. Yet, human behavior is vital to the social system behavior. While simpler models might be more elegant and be useful for prediction, they do not offer the richness and mechanism that might accompany complex models.

This approach emphasizes the need to make models as descriptive as possible and accepts simplification only when evidence justifies it. We too deviate from traditional modelers on one issue. Unlike more tradition-bound research communities such as econometrics and game theory, we do depart from this prevailing paradigm of KISS because we see no convincing methodological or theoretical reasons for blindly adhering to it. This does not mean the models have to be unnecessarily complex either. Complexity of the model must be driven by purpose.

2.2   Introduction to Framework

Let us now introduce our model that can provide a social systemic basis for risk assessment. This introduction should be misconstrued as detailed description of the models; later is not the focus of this paper. It must also be noted that the models introduced in this paper are defaults and one can swap in other models without affecting how the actors think through their resource-based, ethno-cultural conflicts.

Game: The vast majority of conflicts throughout history ultimately centers on the control of resources available to a group and its members (let us refer to this as "greed") and 'settling scores' (let us call this "grievance"). Such conflicts contribute to political instability and hence political risk. We have created a social system model based on this fairly universal class of leader-follower games from our model. Specifically, the socio-cultural game centers on agents who belong to one or more groups and their affinities to the norms, sacred values, and inter-relational practices (e.g., social and communicational rituals) of those groups.

We have constructed, using a framework named CountrySim/ FactionSim, a country based on multi-resolution agent based approach. FactionSim captures globally recurring socio-cultural "game" that focuses upon inter-group competition for control of resources (e.g, Security/Economic/Political Tanks) and settling scores, as mentioned above.

This environment facilitates the codification of alternative theories of factional interaction and the evaluation of policy alternatives. This tool allows conflict scenarios to be established in which the factional leader and follower agents all run autonomously and are free to employ their micro-decision making as the situation requires.

This model has a virtual recreation of the significant agents (leaders, followers, and agency ministers), factions, institutions, and resource constraints affecting a given country and its instabilities.

These influential and important actors, who are required to deliberate over and make key decisions, are represented through cognitively detailed agents. The actors in this category include leaders at various levels and their archetypical followers.

In order to save computational resources, we represent less salient actors (those who do not have to deliberate over key decisions) through a larger population model consisting of dummer agents making decisions on a simplistic scale of support vs opposition.

A diagrammatic representation of an example CountrySim model is given in Figure 1 (forthcoming later in the chapter). For additional details, see Silverman et.al. (2010, 2007a, 2008a).

Cognitively Detailed Agents: We would describe the primary agents that occupy our world as cognitively very detailed, endowed with values including short term goals, long terms preferences, standards of behaviour and personality. The environment provides contexts. According to Gibsonian (1979) ecological psychology, these contexts carry and make decisions available for consideration. These agents make decisions based on a minimum of two set of factors:

- the system of values that an agent employs to evaluate the decision choices and
- the contexts that are associated with choices.

$$\text{Decision Utility} = f(\text{Values, Contexts}) \qquad (1)$$

The values guide decision choices, and in our case, have been arranged hierarchically or as a network. The contexts sway the agent decisions by providing additional and context specific utility to the decisions evaluated. The contexts are broken up into micro-contexts. Each micro-context just deals with one dimension of the context (for example, relationship between the perceiver and target and so on).

With a given set of values, an agent evaluates the perceived state of the world and the choices it offers under a number of micro-contexts, and appraises which of its values are satisfied or violated. This in turn activates emotional arousals, which finally sums up as utility for decisions.

Value Trees: The agent's cultural values and personality traits are represented through Goal, Standards and Preference (GSP) trees. These are multi-attribute value structures where each tree node is weighted with Bayesian probabilities or importance weights.

A Preference Tree is one's long term desires for world situations and relations (e.g., no weapons of mass destruction, stop global warming, etc.) that may or may not be achieved in the scope of a scenario.

The Standards Tree defines the methods an agent is willing to take to attain his/her preferences, and what code that others should live by as well. The Standard Tree nodes that we use merge several best-of-breed personality and culture profiling instruments such as, among others, Hermann traits governing personal and cultural norms (Hermann, 2005), standards from the GLOBE study (House et. al., 2004), top-level guidelines related to Economic and Military Doctrine, and sensitivity to life (humanitarianism). Personal, cultural, and social conventions render inappropriate the purely Machiavellian action choices ("One shouldn't destroy a weak ally simply because they are currently useless"). It is within these sets of guidelines that many of the pitfalls associated with shortsighted Artificial Intelligence (AI) can be sidestepped. Standards (and preferences) allow for the expression of strategic mindsets.

Finally, the Goal Tree holds short term needs the agent seeks to satisfy each turn (e.g., vulnerability avoidance, power, rest, etc.), and hence drive progress toward preferences. In the Machiavellian (Machiavelli, 1965 and 1988) and Hermann-profiled (Hermann, 2005) world of leaders, the Goal Tree reduces to the duality of growing/developing versus protecting the resources in one's constituency. Expressing goals in terms of power and vulnerability provides a high-fidelity means of evaluating the short-term consequences of actions. For non-leader agents (or followers), the Goal Tree also includes traits covering basic Maslovian type needs.

Decision Making Loop: With a unifying architecture, different subsystems are connected. For each agent, PMFserv operates what is sometimes known as an observe, orient, decide, and act (OODA) loop. PMFserv runs the agents perception (observe) and then orients all the sub-systems such as physiological and mental stress receptors to determine levels of fatigues and related stressors, grievances, tension buildup, impact of rumors and speech acts, emotions, and various mobilizations and social relationship changes since the last tick of the simulator clock. Once all these modules and their parameters are oriented to the current stimuli/inputs, the decision-making/cognition module runs a best response algorithm to try to determine or decide what to do next. The algorithm it runs is determined by its stress and emotional levels. In optimal times, it is in vigilant mode and runs an expected subjective utility algorithm that re-invokes all the other modules to assess what impact each potential next step might have on its internal parameters. Under conditions of excessive stress or boredom (limited arousal), the agent's decision making sub-optimal.

Groups/ Factions: The agents belong to factions, which have resources, hierarchies of leadership, followers. The factions that agents belong to, as well as the agents themselves, maintain dynamic relationships with each other. The relationships evolve, or get modified, based on the events that unfold, blames that are attributed etc.

The resources of each group serve as barometers of the health of that aspect of the group's assets – (1) political goods available to the members (jobs, money, foodstuffs, training, healthcare etc.); (2) rule of law applied in the group as well as level and type of security available to impose will on other groups; and (3) popularity and support for the leadership as voted by its members.

Economy and Institution Models: As in the real world, institutions in the virtual world provide public goods services, albeit imperfectly owing to being burdened with institutional corruption and discrimination.

The economic system currently in FactionSim is a mixture of neoclassical and institutional political economy theories. The institutional models describe how different public goods are distributed (often imperfectly) among competing (or cooperative) factions. The public goods themselves are tied to the amount of resources for the institutional functions, including the level of inefficiency and corruption.

Institutions are used as a mediating force which control the efficiency of certain services and are able to be influenced by groups within a given scenario to shift the equitableness of their service provisions. Political sway may be applied to alter the functioning of the institution, embedding it in a larger political-economy system inhabited by groups and their members. However, the followers of each group represent demographics on the order of millions of people. To handle the economic production of each smaller demographic, a stylized Solow growth model is employed: Solow (1956).

Each follower's exogenous Solow growth is embedded inside a political economy which endogenizes the Solow model parameters. Some parameters remain exogenous, such as savings rate- which is kept constant through time. As savings rates are modeled after the actual demographics in question and the time frame is usually only a few years, fixing the parameter seems reasonable.

Each follower demographic's production depends on their constituency size, capital, education, health, employment level, legal protections, access to basic resources (water, etc), and level of government repression. These factors parameterize the Solow-type function, in combination with a factor representing technology and exogenous factors, to provide a specific follower's economic output. The economic output of followers is split into consumption, contribution, and savings. Consumption is lost, for the purposes of this model. Savings are applied to capital, to offset depreciation. Contribution represents taxation, tithing, volunteering, and other methods of contributing to group coffers. Both followers and groups have contributions, with groups contributing to any supergroups they belong to. Contributions, transfers and spoils of other actions (e.g. attacks) are the primary source of growing groups' economy resources.

The unit of interaction is the institution as a whole- defined by the interactions between it with groups in the scenario. An institutions' primary function is to convert funding into services for groups. Groups in turn, provide service to members. In turn, each group has a level of influence over the institution- which it leverages to change the service distribution. Influence can be used to increase favoritism (for one's own group, for example) but it can also be used to attempt to promote fairness.

Groups, including the government, provide funding and infrastructure usage rights. Institutions are controlled by one or more dominant groups and there can be multiple competing institutions offering the same services. The distribution of services is represented as a preferred allotment (as a fraction of the total) towards each group. Institutions also are endowed with a certain level of efficiency, reflecting dollars lost in administration or misuse. The types of institutions currently modeled are public works, health, education, legal protections, and elections. For example, public works provide basic needs, such as water and sanitation; legal protections represent the law enforcement and courts that enforce laws; and the electoral institution establishes the process by which elections are performed, and handles vote counting and announcement of a winner.

Synthesis of Theories: Our modelling framework and the software were developed over the past ten years at the University of Pennsylvania as an architecture to synthesize many best-of-breed models and best practice theories of human behavior modeling. A performance moderator function (PMF) is a micro-model covering how human performance (e.g., perception, memory, or decision-making) might vary as a function of a single factor (e.g., event stress, time pressure, grievance, and so on.).

PMFserv synthesizes dozens of best-of-breed PMFs within a unifying mind-body framework and thereby offers a family of models where micro-decisions lead to the emergence of macro-behaviors within an individual. None of these PMFs are "home-grown"; instead they are culled from the literature of the behavioral sciences.

These PMFs are synthesized according to the inter-relationships between the parts and with each subsystem treated as a system in itself. Elsewhere we have discussed how the unifying architecture and how different subsystems are connected (Silverman et.al. 2009a, 2010a). Some of the salient models currently in the library are listed in the following Table.

Table 1. Examples of Sub-Models and Theories in PMFServ Library

Layer	Functionality	Best of Breed Theories
Cognitive Agent	• Leader Personalities • Follower Archetypes/Norms • Agency Ministers (Corruptible) • External Actors	• OODA Loop (Boyd), Sense/Think/Act (Newell & Simon) • Perception, Focus of Attention (Gibson, Hammond) • Stress/Five Coping Styles (Janis-Mann), Bounded Rationality (Simon) • Personality Traits, Cultural Values, Religious Values, Tribal Norms (Hermann, Hofstede, House:UN Globe Study) • Affective Reasoning, Cognitive Appraisal, and Relationship Management (Ortony) • SEU, Multiattribute Utility Functions (Kahneman, Keeney & Raiffa)
Socio Politico Group	• Factions, Factional Roles • Government, Bureaucracy • Institutional Agencies (Security, Utilities, Elections, etc.) • Resources & Economy • Alliances, Coalitions, Votes • Battle Simulator	• InGroup Membership, Loyalty, OutGroup Bias (Eidelson, Hirshman) • Motivational Congruence, Mobilization, Grievance (Wohl, Collier) • Actual vs. Perceived Group Power/Vulnerability (Machiavelli, Eidelson) • Institutions, Public Goods (Rotberg, Collier, Fearon/Laiton) • Developmental Econ (Lewis/LRF, Solow, Harrod-Domar)
Population Grid	• Spatial layout: groups& regime • Message Filtering/ Propagation • Voting	• Regime spread, Territorial ownership, acquisition, maintenance • Cultural maps, group spatial distribution, segmentation, infrastructure, roads/ports • Line of sight and hearing, spread of news (spin) • Temporality – real time vs. faster than realtime

Metrics Designer and Calculator: Finally, the description will not be complete without saying a few words about the metric system (we call it SAMA). In order to measure, calculate and summarize the performance of the virtual world, there is a layer of standalone metric system sits a top of the database that tracks and stores output and can use lower level parameters to calculate higher level abstract parameters (e.g. EOIs). The metric system is hierarchically organized, showing high level summary metrics that the users deem worth tracking (e.g., equilibrium shifts) computed from combining different summary indicators and trends.

The above description of agents and their world is an oversimplification, but might serve the purpose of introducing the model. In-depth introduction to the framework, including the workings of the modules and the modelling methodology, have been published earlier (Silverman 2009, 2008, 2007a, 2006,). Elsewhere in other publications (Silverman et.al., 2006a, 2007a,b, 2009, 2010a), we have also discussed how these different functions are synthesized, review how agents are profiled, and how their reasoning works to make decisions etc.

Using PMFserv agents and FactionSim, we have developed a comprehensive, integrated, and validated agent-based country modeling generator to monitor, assess, and forecast national and sub-national crises to support decisions about allocating economic, political, and military resources to mitigate them: see Silverman et al. (2009).

With this framework, our collection of country models, CountrySim, can best be described as a set of complex agent-based models that use hierarchically-organized and cognitive-affective agents whose actions and interactions are constrained by various economic, political, and institutional factors. It is hierarchically organized in the sense that the underlying FactionSim framework consists of a country's competing factions, each with its own leader and follower agents. It is cognitive-affective in the sense that all agents are 'deep' PMFserv agents with individually tailored and multi-attribute utility functions that guide a realistic decision-making mechanism. Country-Sim, despite its apparent complexity, is an agent-based model that aims to show how individual agents interact to generate emergent macro-level outcomes.

Even though such models can not be solved mathematically, we can find solutions through validated simulation models with deep agents. If one could find clusters of parameters that pertain to a corresponding game model, we can also start talking about correspondence between game theoretic models and cognitively deep simulation models. There is room for a lot of synergy.

2.3   Country Modeling Case

Using the framework discussed earlier, CountrySim/ FactionSim/ PMFserv, we have constructed, a country based on multi-resolution agent based approach. This model has a virtual recreation of the significant agents (leaders, followers, and agency ministers), factions, institutions, and resource constraints affecting a given country and its instabilities.

Currently four country models are complete (Bangladesh, Sri Lanka, Thailand, and Vietnam) and seven more are underway (China, India, Indonesia, Japan, North Korea, Russia, and South Korea). We are currently populating our CountrySim framework with information about the domestic politics and economics of these countries using social science databases, expert surveys, and automated data extraction technologies, and, in so doing, building additional virtual countries and running simulations to monitor and forecast instabilities and experiment with actions that influence the direction of future political and economic developments. An important part of CountrySim is its end-to-end transparency and drill-down capability – from the front end model elicitation (web interview, database scraping) to the backend metrics views and drill down through indicators to events and even to the ability to query the agents involved in the events.

It is a tool that allows conflict scenarios to be established in which the factional leader and follower agents all run autonomously and are free to employ their microdecision making as the situation requires. A diagrammatic representation of an example CountrySim model is given in Figure 1. For additional details, see Silverman et.al. (2010, 2007a, 2008a).

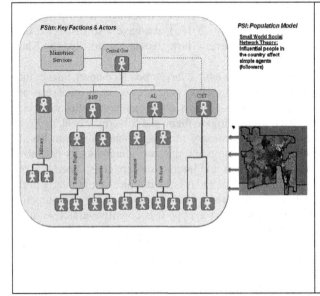

For a given state being modeled, CountrySim uses FactionSim (and PMFserv) typically to profile 10s of significant ethno-political groups and a few dozen named leader agents, ministers, and follower archetypes. These cognitively detailed agents, factions, and institutions may be used alone or atop of another agent model that includes 10,000s of lightly detailed agents in population automata.

Fig. 1. Overview of the Components of a CountrySim Model: Bangladesh (Silverman et.al. 2010)

2.4   Modeling Methodology

In recent years, modeling methodologies have been developed that help to construct models, integrate heterogeneous models, elicit knowledge from diverse sources, and also test, verify, and validate models. A diagrammatic representation of the process is given in Figure 1.

These models are knowledge based systems, and to a significant extent the modeling activity involves eliciting knowledge from subject matter experts as well as extracting knowledge from other sources such as data bases and event data, consolidating the information to build a model of the social system. We designed and tested the Knowledge Engineering based model building process (KE process) to satisfy the following functional requirements: (1) integrate individual social science models, (2) gather and systematically transform empirical evidence, tacit knowledge, and expert knowledge into data for modeling; (3) construct the model with the aim of reducing human errors and cognitive biases (e.g. confirmation bias) in constructing the model; and (4) verify and validate the model as a whole, (4) run model and explore the decision space, and (5) maintain the knowledge base over time.

We have designed an extensive a web interview which elicit subject matter expert inputs to construct the country models. One such model and its output has been shown below. Later, we triangulate these against other sources of inputs such as databases such as World Value Survey and events generated by automatic and manual scraping of webs and documents.

The details of the process are beyond the scope of this paper, but can be found elsewhere (see for example, Silverman & Bharathy, 2005; and Silverman, Bharathy and Kim, 2009). We recap the salient features briefly here.

Data Collection and Model Construction: These models are knowledge based systems, and to a significant extent the modeling activity involves eliciting knowledge from subject matter experts as well as extracting knowledge from other sources such as data bases and event data, consolidating the information to build a model of the social system. We have employed web newsfeeds, country databases, and SME interviewing.

We assembled a compendium of 45 country and social science databases (eg., CIA Factbook, World Values Survey, Global Barometer Survey, etc.), and some of the parameters pertinent to our model (e.g. population level and economic parameters) are available in the databases. Likewise, web newsfeeds provide supplementary material on the events of interest in the target countries. However, most important source of information is Subject Matter Experts (SMEs). In order to elicit knowledge from country experts, we have designed an extensive survey questionnaire containing structured questions. We supplement the structured survey with open ended interviews to capture the richness and intricacies. However, for our purposes, administering a structured, self-explanatory web survey tailored to elicit exactly the information we need would in most cases be preferable to conducting unstructured, open-ended interviews (partly because these interviews would elicit a wealth of information that would then need to be sorted and coded).

The input data, obtained from multiple sources, tend to be incomplete, inconsistent and noisy. Therefore, a process is required to integrate and bring all the information together. We employ a process centered around differential diagnosis.

This design is also based on the fact that directly usable numerical data are limited and one has to work with qualitative, empirical materials. Therefore, in the course of constructing these models, there is the risk of contamination by cognitive biases and human error.

The burden of this integrative modeling process is to systematically transform empirical evidence, tacit knowledge and expert knowledge from diverse sources into data for modeling; to reduce, if not eliminate, the human errors and cognitive biases (for example, for confirming evidence); to ensure that the uncertainties in the input parameters are addressed; and to verify and validate the model as a whole, and the knowledge base in particular. For lack of a better term, the process has been conveniently referred to as a Knowledge Engineering (KE) process due to extensive involvement of KE techniques and construction of the knowledge models. Such systematic approach increase confidence in the models. Details of data sources, their limitations and rationale for selecting and combining data have been discussed in our paper (Silverman, Bharathy and Kim 2009).

Validation: In all modeling activities, in order to be confident that the models are performing adequately (for the purpose), we needed to carry out:

- Verification: Determining, using training starting conditions that the model replicates the training data,
- Validation: Determining, using test starting conditions, the model replicates the test data independent of the model building training data, and
- Sensitivity Analysis: Evaluating the sensitivity of the model to changes in the parameters.

Having built the model, one must test, verify and validate the model. Our approach to validity assessment is to consider the entire life cycle and assess the validity under four broad dimensions of (1) methodological validity, (2) internal validity, (3) external validity, and (4) qualitative, causal and narrative validity. In the ensuing section, we provide a summary of validation. Elsewhere (Bharathy and Silverman, 2010; Bharathy and Silverman, 2012), we elaborate on selected validation techniques that we have employed in the past. We recommend a triangulation of multiple validation techniques, including methodological soundness, qualitative validation techniques such as face validation by experts and narrative validation, and formal validation tests including correspondence testing.

As a social system built primarily of cognitively detailed agents (such as PMF Serv based CountrySim), our models provide multiple levels of correspondence. At observable levels, the models are evaluated for correspondence in behaviors (decisions agents make) and measurable parameters (e.g. GDP, public goods service levels received). They are also evaluated at higher levels of abstractions where aggregated and abstract states of the world (developmental metrics, conflict metrics such as rebellion, insurgency) are compared. During the validation exercise, we attempt to re-create the behavior of the reality using fresh set of empirical evidence hitherto unused in model construction.

Below, we very briefly describe some example validation tests, namely external validity at micro and macro levels.

Micro-Level Validation: In carrying out a micro-level validation process, we primarily aim to create correspondence at the level of agent decisions or other lower level parameters such institutional parameters, socio-economic indicators etc. The intention is to calibrate the model with some training data, and then see if it recreates a test set (actually validation). Examples of this include the mutual entropy and chi-squared pairwise tests, which have been described in Bharathy and Silverman (2012). The mutual entropy correspondence was between the decisions of real and simulated leaders. The leader decisions were classified into categories (e.g. positive, neutral, negative); we were able to calculate a mutual information or mutual entropy (M) statistic between the real and simulated base cases. The mutual entropy values were found to be less than 0.05 (at least an order of magnitude smaller than the mutual entropy of 1.0), indicating reasonable degree of correlation between real and simulated data. Several historical correspondence tests indicate that PMFserv mimics decisions of the real actors/population with a correlation of approximately 80% (see Silverman et al. 2009b, 2007b, 2010).

Macro-Level Validation: Let us summarize Macro-Level and Cross Model Validation test that also directly deals with probability estimates. The direct outputs from the CountrySim model include decisions by agents, levels of emotions, resources, relationships between factions, membership of agents in different factions etc. These parameters are tracked over time and recorded in a database. Based on these, we have defined aggregate metrics or summary outputs of instability called Events of Interest (EOIs). These are, namely: Rebellion, Insurgency, Domestic Political Crisis, Inter-Group Violence and State Repression.

We will describe EOI composition later, but it suffices to say that these EOIs are composed of weighted aggregations of model level outputs.

The indicators we used are items computed by the socio-cognitive models and which can be interpreted as a function of the EOI they characterize. Forthcoming Figures of hierarchical indicator composition (Figure 4) and example indicators (Figure 7) show how model outputs are summarized into indicators that in turn are used to compute EOIs.

During the training period, using the weights on the arcs of the tree, the occurrence of EOIs in the simulated world can be tuned against the occurrence of EOI in the real world. The weights are then employed to make out-of-sample predictions in the test period. The weights tend to be invariant across similar countries (in our case, all countries).

In order to assess Analytical Adequacy, we ask whether the collection of models assembled and implemented thus far satisfy various types of correspondence tests and historic recreation tests. This will often entail backcasts on a set of historic test data with-held during model training and tuning. And to avoid the problem of over-fitting to a single test sample, we always need to examine if the models work across samples.

Here we applied them to models of several States (namely Bangladesh, Sri Lanka, Thailand and Vietnam) and Groups, People and across different types of metrics of interest (different EOIs). The following sample results were drawn from one of our previous paper (Silverman et.al., 2010) and illustrates one EOI for one Country.

Having constructed high level aggregate EOIs, we compared them to Ground Truths of EOIs coded from real data by subject matter experts. Having tested and verified the model over the period of 1998-2003, we ran the model for subsequent 3 years (of 2004 through 2006) and made predictions. The predictions were benchmarked against the Ground Truth consisting of real world EOI for the same interval.

Although we generate and display the multiple futures (from multiple runs), in metrics and calculations, we only employ the mean values across alternative histories for validation. We cast mean likelihood estimates from multiple runs into a binary prediction by employing threshold systems.

In order to get a quantitative relationship between CountrySim and Ground Truth forecasts, we make use of a Relative Operating Characteristic (ROC) curve (Figure 2). The ROC plots the relationship between the true positive rate (sensitivity or recall) on the vertical and the false positive rate (1-specificity) on the horizontal. Any predictive instrument that performs along the diagonal is no better than chance or coin flipping. The ideal predictive instrument sits along the y-axis.

The predictions were benchmarked against the Ground Truth consisting of real world EOI for the same interval. Below, we will see that based on this Ground Truth benchmark, our metrics such as precision, recall and accuracy are mostly in the range of 65-95% for multi-country, multi-year study (Bharathy and Silverman, 2012).

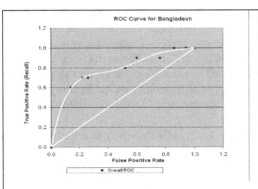

Fig. 2. Relative Operating Characteristic (ROC) Curve for CountrySim (Bangladesh) (Courtesy: Our paper Silverman et.al. 2010)

Definitions

■ Accuracy = (TP+TN)/(P+N)
■ Precision = TP/(TP+FP)
■ Recall/TP rate = TP/P
■ False Positive (FP) Rate = FP/N

Where T: True, F: False, P: Total Positives and N: Total Negatives, TP: True Positives, TN: True Negative, FP: False Positives and FN: False Negatives. ROC curve shown on the left describes the relationship between Recall and FP Rate as FP Rate is varied.

Table 2a. EOI Summary Metrics

Metric for Bangladesh	Accuracy	Precision	Recall
Mean– with two thresholds at 0.65-0.35	87%	66%	81%

Table 2b. Confusion Matrix

		True Class	
		+ve	-ve
Hypothesized Class	Y	TP	FP
	N	FN	TN

This curve well above the diagonal shows that CountrySim largely agrees with the Ground Truth. In fact its accuracy measured relative to Ground Truth is 80+%, while its precision (which is lower at 66%) and recall were listed in the Table 2a. While these would be less than luster results for any physical system, for agent-based models of bottom up social science processes, these are useful results. They are useful both since they significantly beat coin tossing and since these type of models also afford the analyst ways to drill down to try and explore casual factors as we will explain forthcoming subsections. In this case, we employed backcasts with set of data independent of model construction, but eventually, one should move to forecasts and to tracking the actual outcomes to verify the forecast quality.

It is also worthwhile mentioning that accuracy does not distinguish between the types of errors it makes (False Positive versus False Negatives). On the other hand, precision and recall do not stand alone and require to be combined with accuracy. Generally speaking, ROC curve is a comprehensive measure. Yet, there are times when ROC Analysis and Precision could yield contradictory results (e.g. it gives majority class an unfair advantage when proportion of classes are significantly unbalanced). The implication of all these is that one must understand the data and the domain it pertains to before carrying out analysis. That is, there is no substitute for qualitative domain knowledge.

Additional details on the validation of our models, along with examples, have been discussed in our paper (Bharathy and Silverman, 2010). We subscribe to the view that no model can faithfully represent the reality, but detailed, mechanism based models are useful in learning about the system and bringing about a qualitative jump in understanding of the system it tries to model.

3   Social System Model Contributing to Risk Framework

The previous section briefly described our modeling framework, introduced the modeling methodology and set up a case of country study. In this section, using this framework, we will illustrate the risk management framework through this case.

The cardinal question is how we can assess the political risks posed by a social system such as a country through a socio-cognitive modeling framework. In doing so, we will adopt the common risk management structure recommended by a number of coveted risk management standards and organizations (NRC-NAS, 1983; ANS/NZS 4360: 1995). Specifically, we will describe the process along the following steps:

- Risk Conception and Hazard Identification, illustrating conceptualizing and formulating the problem and identifying hazards and factors that contribute to risk are described in the first sub-section,
- Risk Assessment, assessing Probabilities and Consequences of hazards, and Evaluating and Integrating Risks are described in the second sub-section,
- Risk Treatment, illustrating an intervention for mitigating risks, monitoring ongoing process to detect signals surrounding risks drivers and inform of new risks potential, and communication and perception of risk are described in the last sub-section.

3.1   Risk Conception and Hazard Identification

In this stage of the risk assessment, we conceptualize and formulate the problem and identify hazards and factors that contribute to the one, perhaps arguably the most, significant aspect of political risk, the one that arises from the instability.

Events can have negative impact, positive impact, or both. Typically, events with a negative impact are referred to as hazards. Hazard identification involves identifying internal and external events potentially posing a risk or affecting achievement of the objectives, which in this case is healthy functioning of the countries indicated by the absence of instability events.

Pertinent to the scope of political risk assessment, we focused on the external factors such as instability and violence, illustrating how one could gain insight about the risk of such events from the country modeling case study.

Conceptualization and hazard identification are arguably the most important parts of risk management, not only because everything else follows from this, but also owing to the influence of these steps on the management of risk.

When assessing risk through a social system model, two steps of conceptualization and hazard identification are inextricably linked. Therefore, we have discussed them together, breaking from the tradition in a number of mechanism based risk studies, which separate the two steps.

During the conceptual stage, we began by generating an exhaustive list of factors that contribute to instability and disruption. However, we organized these factors into a conceptual model through influence diagrams. In obtaining inputs for the model, we employed surveys and interviews with stakeholders and subject matter experts, and collected information on the system and environment, and established the set of hazards. Once the model was built, like Hazop studies (Kletz, 2006) frequently employed in hazard assessment, we traversed the model and posed systematic questions of what-ifs. We indentified the following summary hazards:

Our central assumption is that instability in a country could be caused by several (interacting and inter-related) factors, including structural and behavioral factors, and the timing of conflict is influenced by frictional or inertial factors and triggers. Structural and Institutional factors include the Institutional/ Regulatory capacity (and vulnerabilities) of a country along social, technical, economic, environmental, and political (STEEP) dimensions. Institutional factors describe the country's political, social, and other institutions that can influence stability. Behavioral factors include the characteristics (personalities, norms, culture, grievances, etc) that motivate a country's leaders, factions, and general population. Other inputs include triggers or shocks to the system such as a tsunami or other key events. Inertial/ Frictional factors are result of inertia or underlying cost to change. Emergent factors on the other hand are due to system level effects. Finally, there are trigger certain factors/ events/ actions that would push the states into conflict. E.g. economic recession, change in the distribution of power, events of violence etc that could accelerate the conflict and instability. Such an abstract view of the conflict can be represented through the fault tree shown in Figure 3a.

Frequently, policymakers focus on symptomatic issues and the triggers of conflict, in an attempt to control a complex system; that is not an easy task to say the least. While these individual, symptomatic trigger factors are helpful in monitoring conflicts (especially lagging or coincident indicators), they are difficult to control per se. Besides, many triggers could bring about an avalanche when the system has precursors.

On the other hand, we approach it based on traditional model of risk, as described in the classic Swiss Cheese Model[3] (Reason, 1990). A system with inherent weaknesses is a precursor to conflict, and is an accident (instability event) waiting to happen; it does do little good to focus on symptomatic solutions. It is not the event itself, but it's the conditions under which they occur that really matters. A significant conflict is a result of several failures in the system, as shown in the fault tree of rebellion in Figure 3b. Let us look at the salient features of this fault tree. The tree identifies several factors (failures in the positive sense) that might lead to conflict. For the sake of simplicity, the likelihoods of these factors are probabilistically combined through "AND" and "OR" gates to estimate the propagation of probabilities of conflict. The escalation of the conflict proceeds through several "gates" and requires multiple factors to have failed in the system. It [fault tree] identifies a broad-brush leadership and institutional failure as contributing to multitude of causes.

---

[3] According to Swiss Cheese model, a system's defenses against failure are modeled as "a series of barriers, represented as slices of Swiss cheese". Accidents occur when the system as a whole fails to provide (frequently more than one) these barriers.

Let us start with the event(s) that a leader of a country discriminating against a minority group has been occurring frequently. Typically, in a nation with mature institutions, there will be institutional mechanisms to address and redress an issue such as discrimination, should it be identified. On the other hand, in a country with corrupt and weak institutions, adequate balancing mechanism may not exist. However, poor institutional framework alone may not be enough to start a separatist conflict.

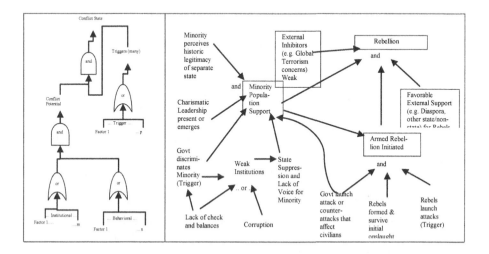

Fig. 3a. Generic and Abstract Fault Tree of Conflict

Fig. 3b. Fault Tree[4] (Network) Representation of Rebellion with Grievance[5] Motivations

In the absence of voice and equality, the members of the discriminated groups might become disenchanted with the leader. However, for them to mobilize themselves and act as collective, they will need a leadership or nucleus of their own. Such competing or opposition leader can politicize and mobilize the masses to overcome the inertia of collective action problem, and take collective actions.

Some of the other factors that might contribute to mobilization by public to support an event of interest such as rebellion (separatism) include perceived legitimacy of separation, acceptability of costs and benefits including status quo.

An armed movement may be formed, but it has to not only survive the initial onslaught by the government, but also need to have public support (from the minority).

---

[4] It must be noted that fault trees tend to assume that each of the leaf nodes (each branch) is independent. In reality, this is often violated. E.g. node pertaining to institutional lack of balance would contribute to more than one parent nodes, making it a network. One may circumvent such difficulties through the use of conditional probabilities based on the those nodes (or in general using Bayesian statistics).

[5] The tree can be expanded to include greed motivations too. See below text for grievance-greed motivations of conflict.

There must also be behavioral incidents such as attack by rebels on the government and that by that should trigger the conflict.

If the current level of security is threatened by a government bombing civilians, then joining a violent separatist movement is not considered to increase the risk significantly.

When a group or section of the population with ties to each other (and less ties to others) regard the system as illegitimate, and when expressing/ voicing grievances result in serious backlash from the incumbent leadership, this could lead to attempts in the form of exits leading to separatist wars. When leadership unleashes violence on the members, this can quell the dissident action, but does not quench the grievances. Once conflict finds a way to begin again, the vicious cycle may resume.

That does not answer the question why the leaders who strive to maximize their power and resources (reduce vulnerabilities), directly or indirectly supply various public and private goods, gain legitimacy or engage in abusive or discriminatory relationship with the members of the group as well as other groups. For this, one has to look at how the decisions are being made by individuals and groups. That is where our model particularly pitches in that other top down systems would find it difficult to do.

In addition to the indirect consequences mentioned above, the act of discriminating another group would have direct consequences to the leader such as decrease in relationship with this group, and perturbing the social or economic equality between groups; it may still pay off in terms of political gains in the leaders' own constituency. The leaders acting myopically compete to win the affections of one group, deliberately or inadvertently discriminating against another group.

The relationship between institutions and conflicts is also significant. Weak, corrupt institutions or discriminatory policies are precursors to conflict. On the other hand, conflicts also weaken the institutional systems, resulting in a vicious cycle. When institutions perform their function well, it strengthens the legitimacy of the leadership and increases the political strength, resulting in a virtuous cycle.

The triggers are certain factors/ events/ actions such as economic hardship (in the extreme cases collapse), change in the distribution of power, events of violence etc that could accelerate the conflict and instability. While these triggers are helpful in monitoring conflicts to a limited extent (they are also difficult to predict or control per se), multitude of such triggers could bring about an avalanche when the system has precursors. Therefore, one must pay attention to the system with weaknesses and precursors of vulnerability, for accidents are waiting to happen in such vulnerable systems (not a matter of whether, but when).

Thus, we briefly described one thread or sequence of events pertaining to rebellion in the fault-tree. It is too naïve to assume that a fault-tree such as above depicts the conflict. Typically, in a country model, there are several such sequences of escalating events that lead to conflict. Frequently, these sequences can be competing, and at times seemingly inconsistent.

According to some key theories, a conflict can be generated and sustained, for example, by scarcity of resources, inequitable and mal-distribution of otherwise abundant resources, as well as the speed of access to resources, grievances accrued due to various resource based and non-resource based conflicts in the past. In particular, the "greed vs. grievance" argument has become important over time.

Many researchers including Collier and Hoeffler (1998, 2002), Bates et al. (2002), Fearon and Laitin (2003), Regan and Norton (2005), Elbadawi and Sambanis (2000, 2003), have explained conflict using the economic factors. Specifically, Collier and Hoeffler (2002) proposed the 'greed' based explanation that internal wars may be motivated more by opportunity than 'grievance', and incentivized by the availability of resources that could be 'conquered'/ 'looted'. Several statistical modelers, for example Imai and Weinstein (2000), Morrison and Stevenson (1971) and Jon-A-Pin, (2008b), determined that political instability and economic growth are negatively correlated.

Several other studies attribute non-instrumental issues, such as intergroup discrimination etc., occur even in the absence of a realistic pretext for the behavior (Billig & Tajfel, 1973; Rabbie & Horwitz, 1969; Rabbie & Wilkens, 1971; Tajfel, Billig, Bundy, & Flament, 1971). Similarly, researchers observe that "The mere categorization of persons into groups on the basis of an arbitrary task is sufficient to cause discrimination in favor of members of one's own group (ingroup) at the expense of members of the other group (outgroup)". Atran (2004) distinguishes the non-instrumental "value rationality of religions and transcendent political ideologies" from the "instrumental rationality of realpolitik and the marketplace". Frequently, virtuous (e.g. altruism), cultural (including for example maintaining honor, exacting revenge) or religious behavior are their own reward and are not subject to trade-off.

Postulating that one or relatively handful of factors[6] accounting for a conflict (e.g. 'greed' vs. 'grievance'), seem simple and elegant worldview, but the truth is messier.

In situations such as the above, each of the fine, but single dimensional, views can be limiting to really understanding the complex system as a whole, and it is helpful to permit competing mechanisms to operate simultaneously, as we do. In our models, for example, greed and grievance mechanisms coexist and depending on the context one or other might predominate or be subservient to some other mechanism at play. On a blank slate, it is not possible to make predictions, even for many experts, which of these conflicts would arise. However, subject matter experts are capable of defining the conditions that prevail on the ground. This is also where an agent based model with detailed agents is helpful, as it permits bottom up design of the system and then later observe the mechanisms operating. A model can also help structure a hazard identification session.

Measuring Conflict and Hierarchically Organizing Hazards/ Factors: In order to formulate a model of hazards, we hierarchically organize the pertinent events, indicators and other factors. At the very top level, the key hazards (or super-hazards) could be conceived as the Events of Interest (EOIs) themselves. Since our intention is to model instability, we have defined instability in five different dimensions, namely Rebellion, Insurgency, Domestic Political Crisis, Inter-Group Violence and State

---

[6] Such views demand various simplifications for mathematical elegance. For example, some researchers attempt to cast non-instrumental values as instrumental values when justifying "greed" based mechanism.

Repression. EOIs reveal a high-level snapshot of the state of the conflict. Specifically, CountrySim generates several EOI scores important to instability. The definitions of these EOIs are given below (O'Brien, 2010).

> "*Rebellion*: *Organized opposition whose objective is to seek autonomy or independence.*
> *Insurgency*: *Organized opposition by more than one group/ faction, whose objective is to usurp power or change regime by overthrowing the central government by extra legal means.*
> *Domestic Political Crisis*: *Significant opposition to the government, but not to the level of rebellion or insurgency (e.g. power struggle between two political factions involving disruptive strikes or violent clashes between supporters).*
> *Inter-Group Violence*: *Violence between ethnic or religious groups that are not specifically directed against the government.*
> *State Repression*: *Use of government power to suppress sources of domestic opposition.*"

In this subsection, we describe our EOI Framework. This EOI Framework has its theoretical basis in a premise that conflict can be measured through a composition of indicators, which include both behavioral and structural or institutional factors (Covey, Dziedzic and Hawley, 2005; Baker, 2003)[7]. Measures of performance of the virtual world will inevitably need to be hierarchical showing high level summary metrics that the users deem worth tracking (e.g., equilibrium shifts) computed from combining different summary indicators and trends. These indicators in turn will summarize the events that transpired and transactions that occurred in the agent interactions.

In our framework, in order to measure if these EOIs are happening in a real or simulated world, the EOIs are composed of, and are calculated from, a set of lower level indicators. Unlike EOIs which are more abstract metrics, indicators are quantifiable and tangible measurements that reflect the EOIs. These indicators tend to constitute the sub-level of hazards and frequently provide leading indicators; Occurrence of these indicators are associated with, and portend, the occurrence of the super hazards. Typically indicators are count up events of that type, or averaged values of the parameter as the case may be, that arise across time in either the real or simulated world. For example, in order to determine the occurrence of EOI Rebellion in the simulated world, we would look at the indicators, which are direct model outputs.

By definition, the indicators are causally related to the EOI they characterize, which makes them relevant as predictors. For instance, three of the leading indicators of rebellion were: (a) claims of discrimination made by followers (members) of an out-group, (b) low intensity military attacks on an out-group by the Central Government (or state apparatus), and (c) number of high intensity attacks on the Central Government (or state apparatus) by out -group or vice versa. Likewise, two of the leading indicators of an Insurgency are: (a) the extent of mobilization among dissident

---

[7] Another framework that was developed when we developed at the same time as ours is MPICE Framework (Dziedzic, Sotirin, and Agoglia, 2008), which has a framework for measuring progress in conflict environments, while the EOIs that we designed are for measuring the conflict itself.

in-group against the Central Government (or state apparatus) and (b) the extent of corruption at the highest levels of the government. One may continue the drill down to find lower levels of hazards or indicators of conflict.

An example hierarchy has been in the Figure 4. Also, the Figure 7 shows that the insurgency EOI indicators. In general, we found it useful to use about 1-2 dozen indicators per EOI to adequately count and track events coming from the simulated world.

One of the uses of a model is in designing the metrics themselves. The metrics to measure and track are not always very clear upfront, and often need to be clarified through the iterative process of trial and error[8]. For us, the metrics layer is located, and is defined, atop of the standard model of the social system. Such a system offers twin benefits: Firstly, the models give a concrete framework and structure against which a metric system could be conceptualized. Secondly, what-if-experiments (for example, through Monte Carlo simulations) can be carried out with different indicators to identify those indicators that are policy relevant and those that the model is sensitive to, thereby, iteratively designing the metrics without incurring the cost of actual program. This works in both the ways. Not only the model will help clarify the metrics, but metrics will also help anchor the model to reality.

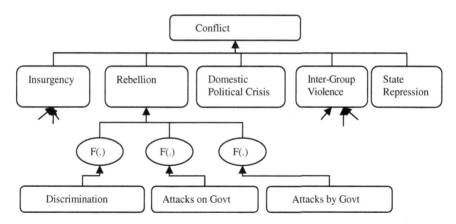

Fig. 4. Hierarchical Design of Metrics of EOIs

3.2 Risk Assessment

The risk was assessed as a combination of the hazard likelihood and the consequences, consistent with the approach recommended in the Australia / New Zealand Standard AS/NZS 4360:1999 (Standards Australia, 1999). In essence, the assessment stage involved calibrating and wherever possible, estimating likelihood and impact of risks.

---

[8] In some case, there are some well known metrics and indicators to follow in any given region, while in other regards this is an open research question and design inquirer will tend to research and evolve his/ her own set. The selection of indicators should take into account not just their statistical performance, which is correlation (as the saying goes, does not imply causation), but as far as possible mechanistic understanding of causality.

As risk analysts, we would also design the assessment process to be amenable to gradual drill down, and carry out the assessments commensurate with the level of risk, level of response action desired, as well as the scope of the project. The actual depth covered by the assessment should vary depending on the nature of activities and potential for harm from hazards identified and ranked.

In a typical risk assessment that is not supported by a virtual world model, one would carry out a screening analysis for selecting the priority risks. Even as modelers, while constructing the models, we had systematically, but implicitly, identified EOIs and factors that would contribute to hazards and assessed the potential hazards and the associated risks. This activity should be regarded as the preliminary analysis, where identified and accounted for all hazards.

So, in the interest of space, we do not described preliminary assessment any further, except to say that model construction process itself is an assessment (albeit implicit) commensurate with the level of risk, level of action desired, as well as the scope of the project.

Special attention was paid to relevant scenarios that may lead to a significant exposure to risk. In the following sub-section, we will look at snippets of detailed assessment. Detailed assessment involved calibrating and wherever possible, creating probability distributions for all identified risks and estimating their consequences.

In the previous sections, we presented our framework, model, summarized model construction process, and provided a brief demonstration of validation. Now, we would like to deploy this validated model to demonstrate risk assessment.

Estimating Probabilities: The direct (or default or base) outputs from the Country-Sim model include decisions by agents, levels of emotions, resources, relationships between factions, membership of agents in different factions etc. These parameters are tracked over time and recorded in the database. Since our intention is to model instability in selected countries, we defined aggregate metrics or summary outputs of instability from default model outputs.

As mentioned earlier in validation, high level aggregate EOIs were compared to Ground Truths (of EOIs). Having constructed and validated the model, we employ the same to make predictions. Below, we show probabilities generated by the model.

Likelihood estimates for each EOI is collected directly from the model output, and therefore, is straight forward. From the multiple runs carried out, distributions of the probabilities were estimated. Examples are shown in Figure 5.

Our prediction of likelihood of Domestic Political Crisis (DPC) matches with the Ground Truth values, but this should be consumed with a pinch of salt. There have been several local challenges to incumbent regime (government) that could be described as DPC. However, these events could also be coded as incipient stages of insurgency, and this ambiguity must be recognized.

Inter-Group Violence has limited and sporadic occurrence. During the latter part of 2005, in Bangladesh, the violent activities by religious extremists such as the JMJB group (Bangla Bhai) launched a terror campaign against other factions.

As to be expected, in Bangladesh where the treaty with the separatist group still holds albeit with some bumps, rebellion has a very low likelihood of occurrence in both real as well as simulated outputs.

We see the insurgency EOI (coup d'état) for Bangladesh is non-zero and rising.

On a tangential note, there is reasonable degree of visual correlation between our prediction of rebellion, domestic political crisis, inter-group violence and that of the respective Ground Truths (we will, however, defer discussing insurgency till later). Actual validation has been discussed separately under the Model and Framework section.

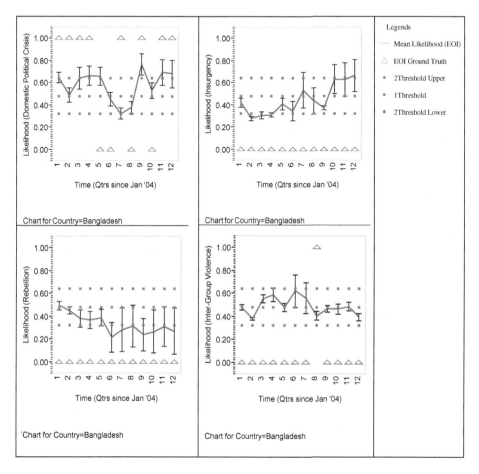

Fig. 5. A Sample of Quarterly Forecasts from our paper (Silverman et al. 2010) for Bangladesh. Each error bar is constructed using a 95% confidence interval of the mean.

Estimating Consequences: Estimating consequences are a little more complicated. Any conflict will have its impact in multiple dimensions, including economic/financial, security, social, cultural and psychological and so on. Consequences are also dependent on the party for whom the assessment is being carried out. For example, an MNC intending to invest in a country will be concerned with Value at Risk or other similar financial measure. On other hand, the individuals living in the country (including the leadership of the country) might be interested in personal, professional and societal consequences. In addition, the leadership of the country or its factions may also focus on the socio-economic consequences.

Estimation of consequences is also more subjective and requires SME or stakeholder input, which could be obtained through such instrument as a survey. We will not be adopting this approach, except to highlight this is an area that requires further exploration and research.

Instead, we will illustrate the risk assessment process through just one dimension of the consequences, namely economic impact of instability at the country level, rather than focusing on the vantage point of various actors and their agendas.

Many researchers have tested the hypothesis that instability in a country negatively affects the economic growth. This is conceivable because instability reduces the availability of capital stock (both human and economic capital), and increases depreciation through destruction and wastage of resources. It is expected to drive investments (and other resources) away from the country, thereby reducing domestic investments. The productivity will also be lower, as workers in the affected areas will be forced to divide their attention between safety-security and productivity issues. At the macro economic levels, instability will also be associated with domestic deficits, as government borrows and pays interest to sustain the war and economy. These seem obvious. Many researchers have studied these relationships through statistical analyses and other empirical techniques, but it is difficult to find studies that clearly outline the outcome.

Some researchers such as (De Haan, 2007), point to methodological shortcomings including measurement errors, treatment of time dimension and robustness of estimation in the present state of such estimation studies. One of the reasons for measurement errors is the treatment of political instability as a single dimension (Alesina et.al., 1996; Gupta, 1980; Perotti, 1996).

Some of the above studies employing principal component analysis have tried to identify different dimensions of political instability and settle for relatively few independent dimensions of instability. One such study, for example, (Morrison and Stevenson, 1971) settle for three dimensions, namely: communal instability, elite instability and political turmoil. Jon-A-Pin (2008b), employing exploratory factor analysis, has relaxed the hitherto imposed requirement that these dimensions have to be uncorrelated. We take the significant factor loadings in Jon-A-Pin (2008b) as a simple measure of relative impact, and this appears to correlate with our own assessment of conflict dimensions.

Imai and Weinstein (2000) fitted regression around the effect of civil war for the economic growth of landlocked, non-oil-exporter African countries in 1980s with average level of initial income, schooling, black-market premium, population grow rate, and ethnic fractionalization. Their regression informed that a widespread civil war may cause reduction in GDP growth rates by about 1.25 % per year [It should also be noted that Imai and Weinstein corrected for any fundamental uncertainty by using the method suggested by King et al. (2000)]. Ima and Weistein's work also help account for geographic spread of the conflict, which is a proxy for the size of the conflict. Their results in Table 3 suggest that wide-spread civil wars are five times more costly than narrowly fought internal conflicts.

Table 3. Simulated First Difference Effects of Geographic Spread of Civil War (Imai and Weinstein, 2000)

Scenario	MSPREAD (from, to)	Estimated First Difference	95% Confidence Interval		Negative First Difference
No war to Minor War	(0, 1)	0.19	-0.05	0.41	94.0%
Minor War to Medium War	(1, 3)	0.39	0.02	0.71	98.2%
No War to Major War	(0, 5)	0.95	0.19	1.69	99.2%

We were able to construct a simple model to estimate approximate consequences of instability. Firstly, we took Imai and Weinstein's estimate as the base value. We hypothesized that specific types of conflict with their contextual factors may modify this further. In general, not all conflicts are equal in terms of consequences. The economic consequences are a function of the conflict type, scale and other contextual factors. For example, geographic spread or fraction of the country affected, human toll, and strength of the insurgent groups etc. might be used to index the consequence of the instability. Given that we are concerned about significant conflicts, we omit the scale of the conflict, but the type of conflict is still significant, for which, previously, there were no breakdowns of economic impacts of conflict based on different instability types. This is where Jon-A-Pin's research (2008b) informs us about the relative contribution of each instability Events of Interest (EOI).

Table 4. Instability Types/ EOIs and Relative Consequences (compiled from Jon-A-Pin 2008b and mapped to our EOIs). Mean percentage effect on GDP has been indicated below. Adverse nature of the consequences is indicated by negative sign.

SNo	EOI [Instability Type from Jon-A-Pin]	PinFactor Mean Effect on GDP	PinFactorDevn Standard Deviation
0	Civil War [Civil War]	-0.67	0.50
1	Rebellion [Separatists' Guerrilla War]	-0.84	0.39
2	Insurgency [Coup]	-0.60	0.88
3	Domestic Political Crisis [Major Govt Crisis]	-0.45	0.75
4	Inter-Group Violence [Ethnic Conflicts]	-0.47	0.77
5	State Repression [None. Estimated]	-0.35	0.77
6	Consolidated [Generic Internal Conflict]	-0.65	0.60

We are able to approximate the economic consequences of various instability events by scaling the base level of economic consequences (of EOI_0: $C_0 = -1.25\%$ Reduction of GDP per annum) estimated by Imai and Weinstein (2000) for civil wars with relative mean effect factor for j^{th} EOI ($PinFactor[EOI_j]$) estimated by Jon-A-Pin (2008b) for types of conflict.

$$Consequence\ (EOI_j) = C_0\ *\ (PinFactor[EOI_j])\ /\ (PinFactor[EOI_0]) \qquad (2)$$

where it is known that $PinFactor[EOI_0]) \neq 0$ and C_0 refers to base consequences (i.e. for EOI_0]

We have assumed that the economic consequences can be approximated by normal distribution. This makes sense given that consequences are aggregates of numerous consequences at a lower level (law of large numbers and central limit theorem are common knowledge) even if we could not say much about the individual effects themselves. Typically, a hazard may also entail both positive and negative consequences. There is also some face validity and intuitive rationale to the above table of effects. For example, both Rebellion and Insurgency are nationally significant phenomena and tend to be associated with larger consequences; however, insurgency is associated with larger uncertainties in economic consequences; the direct economic costs of Domestic Political Crisis and Inter-Group Violence tend to be limited, although grievances of the followers and political consequences (especially the long term consequences) for the leaders are still something to reckon with. The state repression has long term adverse consequences to the performance of the economy, but at the onset or early stages of this EOI (when model prediction matters), the consequences are limited.

Risk Evaluation and Integration

Evaluation: The estimation of risk is obtained as a combination of likelihood and consequences. In this case, for the sake of simplicity, we combine as a product, similar to expectation calculations. However, we do not collapse it into a single number by aggregating, but retain the distributional nature of risk through Monte Carlo sampling and simulation to account for the distribution of inputs. We present our assessment of risk from economic perspective (see Figure 6). In Figure 6, x-axis shows the percentage effect on the GDP and y-axis shows the probabilities of these effects.

Although conflicts in general predominantly pose negative financial risks (losses) (consequences modeled as normal distributions with negative mean for simplicity), some marginal positive risk can also be seen. Risk events such as rebellion (separatist conflict) pose largely downside risks, while ambiguous events such as state repression, insurgency (coup), domestic political crisis etc. are not entirely downsided. For example, shaking up of government as in the case of domestic political crisis or a peaceful coup may not necessarily result in negative consequences. Therefore, the risks estimated for a generic conflict also shows both positive and negative risks. This is also not inconsistent with the literature.

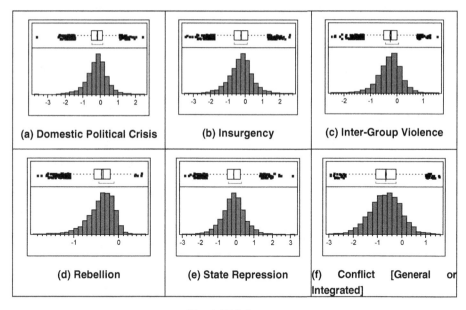

Fig. 6. Risk Outputs

Integration of Risks: In most risk projects, we encounter diverse risks. Integration of individual risks could be carried out at different levels of complexity and coupling. The simplest form of integration of risk is through weighted aggregation. While simple, it may or may not take the interaction effects between the risks into account. Alternatively, one could aggregate statistically, reflecting correlations and portfolio effects. An interesting method of combining risks at a more fundamental level is through developing a model that integrates risks.

In the case that we illustrated, we have not said anything about the independence of the top level EOIs. In reality, this may not be the case. The consolidated generic conflict is an aggregation of all conflicts and its effects.

For the sake of simplicity, the probabilities of hazards at EOI level are integrated as though they are independent events and the consequences are obtained from literature as other EOIs. However, there is more to it than that meets the eye. A significant proportion of hazards, and hence risks, already come integrated through the modeling efforts itself. While this level of integration is not illustrated at the level of EOIs, it is fundamental to the outputs below the level of EOIs. At the level of indicators, such individual hazards are combined and a consolidated risk is presented for each EOI.

Explanation of Risk and Drilldown: Earlier, in Metrics section, we described how the metrics of conflict are put together in a hierarchical fashion. In the model framework, we also described how the model itself is hierarchically organized. Such structure opens up some avenues for studying the model.

One of the advantages of using a hierarchically organized, mechanistic model such as socio-cognitive model is the ability to drill down, explore and explain the behavior, both when the model predicts correctly to provide additional information, but also when the model disagrees with reality. As an example, let us begin drilling into one of those EOIs to understand them more fully.

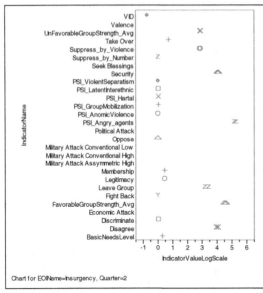

The Figure 7 shows high degrees of disagreeing and angry agents wanting to leave the government group, people mobilizing against government (unfavorable group strength building up against government), and high military buildup. We also see that Take Over event (act of coup or hostile takeover) is non-zero which implies that a group of agents have mobilized (in this case, military) to take over the government. In fact, in the real world, it is a military take over that happens in the near future, as described earlier.

Chart for EOIName=Insurgency, Quarter=2

Fig. 7. Insurgency Indicators

Reader might recall that in predicting insurgency (coup d'état) (Figure 5, top right), our model seems to show rising potential of insurgency while ground truth reports none (which is the official version of the story). That seems to be an incorrect prediction, right? Actually, there was an extraordinary change of government between Jan 11th -12th, 2007 (just after the quarter 12, just off the chart) following political turmoil. Political analysts (Economist, 2007) allege that it was a coup by the military, but the military and the country fearing international repercussions (e.g. losing peace keeping role, sanctions, aid withholding), never declared it as one. It seems that while the model does not get the timing of the insurgency right (off by a quarter), it captures the underlying mechanism (as military coup) and the leading indicators such as perception of escalating corruption, political crisis, declining legitimacy of the government (violent protests) and impending and consequent military take over. In fact, the model reports that the likelihood of insurgency as rising and does not say anything about actual occurrence.

Specifically, the reader will recall that EOI scores are computed by combining together a number of indicator variables which in turn count up the kinds of events that occurred in the model. We see that Insurgency may be explained by noting that indicators are substantially above zero and one is below. This is a logarithmic scale, so the elevated levels are quite high.

Some useful techniques for resolving such ambiguities and adding richness to model are tracing the results back to their origins (in this case, indicators, but could go deeper into sub-indicators or follow them temporally and causally), developing a story from the model outputs and obtaining qualitative feedback from subject matter experts. In presence of a virtual world, one might seek to drill into some other variable to verify that the discrepancy is valid (a form of triangulation or differential diagnosis). After all, observed reality may not always be correct in the world of partial information. Below, we present a snapshot of indicators pertinent to Insurgency.

In addition to direct drill down, an agent world such as ours provides new opportunities – both to include novel indicators, and to interrogate the simulated actors directly. We have made some limited progress towards conversational agents. These agents, demonstrated in preliminary form, can explain how they feel about the current state of the world, about their own condition, about the groups in the region and their leaders' actions, and why they took various actions and how they felt about doing so.

These are work in progress and there is room for improvement, but we have found that our descriptive approach to agent-based modeling, based on synthesis of models, is amenable to drill down and explanation.

3.3   Treatment of Risk

Risk management theory recommends (in that order) elimination, mitigation, transfer or acceptance strategies for the treatment of significant risk categories. Most commonly used strategies, however, are mitigation and transfer.

In the case of country risk assessment, our focus was on evaluating the risk due to various political events. Managing, especially treating, political risk is not an easy task to say the least. As in consequence assessment, management of risk will depend on whom the activity is being carried out for.

For a neutral outside agency or a political decision maker in the leadership position of the country, the concern might be in making interventions or changes that mitigate, if not eliminate, the hazards (and hence risks) in the first place. Ideally, this translating into finding ways to reduce the potential for conflict, and might imply taking steps to address the root causes of the conflict. However, with a validated social system model, one could explore what-if alternatives in a MonteCarlo fashion. It is also possible to carry out experiments to determine the effectiveness of policy intervention.

Intervention and Risk Mitigation: CountrySim's experiment dashboard and Monte Carlo engine permit us to design experiments and explore sensitivities surrounding those starting settings. We do frequently carry out such experimentation and exploration as part of modeling activity. However, describing a full experiment requires a series of full-fledged policy papers and is beyond the scope of this paper. Instead, we provide a snippet below.

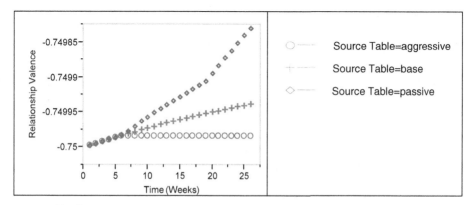

Fig. 8. Effects of Interventions by a Third Party, External (International) Actor

In the Figure 8, we see one such what-if exploration under three different policy intervention strategies, namely aggressive, neutral (no-intervention) and passive interventions, by an external third party international actor. The figure describes the effects course of relationships between two protagonists to the conflict, namely the government and the insurgents.

With aggressive intervention against insurgents by external actor, the relationship between insurgents and government remain the same at the negative value. Without external intervention, this relationship improves marginally. With passive interventional strategy (involving dialogue, exploration and conflict resolution) adopted by international actor towards insurgents, the relationship between insurgents and government appears to move in a positive direction. Of course, one must note that the changes are small (scale is zoomed) in all cases, but noticeable.

Likewise, if the risk treatment were to be considered from the perspective of a multinational or inter-governmental body investing in various countries, each of those countries could be treated as a portfolio and management might involve optimizing the portfolio through hedging instruments. An MNC could compare the results in term of total return to ultimately identify the source of value destruction as well as its magnitude.

Monitoring Risk: According to a recent article on McKinsey Quarterly (2009), "Risk-assessment processes typically expose only the most direct threats facing a company and neglect indirect ones that can have an equal or greater impact." This is true of most risks.

On the other hand, it is necessary to scan the environment and monitor for emergent risks that are "lurking" around the corner. By integrating techniques in social system modeling and simulation and risk identification fields, we propose a methodology for establishing a general monitoring system, set up to identify known and unknown and emerging risks at an early stage. This would be part of, but much more 'holistic' and systemic than, the typical emergency warning system. This step of the iterative risk management process entails dynamic monitoring and adjustment. So it is important to develop a monitoring system with early warnings when key assumptions that were previously uncertain become more concrete. We propose that such a system be established for achieving flexibility, robustness, and vigilant tracking and monitoring of key assumptions or parameters in the external environment.

By being able to data-mine the model outputs and linking the signals or leading indicators to the defined EOIs enables one to monitor the hazard event. We recommend the following outline methodology to increase the chances of identifying risks that have hitherto not manifested:

- setting up an expert network,
- organizing and developing a system (knowledge elicitation instruments such as survey, checklists) and software for both displaying instruments and models to experts,
- monitoring the environment, including:
  o continuously gauging the environment, using specific indicators and updating the evolution of the strategies,
  o continuously gauging the evolution and handling of risk, and hence updating the risk management practices, and
  o periodically updating the model with new data from SMEs.

- periodic or ad hoc knowledge elicitation, dialogue and discussion as in:
  - o periodic scanning of conceptual models to prime analysts,
  - o simulating social system model to generate outputs, alternatives and drill down information,
  - o periodic and ad hoc interviewing of expert networks, again primed by not only the output of the model, but also drill-down into the model and narrative,
  - o dialogue among experts to reconcile reasons for discrepancies among themselves and between the model and experts.

The key part relevant to this paper is prompting the expert networks with outputs, drill down into the model, then creating a dialogue where potential risks could be discussed.

This is could also be aided using such techniques as brain storming, by traversing the conceptual influence diagrams; we frequently prompt domain experts to identify set of signals that can be indicative of a particular event or chain of events unfolding. However, introducing a dynamic complex system model (especially virtual worlds) to experts tends to result in a level of richness (in the information elicited from SMEs) that cannot be captured by brainstorming, influence diagrams or statistics alone.

Risk Communication and Perception: In risk communication and perception, a social system model is useful in providing a framework around which both communication and potentially subsequence perception can take place. Risk is an abstract concept. Like the fabled "four blind men perceiving an elephant in parts", experts and non-experts make tacit assumptions around conception, assessment, management, communication and perception of risk, resulting in competing models of risk. This is one of the causes of much miscommunication and "misperception".

The communication process could be greatly standardized, if risk can be understood in a more concrete terms. Unfortunately, risk is about the future, unmanifest hazards, therefore, will always remain abstract. However, a validated social system model can bridge this gap to some extent, as it can simulate both the perception of risk using cognitively transparent agent processes as well as assessment of risk using a virtual world. In time, we believe that it will go a long way in coming to grips with risk through a common framework.

4   Conclusions

In this paper, we presented a case for risk assessment through a framework that employs rich descriptive modeling of social system.

We introduced an agent based (or primarily agent based) social system model where the agents are cognitively deep and come equipped with values (goals, standards, preferences, cultural and ethical values, personality). The agents belong to factions, which have resources, hierarchies of leadership, followers. The factions that agents belong to, as well as the agents themselves, maintain dynamic relationships with each other. The relationships evolve, or get modified, based on the events that unfold, blames that are attributed etc. As in the real world, the virtual world has processes such as institutional processes, which function, albeit imperfectly, to distribute social/ public goods. Institutions are imperfect owing to being burdened with institutional corruption, discrimination and inefficiency. This environment captures a globally recurring socio-cultural "game" that focuses upon inter-group competition for control of resources (e.g, Security/Economic/Political Tanks), and facilitates the codification of alternative theories of factional interaction and the evaluation of policy alternatives. We also outlined some criteria that the models of this nature satisfy and presented methodology for constructing and validating models. As a social system built primarily of cognitively detailed agents, our model is amenable to providing multiple levels of correspondence (micro, macro). At observable (micro) and abstracted (macro) levels, we showed correspondence in behaviors (e.g. decisions agents make and events of interest respectively). Our rigorous approach to model construction and validation give confidence in the model.

Having validated a complex agent based model (such as the one described earlier), one can exploit the richness behind the model. In line with the social mechanism advocacy (Hedström and Swedberg 1998; Mahoney 2001; Bunge 1997; Boudon, 1998), agent based model would be a tool to help expose the mechanism behind the models, especially addressing the question of causal equivalence through qualitative and more holistic data.

Based on this framework, we have very briefly illustrated how such a social system model could be employed to potentially identify and assess risk in a country, making them transparent to policy makers and business leadership.

It is our hypothesis that a descriptive, virtual world with cognitively detailed agents can also assist with understanding concepts and manifestations of cause, effects, risk and other system level relationships that are otherwise difficult to grasp. So far, social system models have only had been tested in limited applications to make sweeping statements about its benefits. Causal effects, for example, is not yet well established. While there are encouraging preliminary findings, including limited validated studies, it needs to be explored further as a hypothesis.

On the other hand, use of social system models to provide concrete framework and structure to otherwise abstract problems and concepts is a definitive benefit. Here, social system models will act as a knowledge acquisition, structuring and communication framework. For example, social system models can be employed to clarify design of metrics and data acquisition framework. Besides, what-if-experiments can be carried out with different indicators to iteratively design the metrics without incurring the cost of actual program.

Besides, our framework is open and is based on synthesis of best-of-breed social and behavioral theories. This is pivotal as the social science is inherently fragmented into sub-disciplines and is still evolving; Facilitating continuous learning and improvement in our framework is essential to bridging the gaps between theories and in turn improving the model (and vice-versa). In our framework, as the science of social systems advance with time, these sub-models (or PMFs in our case), which are replaceable, will be either refined or replaced with new best of breed models. Thus the entire system of models is amenable to continuous improvement, ultimately bringing about an evolutionary growth in virtual social systems.

However, this is not all without challenges. In other papers, we discuss these challenges in detail (Silverman, 2010; Bharathy and Silverman, 2012). Here, it suffices to say that social system models are complex, with imprecise, incomplete and inconsistent theories, and is burdened with large feature spaces. It is a paradigm shift to work with these models. The progress is limited by various concerns including limitations in resources and slow rate of diffusion of these new ideas. Besides, complex models come at a price. There is need for investing time and resources to make model construction efficient and economical for commercial use. Multi-dimensional approach to validation combined with emphasis on out-of-sample comparisons (common knowledge) as well as qualitative insights, including extensive use of domain knowledge in the model construction process, increase confidence in such complex models.

The risk management approach that we demonstrated could be regarded as design inquiry approach to learning about social systems through deliberate experimentation with a validated social system model. In summary, such an approach to risk identification, assessment and management in political domains could complement the traditional techniques, both the qualitative and the quantitative, typically employed for studying political risks.

Acknowledgement. This research was supported by Becks Fellowship, various government agencies, including an advanced research branch of DoD, a COCOM, and the US Army. The authors alone are responsible for any opinions, unsubstantiated claims, or errors made in this manuscript.

Several past and present members of the Ackoff Collaboratory for Advancement of Systems Approach (ACASA) at the University of Pennsylvania contributed to development of software framework described in this paper. They, in the chronological order of getting involved, are: Barry Silverman, Kevin O'Brien, Jason Cornwell, Michael Johnes, Ransom Weaver, Gnana Bharathy, Ben Nye, Evan Sandhaus, Mark Roddy, Kevin Knight, Aline Normoyle, Mjumbe Poe, Deepthi Chandasekeran, Nathan Weyer, Dave Pietricola, Ceyhun Eksin and Jeff Kim.

References

1. Alesina, A., Ozler, S., Roubini, N., Swagel, P.: Political instability and economic growth. Journal of Economic Growth 1, 189–211 (1996)
2. Alesina, A., Tabellini, G.: A Positive Theory of Fiscal Deficits and Government Debt. Review of Economic Studies 57(3), 403–414 (1990)
3. Allen, V.L., Wilder, D.A.: Categorization, Belief Similarity, and Intergroup Discrimination. Journal of Personality and Social Psychology 32(6), 971–977 (1975)
4. Atran, S.: In Gods We Trust: The Evolutionary Landscape of Religion. Oxford University Press (2004)
5. Barro, R.J., Sala i Martin, X.: Economic Growth. MIT Press, Cambridge (1999)
6. Barro, R.J.: Economic Growth in a Cross Section of Countries. Quarterly Journal of Economics 106(2), 407–443 (1991)
7. Bates, R., Greif, A., Singh, S.: Organizing Violence. Journal of Conflict Resolution 46(5), 599–628 (2002)
8. Bharathy, G., Silverman, B.: Challenges in Building and Validating Agent Based Social Systems Models. In: Simulation: Tran. of the Society for Modeling and Simulation International (accepted, 2012)

9. Boudon, R.: Social Mechanisms without Black Boxes. In: Hedstrom, P., Swedberg, R. (eds.) Social Mechanisms: An Analytical Approach to Social Theory, pp. 172–203. Cambridge University Press, Cambridge (1998)

10. Bunge, M.: Mechanisms and explanation. Philosophy of the Social Sciences 27(4), 410–465 (1997)

11. Collier, P., Hoeffler, A.: On the Incidence of Civil War in Africa. Journal of Conflict Resolution 46(1), 13–28 (2002)

12. Collier, P., Hoeffler, A.: On the Economic Causes of Civil War. Oxford Economic Papers 50(4), 563–573 (1998)

13. Collier, P., Hoeffler, A.: Greed and Grievance in Civil War. Center for the Study of African Economies, Oxford (2001)

14. Collier, P.: On the Economic Consequences of Civil War. Oxford Economic Paper 51(1), 168–183 (1999)

15. Davis, P.K.: Representing Social Science Knowledge Analytically. In: Davis, P.K., Cragin, K. (eds.) Social Science for Counterterrorism: Putting the Pieces Together. RAND Corporation, Santa Monica (2009b),
http://www.rand.org/pubs/monographs/MG849/ (as of March 22, 2010)

16. De Haan, J.: Political institutions and economic growth reconsidered. Public Choice 127, 281–292 (2007)

17. De Marchi, S.: Computational and Mathematical Modeling in the Social Sciences, Cambridge (2005)

18. Dziedzic, M., Sotirin, B., Agoglia, J. (eds.): Measuring Progress in conflict Environments (MPICE) – A Metrics Framework for Assessing Conflict Transformation and Stabilization, Defense Technical Information Catalog. United States Institute of Peace (August 2008)

19. Easterly, W., Levine, R.: Africa's Growth Tragedy: Policies and Ethnic Divisions. Quarterly Journal of Economics 112(4), 1203–1250 (1997)

20. Economist, The coup that dare not speak its name (January 18, 2007),
http://www.economist.com/world/asia/displaystory.cfm?story_id=8560006 (accessed June 14, 2009)

21. Edmonds, B., Moss, S.J.: From KISS to KIDS – An Anti-simplistic Modelling Approach. In: Davidsson, P., Logan, B., Takadama, K. (eds.) MABS 2004. LNCS (LNAI), vol. 3415, pp. 130–144. Springer, Heidelberg (2005),
http://bruce.edmonds.name/kiss2kids/kiss2kids.pdf

22. Elbadawi, I., Sambanis, N.: Why Are There So Many Civil Wars in Africa? Understanding and Preventing Violent Conflict. Journal of African Economies 9(3), 244–269 (2000)

23. Elbadawi, I., Sambanis, N.: Now Much War Will We See? Explaining the Prevalence of Civil War. Journal of Conflict Resolution 46(3), 307–344 (2003)

24. EPA (Environ. Prot. Agency), Guidance for Performing Aggregate and Risk Assessments. Off. Pestic. Programs, Washington, DC (1999)

25. Fatehi, S., Safizadeh, K., Hossein, M.: The Association between Political Instability and Flow of Foreign Direct Investment. Management International Review 29(4), 4–13 (1989)

26. Fearon, J., Laitin, D.: Ethnicity, Insurgency, and Civil War. American Political Science Review 97(1), 75–90 (2003)

27. Fitzpatrick, M.: The definition and assessment of political risk in international business: A review of the literature. Academy of Management Review 8(2), 249–254 (1983)

28. Freedman, D.H.: Why Economic Models Are Always Wrong, Scientific American, October 26 (2011), http://www.scientificamerican.com/article.cfm?id=finance-why-economic-models-are-always-wrong

29. Goldstone, J.A.: Toward a Fourth Generation of Revolutionary Theory. Annual Review of Political Science 4, 139–187 (2001)
30. Green, R.T., Smith, C.H.: Multinational Profitability as a Function of Political Instability. Management International Review 6, 23–29 (1972)
31. Gupta, D.: The Economics of Political Violence. Praeger, New York (1980)
32. Hedstrom, P., Swedberg, R.: Social Mechanisms; An Introductory Essay. In: Hedstrom, P., Swedberg, R. (eds.) Social Mechanisms: An Analytical Approach to Social Theory, pp. 172–203. Cambridge University Press, Cambridge (1998)
33. Hedström, P., Swedberg, R.: Social mechanisms. Acta Sociologica 39(3), 281–308 (1996)
34. Hubbard, D.: How to Measure Anything: Finding the Value of Intangibles in Business, p. 46. John Wiley & Sons (2007)
35. Hubbard, D.: The Failure of Risk Management: Why It's Broken and How to Fix It. John Wiley & Sons (2009)
36. Imai, K., Weinstein, J.: Measuring the Economic Impact of Civil War, CID Working Paper No. 51, Center for International Development, Harvard University (June 2000), http://www.cid.harvard.edu/cidwp/051.pdf (accessed: May 12, 2010)
37. Jakobsen, J.: Old problems remain, new ones crop up: Political risk in the 21st century. Business Horizons 53, 481–490 (2010)
38. Johnson, B.L., Reisa, J.J.: Essays in commemoration of the 20th anniversary of the National Research Council's risk assessment in the federal government: managing the process. Hum. Ecol. Risk Assess. 9, 1093–1099 (2003)
39. Jong-A-Pin, R.: Essays on Political Instability: Measurement, Cause and Consequences, Dissertation, University of Groningen (2008b)
40. Jong-A-Pin, R.: On the measurement of political instability and its impact on economic growth, SOM research report, 06C05, University of Groningen, Research Institute Systems Organizations and Management, SOM (2006)
41. Jong-A-Pin, R., De Haan, J.: Growth accelerations and regime changes: a correction. Economic Journal Watch 5, 51–58 (2008a)
42. Kaplan, S., Haimes, Y.Y., Garrick, B.J.: Fitting hierarchical holographic modeling into the theory of scenario structuring and a resulting refinement to the quantitative definition of risk. Risk Anal. 21, 807–819 (2001)
43. Keey, R.B.: Management of Engineering Risk (for Petroleum Transportation Study Project), Centre for Advanced Engineering, University of Canterbury (April 2000)
44. King, G., Honaker, J., Joseph, A., Scheve, K.: Analyzing Incomplete Political Science Data: An Alternative Algorithm for Multiple Imputation. Harvard University (2000) (unpublished manuscript)
45. King, G., Tomz, M., Wittenberg, J.: Making the Most of Statistical Analysis: Improving Interpretation and Presentation. American Journal of Political Science 44(2), 341–355 (2000)
46. Kletz, T.: Hazop and Hazan, 4th edn. Taylor & Francis (2006)
47. Kobrin, S.J.: Political Risk: A Review and Reconsideration. Journal of International Business Studies 10(1), 67–80 (1979)
48. Lindley, D.: The Philosophy of Statistics. The Statistician (2000)
49. Loree, D.W., Guisinger, S.E.: Policy and Non-Policy Determinants of U.S. Equity Foreign Direct Investment. Journal of International Business Studies 26(2), 281–299 (1995)
50. Mahoney, J.: Beyond correlational analysis: Recent innovations in theory and method. Sociological Forum 16(3), 575–593 (2001)

51. McKinsey Quarterly: Risk: Seeing around the corners (October 2009),
    http://www.mckinseyquarterly.com/Strategy/
    Strategy_in_Practice/Risk_Seeing_around_the_corners_2445
    (accessed: August 2010)
52. Morrison, D., Stevensson, H.: Political instability in independent black Africa. Journal of
    Conflict Resolution 15, 347–368 (1971)
53. Natl. Res. Council [NRC-NAS]. Risk Assessment in the Federal Government: Managing
    the Process. Natl. Acad. Press, Washington, DC (1983)
54. Nehrt, L.C.: The Political Environment for Foreign Investment. Praeger Publishers, NY
    (1970)
55. Noricks, D.M.E.: The Root Causes of Terrorism. In: Davis, P.K., Cragin, K. (eds.) Social
    Science for Counterterrorism; Putting the Pieces Together. RAND Corporation, Santa
    Monica (2009), http://www.rand.org/pubs/monographs/MG849/
    (as of March 22, 2010)
56. O'Brien, S.P.: Crisis Early Warning and Decision Support: Contemporary Approaches and
    Thoughts on Future Research. International Studies Review 12(1), 87–104 (2010)
57. Olibe, K.O., Crumbley, C.L.: Determinants of U.S. Private Foreign Direct Investments in
    OPEC Nations: From Public and Non-Public Policy Perspectives. Journal of Public Bud-
    geting, Accounting & Financial Mgmt. (1), 331–355 (1997)
58. Perotti, R.: Growth, income distribution, and democracy: What does the data say. Journal
    of Economic Growth 1, 149–187 (1996)
59. Reason, J.: Human Error. Cambridge University Press (1990) ISBN: 0521314194
60. Regan, P., Norton, D.: Greed, Grievance, and Mobilization in Civil Wars. Journal of
    Conflict Resolution 49(3), 319–336 (2005)
61. Robock, S.H.: Political Risk: Identification and Assessment. Columbia Journal of World
    Business 6(4), 6 (1971)
62. Rodriguez, R.M., Carter, C.E.: International Financial Management. Prentice-Hall,
    Englewood Cliffs (1976)
63. Shapiro, A.C.: Capital Budgeting for the Multinational Corporation. Financial Manage-
    ment (Financial Management Association) 7(1), 7–16 (1978)
64. Silverman, B.G., Bharathy, G.K.: Modeling the Personality & Cognition of Leaders. In: 14th
    Conference on Behavioral Representations In Modeling and Simulation, SISO (May 2005)
65. Silverman, B.G., Bharathy, G.K., Kim, G.J.: The New Frontier of Agent-Based Modeling
    and Simulation of Social Systems with Country Databases, Newsfeeds, and Expert Sur-
    veys. In: Uhrmacher, A., Weyns, D. (eds.) Agents, Simulation and Applications. Taylor
    and Francis (2009a)
66. Silverman, B.G.: Systems social science: a design inquiry approach for stabilization and
    reconstruction of social systems. J. of Intelligent Decision Technologies 4(1), 51–74
    (2010a)
67. Silverman, B.G., Bharathy, G.K., Eidelson, R., Nye, B.: Modeling Factions for 'Effects
    Based Operations': Part I –Leaders and Followers. Journal of Computational and Mathe-
    matical Organization Theory 13, 379–406 (2007a)
68. Silverman, B.G., Bharathy, G.K., Nye, B., Smith, T.: Modeling Factions for 'Effects Based
    Operations': Part II – Behavioral Game Theory. Journal of Computational and Mathemati-
    cal Organization Theory 14(2), 120–155 (2008a)
69. Silverman, B.G., Bharathy, G.K., Smith, T., Eidelson, R., Johns, M.: Socio-Cultural
    Games for Training and Analysis: The Evolution of Dangerous Ideas. IEEE Systems, Man
    and Cybernetics 37(6), 1113–1130 (2007b)

70. Silverman, B., Bharathy, G.K., Nye, B., Kim, G.J., Roddy, Poe, M.: Simulating State and Sub-State Actors with CountrySim: Synthesizing Theories across the Social Sciences. In: Sokolowski, J., Banks, C. (eds.) Modeling and Simulation Fundamentals: Theoretical Underpinnings and Practical Domains. M. Wiley STM (2010)
71. Silverman, Bharathy, G., Nye, B.: Gaming and Simulating Sub-National Conflicts. In: Argamon, S., Howard, N. (eds.) Computational Methods for Counter-Terrorism. Springer, Heidelberg (2009b)
72. Tetlock, P.: Expert Political Judgment: How Good is It? How Can We Know? Princeton University Press, Princeton (2005)
73. Woodward, D.P., Rolfe, R.J.: The Location of Export-Oriented Foreign Direct Investment in the Carribean Basin. J. of International Business Studies 24(1), 121–144 (1993)
74. Zacharias, G.L., MacMillan, J., Van Hemel, S.B. (eds.): Behavioral Modeling and Simulation: From Individuals to Societies. National Research Council, National Academies Press (2008)
75. Zink, D.W.: The Politicial Risks for Multinational Enterprise in Developing Countries. Praeger, NY (1973)

Chapter 18
An Integrated Intelligent Cooperative Model for Water-Related Risk Management and Resource Scheduling

Yong-Sheng Ding[1,2], Xiao Liang[1], Li-Jun Cheng[1], Wei Wang[1], and Rong-Fang Li[1,3]

[1] College of Information Sciences and Technology,
Donghua University, Shanghai 201620, P.R. China
[2] Engineering Research Center of Digitized Textile & Fashion Technology,
Ministry of Education, Donghua University, Shanghai 201620, P.R. China
[3] Jiangxi Provincial Institute of Water Sciences, Nanchang 330029, Jiangxi, P.R. China

Abstract. Risk management for natural disasters that focuses on early warning, dynamic scheduling and aftermath evaluation, has been one of the key technologies in the field of disaster prevention and mitigation. Some water-related disasters have common characteristics, which are usually generalized and then applied for establishing an integrated model to cope with these disasters. The rapid development of the artificial intelligence (AI) technique makes it possible to enhance the model with intelligent and cooperative features. Such type of model is firstly proposed in this chapter and then derived into two instances, which aim to solve problems in drought evaluation and water scheduling of reservoirs, respectively. The former is based on the radial base function neural network (RBFNN) and the later takes an improved particle swarm optimization (I-PSO) algorithm as its carrier for implementation. Simulation results demonstrate that the first model can make full use of the spatial and time data of the drought and high accuracy of evaluation and classification of the drought severity can therefore be acquired. The second model can distribute the water storage among the reservoirs timely and efficiently, which is of great significance of eliminating the damage of the seasonal droughts and floods occurred in the tributary.

Keywords: integrated intelligent model, water disaster, risk management, water resource scheduling, neural network, particle swarm optimization.

1 Introduction

Risk management is the identification, assessment, and prioritization of risks followed by coordinated and economical application of resources to monitor, control or minimize the probability and/or impact of unfortunate events [1] or to maximize the realization of opportunities. Risks come from uncertainty in financial markets or project failures as well as deliberate attacks from adversaries, especially from frequent natural disasters which may bring great losses in life and property. Effective

J. Lu, L.C. Jain, and G. Zhang: Handbook on Decision Making, ISRL 33, pp. 373–402.
springerlink.com   © Springer-Verlag Berlin Heidelberg 2012

risk management practices are comprehensive in recognizing and evaluating all potential risks. Its goal is less volatility, greater predictability, fewer surprises, and arguably most important the ability to bounce back quickly after a risk event occurs. Disaster risk management has become essentially an urgent problem that any preparedness and mitigation planning have to be taken up. A simple view of risk is that more things can happen than will happen. If we can devise a model of possible outcomes whose probability or emergency methods can be acquired, then we can consider how we will deal with surprises–outcomes that are different from what we expect. We can evaluate the consequences of being wrong in our expectations. In this chapter, the issue that how to construct integrated holistic intelligent models on the early warning and evaluation for disaster risk management will be provided, which combined with practical experiences of leading figures in its specific field. Some of them provide rigorous research results, while others are in-depth reports from the field.

2   The Integrated Model for Risk Management of Disasters

2.1   The Artificial Intelligent Approaches

Intelligent system models take on a very meaningful role in the broad area of information technology for risk management prediction of disasters by analyzing their intricate factors. Many intelligent modeling approaches come from the artificial intelligence (AI), which is a science of intelligence system design. In fact, AI is so prevalent that many people encounter such applications on a daily basis without even being aware of it. The AI systems provide a key component in many computer applications. One of the most ubiquitous uses of AI can be found in network servers that deal with weather and typhoon early warning. The wide range of AI application makes people's lives easier and more convenient.

AI is highly technical and specialized and deeply divided into subfields that often fails to communicate with each other. Subfields have grown up around particular institutions, the work of individual researchers, the solution of specific problems, longstanding differences of opinion about how AI should be done and the application of widely differing tools. The central problems of AI include such traits as reasoning, knowledge, planning, learning, communication, perception and the ability to manipulate objects. But general intelligence (or "strong AI") is among the field's long term goals. Soft computing, a part of AI as supplement and complement to conventional techniques, provide a possibility in integrated intelligent computation for human-machine system design and simulation.

Soft computing techniques, which emphasize gains in understanding system behavior in exchange for unnecessary precision, have proved to be important practical tools for many contemporary problems. The principal constituents of soft computing are fuzzy logic, neurocomputing, evolutionary computing and probabilistic computing, simulation annealing, tabu search approach, swarm intelligence systems (such as) with the later subsuming belief networks, chaotic systems and parts of learning theory [2, 3]. Each of them contributes a reliable methodology which only in a cooperative rather than competitive manner for persuading problems in its field, so these methods open new possibilities of application at the sphere of risk management. The methods enable to solve not only non linear dependencies but multi-criteria and optimization problems. In recent years, the application of soft computing techniques has expanded to fuzzy logic, neuro-computing, evolutionary computing and probabilistic computing in combination, which leads to the emergence of hybrid intelligent systems. Such integrated intelligence model is rapidly growing in real-world applications as detecting highly nonlinear, unknown, or partially known complex systems, or processes. "Integrated" refers to the process of subsystems cooperation in which subsystems need to collaborate, coordinate and synchronize so that the material, energy or information can be exchanged. By this means, the system takes a whole effect or builds a new structure by the subsystems' teamwork toward optimization development. Currently the multi-objective optimization problems, in fact, are integrated processes, which are mostly complex system optimization, such as the deployment of the production lines or allocation of water among different districts. They need to achieve the product optimization on all levels or the global optimization on quality. A classical instance is that the business integrated intelligence helps business people make more informed decisions by providing timely and data-driven answers to their specialized questions. In general, the integrated intelligent model can be realized by soft computing approaches: fuzzy logic, neural networks and swarm intelligence, etc.

With the extensive use of integration of the AI and information technology, the advanced engineering technology is utilized in produce, the emergency response capabilities are improved to the community, local self-governments, urban bodies and the state authorities so that they can prepare leisurely, prevent and respond immediately to natural calamities. So integrated intelligent information systems play a significant role in our life, produce and gain better acceptance both in academia and in industry.

2.2 The AI-Based Integrated Intelligent Cooperative Model for Risk Management of Disasters

Many problems to be solved about natural crisis or disasters have a common structure that it includes a "net" framework with some certain "information points" attached to different places on the net. The net usually consists of some small components in a large entity that has a direct relation to the problem itself. For the drought in a large area, e.g., a city or a province, the droughts in its subordinate administrative areas are the "components", and the corresponding large entity is the drought of the city or province. If the problem refers to something about river, the components will be changed to the tributaries of the river and the facilities on them, if any. The river itself, therefore, becomes the entity in this situation. With these components, the information about the problem can be extracted, quantized and reorganized to abstract units, which are called "information points". Accordingly, the solution of the problem under discussion can also be transferred to a digital form, which is another type of information points. The solving process of the target problem is an approach on how to utilize some information points to acquire others, and the transformation between them is the central issue that the technicians should focus, which is also known as methods. On many occasions, the details of the methods are less important, compared to the information points they may use. Sometimes they cannot even be acquired, especially when the modern AI technology is applied. The AI methods build a black box for the target problem, receive some information points as sources and generate the other points as output. Fig. 1 illustrates such relations between information points of different types and the problem structure they connect to (like a cram lifting its two chelas). It can be observed that there exists a correspondence between the entities in the problem domain and the model domain. The natural factors, aka "components" like rivers, grounds and any other things possibly related to the crisis or disaster can be extracted and organized to the "information points". The aftermath of the crisis or disaster, meanwhile, can also be regarded as a type of components, which is naturally changed to another type of information points. So, the crisis or disaster itself is the complex relations among these components (the "net structure" in Fig. 1), which can be transferred to some mathematical expressions in the model domain. In this chapter, such expressions are the AI approaches.

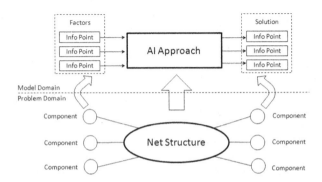

Fig. 1. "The cram": the correspondence between problem domain and model domain of natural crisis or disasters

Fig. 2 shows a classical configuration of an intelligent model based application for risk management business. It is the detailed realization of the mapping relation in Fig. 1 and can be carried out by software designers to form a practical application, e.g. an expert system and so on. It can be divided into three layers according to the different responsibilities they take charge in. The practice layer is the disaster itself. All the data of the disaster in focus can be monitored and collected by researchers and then transferred to its upper layer, namely, the business layer. The business layer is the core of the whole structure that consists of three main components, a database, a knowledge library and an AI/model mechanism built on them. The database receives data from the practical, makes them persistent and finally stored in electronic form. The database is the principle foundation of the AI model which can therefore generalize the desired results through intelligent and cooperative calculation. The output of the AI model can also be regarded as suggestion on how to cope with the disaster in focus, which helps build the knowledge library, together with the original data from the database. The knowledge library takes the data as its long-term memory and provides the practical layer below with instructions which actually come from the AI model. Such instructions can therefore be used by technicians to directly guide the disaster elimination process and control the corresponding equipments (e.g. the artificial rainfall for drought and dam control for reservoirs). The top layer of the structure is the application layer which is mainly the contribution of software engineers. The responsibility of this layer is to provide the users or decision-makers a visual and convenient tool so that the panoramic view of the disaster can be illustrated clearly and the corresponding strategies can be automatically supported. Following the instructions from the users or decision-makers, the application layer organizes all the possible resources and creates certain results with the help of the AI model in the lower layer. Finally, the details of the disaster in focus will be arranged appropriately and expressed on the interface of the decision making applications, together with the analysis and possible solutions.

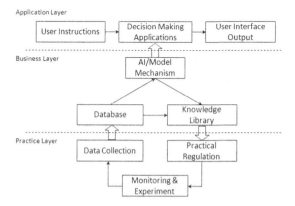

Fig. 2. The basic structure of an intelligent model based application for practical use

3   The Intelligent Classification of Regional Drought Severity Based on Double-Layered Radial Base Function Neural Networks

Risk management for natural disasters, which focuses on early warning, dynamic scheduling and aftermath evaluation, has been one of the key technologies in the field of disaster prevention and mitigation. In most regions of China, the drought is one of the main natural disasters towards agricultural production and farming life. The annual area affected by drought has reached to 21.4 million hm2 and nearly half of these areas are suffered with the loss of cereals as high as 14.2 billion kg [4] and the situation still keeps on worsening with the rapid development of society and economy. The researchers have to not only promote the prevention and mitigation of drought but also establish a series of reasonable and effective drought evaluation and prediction mechanisms, which can therefore lay a foundation for the strategy making and help to make the prevention of drought more precisely.

In this part, a regional drought severity evaluation model based on the radial base function neural network (RBFNN) is proposed to conduct an intelligent evaluation and classification process towards the drought severity of some regions in China. By introducing the size of the affected areas as the initial data, two models based on the RBFNN are firstly applied to the evaluation of target regions and their subregions, respectively. The drought severities of the subregions are then integrated and used for making the evaluation of their root region. The statistically learning and functional approximation feature of the RBFNN helps the proposed model to identify the correct classes for the drought severity. The diversity of drought causing factors in different regions are embodied in the structure of the proposed model, and the differences of drought severities between large regions and their component subregions are also taken into consideration. Simulation results demonstrate that the proposed model can make full use of the spatial and time data of the drought happened in the given regions efficiently, and high accuracy of evaluation and classification toward the drought severity can be acquired.

3.1   Radial Base Function Neural Network

The RBFNN that firstly proposed by J. Moody and C. Darken in 1980s [5] is one of the most prevailing types of artificial neural networks (ANNs). Literatures have proved that the RBFNN with its improved derivatives can make approximation toward any continuous functions in any level of precision. Meanwhile, the RBFNN also has strong ability in extracting features from large amount of data to carry on classifications. It should be noted that the RBFNN shows an impressive performance on analyzing discrete data, which endows this type of ANN the ability to deal with problems containing fewer data provided. The RBFNN is a supervised ANN structure which needs to be trained with some samples and the corresponding teacher signals before put in to practice. The samples and the teacher signals are collected from the problem domain to which the network is about to be applied. By conducting the training process, the features of the problem domain can be recognized by the network, and some possible hidden relations between components of the target problem can also be found and remembered. Such features of RBFNN are more likely to those of another ANN type, the back propagation neural network (BPNN), but the training speed of the RBFNN is far less than the BPNN on most occasions. The classical applications of RBFNN presently include function approximation [6], nonlinear system identification [7], balancing of communication channels [8] and prediction of unstable signals [9].

The basic structure of RBFNN is shown in Fig. 3. A classical RBFNN consists of one input layer, one output layer and a single hidden layer (different from other neural networks that have multiple hidden layers). The input layer consists of some distribution neurons which simply transfer the input signals to the neurons in the hidden layer without any additional calculation. The neurons of the input layer may not be actually treated as standard neurons for they are lack of some mapping functionalities from input to output through a certain relation. On the contrary, the neurons in the hidden layer are real neurons with some classical features of artificial neurons, e.g. the adjustable stimulation functions, weight vectors and threshold vectors. In this layer, the Gaussion function is usually selected as the stimulation function for the neurons. The output layer consists of one or more neurons, and their function is to merge the intermediate results from the hidden layer and finally provide a complete output [10]. In general, the calculation process of an RBFNN can be stated by the following expression,

$$y = f(x_i \mid w1_{ij}, w2_j, r_j), \quad i = 1, 2, \ldots, m, \quad j = 1, 2, \ldots, n \tag{1}$$

where y is the output of the RBFNN, x_i is the input vector with m dimensions, w_1 and w_2 are weight vectors among input layer, hidden layer and output layer, respectively. r_j is the output of neurons in the hidden layer. m and n are the number of input neurons and hidden neurons, respectively. Note that w_1 has double subscript while w_2 has only single, for the mapping between the input layer and the hidden layer is complete and both these layers have more than one neurons. However, the output layer has a single neuron, and the corresponding weight vector w_2 has only one subscript.

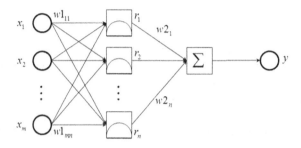

Fig. 3. The basic structure of the RBFNN

Fig. 4 is the diagram illustrating the calculation process of the i-th neuron in the hidden layer. The intermediate input vector for the i-th hidden neuron can be written as

$$k_i = \sqrt{\sum_j (w1_{ji} - x_j)^2 \times b_i}, \quad j = 1, 2, \ldots, m \tag{2}$$

where k_i is the intermediate input vector before the calculation of Gaussian function, b_i is the bias value. Accordingly, the output of the i-th neuron is

$$r_i = \exp(-(k_i)^2) = \exp\left(-\left(\sqrt{\sum_j (w1_{ji} - x_j)^2} \times b_i\right)^2\right) = \exp\left(-(w1_j - X^q \times b_i)^2\right) \tag{3}$$

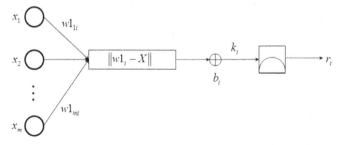

Fig. 4. The working process of a hidden neuron in the RBFNN

The sensitiveness of the radial base function of the i-th hidden neuron can be adjusted by a threshold value b_i, as shown in Fig. 4. However, another parameter C which is called extend constant is more widely used in practice, and several methods can be used to make the transformation between $b1_i$ and C. In the definition of MATLAB (also the software environment to carry out the simulation below), such transformation can be expressed as

$$b_i = 0.8326 / C_i \qquad (4)$$

so the output of the i-th hidden neuron can be rewritten as

$$g_i = \exp\left(\frac{\sqrt{\sum_j (w1_{ji} - x_j)^2} \times 0.8326}{C_i}\right) = \exp\left(-0.8326^2 \times \left(\frac{w1_j - X \times b_i}{C_i}\right)^2\right) \qquad (5)$$

and Eq. (5) actually unveil a fact that the extend constant C determines the range of output vector, compared with the input vector of the hidden neuron. The hidden neuron can respond to the input vector more widely if the value of C is set larger. In that case, better smoothness among hidden neurons can be achieved. The output vector of an RBFNN is calculated by adding the weighted output vectors of each hidden neuron, and different selections of the mapping function for the output neuron can acquire different network output results. Take the linear simulation function in the output layer as an example, and the expression of the output vector can be written as

$$y = \sum_{i=1}^{n} r_i \times w2_i \qquad (6)$$

This result can also be observed in Fig. 3.

The training process of the RBFNN includes two phases, an unsupervised training phase, and then a supervised training phase. In the unsupervised training phase, the weight vectors between the input layer and hidden layer (as $w1$ in Fig. 4) are calculated. In the following supervised phase, the weight vectors between the hidden layer and output layer (as $w2$ in Fig. 4) are calculated. The detailed training methods can be found in many literatures about neural networks [11]. The input data vectors and the corresponding output vectors which act as a foundation of both unsupervised and supervised training should be prepared before the training process begins. It should be noted that the number of hidden neurons is a key factor to the performance of the whole RBFNN, and this number is often set to the same as the dimension of the input vectors in practice. The characteristics above help the RBFNN to achieve high precision for approximating functions, especially those discrete ones.

3.2   The Design of the Intelligent Classification Model Using Double-Layered RBFNN

3.2.1   Design Principle

The drought evaluation is a complicated problem that hasn't be solved on a generally accepted foundation because there are so many causing factors of drought.

Meanwhile, drought occurred in different regions has different forms of damage so that a general standard suitable for all kinds of regions to classify the drought severity levels cannot be easily found. In this paper, a new evaluation model is proposed based on the principles as follows:

Principle 1. The region suffered from drought can be divided into several small regions (called subregion), and drought in each small subregion has its own form by influence diffenent kinds of farmland in different severity levels.

Principle 2. The drought severity of the whole region is a certain kind of combination of drought severities in its subregions.

Principle 1 illustrates each region suffering from drought has its own special condition, and various types of factors are possible to be responsible for the drought. Principle 2 shows the possibility that the drought severity level in a large region may be different from those in its components, but can be influenced by those in its components. Based on these principles, a model that can embody both the two specifications, namely, the variety of drought causing factors and the regional differences is required. Fig. 5 is the overall diagram of an RBFNN-based regional drought severity evaluation model.

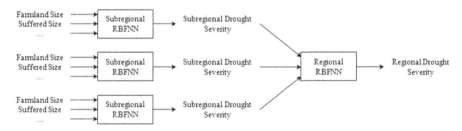

Fig. 5. The general structure of the double-layered RBFNN for drought severity classification. The severities of subregions are firstly calculated by the front RBFNN layer, which are then used for decide the total severity level of a larger region. Note that each layer contains one or more complete RBFNN models.

The drought severity evaluation model in Fig. 5 is based on a double-layered RBFNN foundation. The drought suffered region is separated into several subregions (3-separation case is applied in Fig. 5 for illustration), whose drought severities are firstly evaluated. The evaluation model for each subregion is based on an RBFNN which receives different drought causing factors of its corresponding subregion as input (such as total farmland size, drought suffered size, etc.) and the subregional drought severity as output. The subregional drought evaluation results are then formed another input vector, which is applied to conduct evaluation for their parent region by another RBFNN structure. Finally, the overall regional drought severity can be achieved after this two-step process, and the drought severity levels in the smaller subregions can also be achieved simultaneously.

3.2.2   Parameters and Data Preparation for Modeling

The parameters in the proposed evaluation model mainly consist of the RBFNN size (including the number of neurons in input, hidden and output layers) and the quantized output types of drought severity. The RBFNN size is determined by the number of drought causing factors of its corresponding subregion (for the subregional evaluation), or the number of subregions (for the regional evaluation). The drought severity levels literally should be transferred to digital forms and then quantized to some discrete values between 0 and 1 so that the proposed model can make use of them. The output of the proposed model is also discrete digit, which should be transferred to its literal level description. On account of this, a quantization scheme including five levels is applied to the proposed model as shown in Table. 1. Note that the digits in Table. 1 are all precise values while the output values of the evaluation model may not always fall on such precise levels. So a tolerance Δ_{tol} is introduced to the model, and the range of the output can be written as

$$Y_{real} \in \{R_{real} \mid |Y_{real} - Y_q| \le \Delta_{tol}\} \tag{7}$$

where Y_{real} is the practical value of the model, R_{real} is the collection of Y_{real} adjacent to a certain quantized drought level shown in Table.1, and Y_q is the precise value of this level. The value of Δ_{tol} is determined according to the precision required and often ranges between 0.05 and 0.1. If an output value of the model falls into the range R_{real} of a certain level, which is the same as the practical level, it can be regarded as a successful evaluation. Based on such quantization scheme, the history drought data with evaluation results made manually are firstly taken as teaching signals and sent to the RBFNN model. Then the practical drought data to be evaluated can be provided to the trained model.

Table 1. Quantization scheme with 5-level drought severity

Literal Description	No Drought (NO)	Slight Drought (SL)	Moderate Drought (MD)	Severe Drought (SD)	Giant Drought (GD)
Quantized Level(Y_q)	0.1	0.3	0.5	0.7	0.9

4   The Intelligent Optimization of Multiple Reservoirs Operation Using Improved Particle Swarm Optimization

The shortage of water has become a major problem that takes effects on the industrial and agricultural development in southwestern regions of China. The challenges of water resources management for this region include the trade-off between water demand and supply capacity and the operation of water resources infrastructures (e.g., reservoirs, dams, etc.) for dealing with possible water shortage be sides for flood control. Early water-limiting method may be used to avoid the shortage phenomenon before the wet season through reservoirs adjusting. In this part, a multiple reservoir operation system management model is proposed to tune the distribution of water resources in the southern China by scheduling the volumes of reservoirs reasonably. This system is designed based on an improved particle swarm optimization (I-PSO) algorithm.

4.1   Particle Swarm Optimization

Particle swarm optimization (PSO) is an evolutionary computation technique developed by Kennedy and Eberhart in 1995 [12-13], which is derived from the metaphor of social interaction and communication such as bird flocking. It gets inspiring from the biological populations' behavior characteristics to solve optimization problems as one optimization method on a swarm intelligence basis. The basic PSO mechanism and its various derivations have been widely applied to the multivariate function optimization problems [14], e.g. the complex nonlinear optimization and pattern recognition [15], signal processing [16], robot technology application [17], power resource dispatch [18], scheduling of hydraulic multi-source and multi-reservoir (MSMR) interconnected systems [19-20], etc.

4.1.1   The Basic Principle of PSO

The basic PSO is based on neighborhood principle which roots in the social network structure research, and some terms for PSO are inherited from its origin accordingly. A swarm is defined as one interaction organization and consists of particles each of which represents a potential solution and flies in the solution hyperspace by following the current optimum particles. In the hyperspace determined by the number of dimensions that the particles occupy, the particles keep track of their coordinates which are associated with the best solution (fitness) it has achieved so far. The value of that fitness is also stored. This value is called p_{best}. Another "best" value is also tracked. The "global" version of the particle swarm optimizer keeps track of the particle swarm optimizer keeps track of the overall best value, and its location ,obtained thus far by any particle in the population; this is called g_{best}.

The procedure of PSO consists of, at each time step, changing the velocity (accelerating) of each particle toward the p_{best} and g_{best}. New Acceleration is calculated by the present acceleration, with separate random numbers being generated for particle toward p_{best} and g_{best}. The updating process of the velocity and position are accomplished according to the following equations:

$$v(t+1) = v(t) + c_1 r_1(t)(p_{best} - x(t)) + c_2 r_2(t)(g_{best} - x(t)) \tag{8}$$

$$x(t+1) = x(t) + v(t+1) \tag{9}$$

where $v(t)$ is the particle velocity, $v(t+1)$ is the new particle velocity which will promote optimized process and reflects exchange of social information, $x(t)$ is the current particle position and $x(t+1)$ is a new particle position (solution). Eq. (8) consists of three parts: the first part is the previous velocity of the particle, the second and the third parts are the ones contributing to the change of the velocity of a particle, respectively. Without these two parts, the particles will keep on moving with the current speed in the same direction until they hit the boundary of the hyperspace they belong to. PSO will not find a acceptable solution unless there are acceptable solutions on their"flying" trajectories. But that is a rare case. On the other hand, referring to Eq. (8) without the first part, the "flying" particles' velocities are only determined by their current positions and their best positions in history. The velocity itself is memory less. Supposing one particle is located at the global best position at beginning, and then its velocity will be zero, that is, it will keep the same position until the global best position changed. At the same time, each other particle will be "flying" toward its weighted centroid of its own best position and the global best position of the population. It means that the searching space will shrink with population evolution. So only if the global optimal solution existed within the limit of the initial searching space, then one particle may find it. This indicates that the PSO resembles a local optimal algorithm without the first part of Eq. (8).

Another feature of PSO is that each particle will be "flying" toward its own best position and the global best position of the population simultaneously and a certain degree of trade-off should be met between these two directions, if they are different. In [13], this paper proposes that a recommended choice for constant c1 and c2 is integer 2 since it on average makes the weights for the second part and the third part of the Eq. (8) to be 1.Under this condition, the particles statistically contract swarm to the current global best position until another particle takes over from which time all the particles statistically contract to the new global best position. It can be imaged that the search process for PSO without the first part of the Eq. (8) would be a process where the search space statistically shrinks through the generations. It degrades to a local search algorithm. This can be illuminated more clearly by displaying the "flying" process if its dynamic behaviors can be illustrated by simultaneous simulation on the screen. It's the combination of local search and global search that form a complete PSO mechanism and finally leads to an appropriate solution. The final solution is highly dependent on the initial seeds (population). So it is more likely to exhibit local search ability without the first part. On the other hand, by adding the first part of the Eq. (8), the particles have a tendency to expand to the whole search space, that is, they have the ability to explore the new area. So it more likely has global search ability by adding the first part of the Eq. (8). Both the local search and global search will benefit solving some kinds of problems. Finally, the trade-off between the global and local search should be treated more carefully. There should be different balances between the local search ability and global search ability for different problems [21].

4.1.2   Standard PSO Algorithm

For keeping the balance of the global and local search ability, many technicians endeavor to make modifications on Eq. (8). As mentioned in [14], Shi and Eberhart add a self-adapting inertia weight w in the speed evolution, which can improve the performance of the particle swarm optimizer. The role of the parameter w is to balance the global search and local search. It can be a positive constant or a positive linear or nonlinear function with time. All the procedures represented by Eq. (8) and Eq. (9) with the inertia weight w are generally called the standard particle swarm optimization algorithm (SPSO). It consists of a swarm of particles which are initialized with a population of random candidate solutions. They move iteratively in the D-dimension hyperspace to look for newer solutions where a fitness function can be calculated and acquired a better result according to a certain quality measurement. Each particle has a position represented by a position vector x_i^{iter} (where i is the index of the particle, $iter$ is the iteration time) and a velocity represented by a velocity vector v_i^{iter}. Each particle stores its own best position that it has reached so far in a vector p_{ibest}^{iter}. The best position vector among the swarm so far is then stored in a vector g_{best}^{iter}. At the tth iteration, the update of the velocity from its previous value to the new value is determined by the following Eq. (10). The new position is therefore determined by the sum of the previous position and the new velocity by Eq. (11).

Position: $x_i^{iter} = (x_{i1}^{iter}, x_{i2}^{iter}, ..., x_{iD}^{iter})$, where $x_{ij}^{iter} \in [L_j, U_j]$, L_j and U_j are lower and upper limits of the search space respectively, $1 \le i \le M$, $1 \le j \le D$; M is the populations size.

Velocity: $v_i^{iter} = (v_{i1}^{iter}, v_{i2}^{iter}, ..., v_{iD}^{iter})$, where $v_{ij}^{iter} \in [v_{j\min}, v_{j\max}]$, $v_{j\min}$ and $v_{j\max}$ are the lowest and fastest velocities respectively;

Personal optimal position: $p_{ibest}^{iter} = (p_{i1}^{iter}, p_{i2}^{iter}, ..., p_{iD}^{iter})$;

Global optimal position: $g_{best}^{iter} = (p_{gbest1}^{iter}, p_{gbest2}^{iter}, ..., p_{gbestD}^{iter})$;

Therefore the velocity and position of each particle will be updated according to the following equations:

$$v_{ij}^{iter+1} = wv_{ij}^{iter} + c_1 r_{1j}^{iter}(p_{ibest}^{iter} - x_{ij}^{iter}) + c_2 r_{2j}^{iter}(g_{best}^{iter} - x_{ij}^{iter}) \tag{10}$$

$$x_{ij}^{iter+1} = x_{ij}^{iter} + v_{ij}^{iter+1} \tag{11}$$

where $1 \le i \le M$, $1 \le j \le D$; c_1, c_2 are learning factors, usually we make $c_1 = c_2 = 2$; w is the inertia weight and used to control the trade-off between the global and local exploration ability of the swarm. Random numbers r_{1j}^{iter} and r_{2j}^{iter} are uniformly distributed in [0, 1].

Eq. (10) consists of three parts. wv_{ij}^{iter} is called the "memory" session, which presents particle maintains the ability of previous speed and w can adjust it. $c_1 r_{1j}^{iter}(p_{ibest}^{iter} - x_{ij}^{iter})$ is the "cognition" session, which presents particle itself express. $c_2 r_{2j}^{iter}(g_{best}^{iter} - x_{ij}^{iter})$ can be regarded as the "social" session, which presents the information sharing between particles and particle's ability to learn from the optimal individual.

As the memory session has randomness, expands the search space, and therefore has the tendency of global optimization. The second part and the third part of Eq. (10) represent the individual experience acquired in small groups and the social learning achievements from the whole particle society, respectively. All of them together realize the local and global search mechanism.

Shi and Eberhart propose an adaptive inertia weight as follow [15]:

$$w = w_{max} - iter \times (w_{max} - w_{min}) / iter_{max} \tag{12}$$

where w_{max} and w_{min} are the maximum or minimal inertia weight, respectively; $iter$ is the current iteration time; $iter_{max}$ denotes the maximum iteration number; w_{max} is uniformly distributed in [0.9, 1.2] and w_{min} is usually set to 0.4. Shi and Eberhart have proved that a time decreasing inertia weight as Eq. (12) can bring significant performance improvement to the PSO [21]. It enables the PSO to possess more exploitation ability to not only find a good seed at the beginning but also tine search the local area around the seed.

4.2   An Improved PSO Algorithm for Long Term Optimal Hydraulic Scheduling of Multi-reservoir Systems

4.2.1   An Improved PSO Algorithm

Although the standard PSO brings significant performance improvement to the PSO, but it can easily get trapped in the local optima and convergence rate decreased considerably in the later period of evolution, when solving complex multimodal problems [22]. Many scholars also consider it is an undesired process of diversity loss [23]. Furthermore, when the standard PSO is reaching a near optimal solution, the algorithm stops optimizing, thus the solution accuracy is limited [24]. These weaknesses have restricted wider applications of the PSO [25].

Therefore, accelerating convergence speed and avoiding the local optima have become the two most important and appealing goals in PSO research. In recent years, various attempts have been made to achieve these two goals [22,26-34], such as Eberhart and Shi's linearly decreasing weight (LDW) of PSO, which has obviously effect on raising optimal capability[26]; a fuzzy adaptive turbulent PSO[27]; a new diversity based PSO with Gaussian mutation[28] ,etc. In this development, a diversity-guided[35] method has been employed into the improved PSO algorithm to maintain the level of diversity in the swarm population, thereby maintaining a good balance between the exploration and exploitation phenomena and preventing premature convergence. Thangaraj, Pant and Abraham propose some improved PSO algorithm based on adaptive mutation strategy and use this algorithm on unconstrained test problems and real life constrained problems taken from the field of Electrical Engineering. The numerical results show the competence of the proposed algorithms with respect some other contemporary techniques [29, 36].

The Beta Mutation Particle Swarm Optimization (BMPSO) is a simple and modified version of Particle Swarm Optimization algorithm; it uses Beta distribution to mutate the particle [30]. The BMPSO algorithm has two phases namely attraction phase and mutation phase. The attraction phase is the same as that of the standard PSO, while in the mutation phase the swarm particles position vectors are mutated using Beta Distributed Mutation (BDM) operator. The BDM operator is EP based mutation operator and is defined as

$$x_{ij}^{iter+1} = x_{ij}^{iter} + \sigma_{ij}' \cdot Betarnd_j() \tag{13}$$

where $\sigma'_{ij} = \sigma_{ij} \cdot \exp(\tau N(0,1) + \tau' N_j(0,1))$, $N(0,1)$ denotes a normally distributed random number with mean zero and standard deviation one. $N_j(0,1)$ indicates that a different random number is generated in j-th variables. τ and τ' are set as $1/\sqrt{2n}$ and $1/\sqrt{2\sqrt{n}}$ respectively [37]. $Betarnd_j()$ is a random number generated by beta distribution with parameters less than 1.

The BMPSO starts like the standard PSO i.e. it uses attraction phase (Eq. (10)) for updating velocity vector and uses Eq. (11) for updating position vector. In this phase, the swarm contracts rapidly due to the fast information flow among the particles, as a result, the diversity also decreases consequently the chances of the swarm particles to get trapped in some local region or some suboptimal solution also increases. In BMPSO algorithm, we keep a check on the decreasing value of diversity with the help of a user-defined parameter called d_{low} . When the diversity of population drops below d_{low} , it switches over to the mutation phase, with the hope of increasing the diversity of the swarm population and thereby helping the swarm to escape the possible local regions. This process is repeated until a maximum number iteration is reached or the stopping criterion is reached [36].

The "diversity" measure of the swarm is defined as [37]:

$$f(m)_{diversity} = \frac{1}{m \cdot L} \sum_{i=1}^{m} \sqrt{\sum_{j=1}^{D} (p_{ij} - \overline{p_j})} \tag{14}$$

where m is the particle size of population; L is the length of longest the diagonal in the search space, D is the dimensionality of the problem, p_{ij} is the jth value of the ith particle and $\overline{p_j}$ is the average value of the jth dimension over all particles, i.e.

$\overline{p_j} = \sum_{i=1}^{m} p_{ij} / m$.

From this Eq. (13), the smaller $f(m)_{diversity}$ is, the nearer the particle is to an equal value face, otherwise. If $f(m)_{diversity} < d_{low}$, so the current particle carries on the adaptive mutation as Eq. (13):

4.2.2   Optimal Scheduling of Multi-reservoir System

The reservoirs in the tributary region of a river are generally functioned for flood control, hydropower generation and balance between the water supply capacity from the river and the water demand potential from the residential areas, such as cities and villages nearby. In this situation, an integrated water resources management toward the river and its attachments (reservoirs, etc.) is vital and necessary. In the following paragraphs, we will analyze the entity distribution around the river including reservoirs, irrigation district and residents, etc., and then construct an overall operation planning for the reservoirs to control water to cope with the shift between dry and wet seasons.

The following figure (Fig. 6) shows a generalized network of a river basin in the southern China. This river basin has four reservoirs, three irrigated districts and four major tributaries. It will supply millions of residents' domestic water requirement and maintain eco-environment water requirement, etc. In each year the water demand of the river basin sharply increases through July to September because the agricultural irrigation should be done, which covers about 66 percent (two third) of the total water consumption. If water shortage occurs, both the agricultural production and industrial production will be threatened.

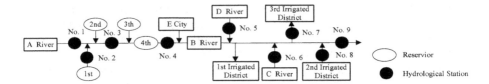

Fig. 6. A generalized network of a river basin

In the instance shown in Fig. 6, the No. 1 and No. 4 reservoir are large-scale reservoir, and the reservoirs No. 2 and No. 3 are middle-scale ones. The difference between large-scale reservoirs and smaller ones means that the former plays a comparatively main role in the water dispatch process, thus the latter takes auxiliary responsibilities. In the term of the relations between these reservoirs and the river, the reservoirs with No. 1, 2, and 3 are parallel connection, and all of them with the No. 2 reservoirs are series connection that means the paralleled reservoirs outflow inflow into water of the No. 2 reservoirs. The whole river basin is usually divided manually into several sections to make the following analysis simpler, and some hydrological stations can be set as control section to monitor the reservoir's storage to keep the balance of supply and demand simultaneously. Fig. 6 shows the reservoirs in space distribution and correlations, which demonstrate clearly the inner relationship of the multi-reservoir systems.

Many scholars have proposed dynamic programming (DP), progressive optimization algorithm (POA), genetic algorithm (GA) etc. various theories to solve the optimal hydraulic scheduling of interconnected system problem [38-40], but many simulation tests demonstrate PSO algorithm can perform better convergence accuracy and convergence rate [41]. Therefore, we can propose the PSO algorithm to solve the water dispatch problem.

For the water dispatch problem, the objective function can be defined as the minimum shortage of water storage:

$$F(x) = \min \sum_s [R_s(t) - D_s(t)] \tag{15}$$

where $R_s(t)$, $D_s(t)$ are the sth control section's water supply capacity (including the recommended reservoirs outflow) and water demand capacity respectively; $1 \leq s \leq S$, S denotes the number of the control section in this river basin.

Applying PSO algorithm to water dispatch problem, we assume that the reservoir outflow as a particle position. In addition, making sure the solution hyperspace dimensionality is elementary. For long term optimal hydraulic scheduling problem, generally choosing one year as operation cycle and set every month as the dispatch period. Other parameters of the PSO respectively are set as follow: $D = 12$, $m = 50$, $iter_{max} = 1000$, $w_{max} = 0.9$, $w_{min} = 0.4$, $c_1 = c_2 = 2$, where D presents the dimensions of the solution hyperspace; m is the particle size of population; $iter_{max}$ is the maximum iterations; w_{max}, w_{min} are respectively the maximum or minimal inertia weight; c_1, c_2 are learning factors of the Eq. (8).

The maximum, minimal velocities can be calculated as the following equations respectively:

$$V_{jmin}^t = \alpha Q_{jmin}^t + (1-\alpha) Q_{jmax}^t \tag{16}$$

$$V_{jmax}^t = \beta Q_{jmin}^t + (1-\beta) Q_{jmax}^t \tag{17}$$

where Q_{jmax}^t, Q_{jmin}^t are respectively the maximum or minimal reservoir outflow. $\alpha = (1 + rand()) / 2$, $\beta = (1 - rand()) / 2$. $1 \le j \le N$, $1 \le t \le D$.

The velocity of each particle represents the transition rate of every reservoir outflow:

$$V_i^{iter} = (v_{i1}^1, v_{i1}^2, ..., v_{i1}^D, v_{i2}^1, v_{i2}^2, ..., v_{i2}^D, ..., v_{ik}^1, v_{ik}^2, ..., v_{ik}^D) \tag{18}$$

The position of each particle illustrates the monthly changes of the reservoir outflow and presents a potential solution:

$$X_i^{iter} = (q_{i1}^1, q_{i1}^2, ..., q_{i1}^D, q_{i2}^1, q_{i2}^2, ..., q_{i2}^D, ..., q_{ik}^1, q_{ik}^2, ..., q_{ik}^D) \tag{19}$$

where $1 \le i \le m$; k is the number of reservoirs.

Based on the PSO algorithm referred above, an objective function specific to the water management is proposed as follows:

$$F(x) = \max(R_1 + R_2 + R_3 - D_3 - D_2 - T \times D_1) \tag{20}$$

where D_1 is the average monthly ecological water requirement of each control section, D_2 is the average monthly off-stream ecological water requirement of each control section, D_3 is the average monthly human life water requirement of each control section, R_1 is the average monthly drain outlet water of each control section, R_2 is the average monthly runoff of each control section, (which may contain the natural runoff volume or the reservoir outflow volume) and R_3 is the average monthly rainfall of each control section. T is a coefficient setting as 2.59×10^6.

Fitness function can be defined as the same as the objective function:

$$Fit(x_i) = R_1 + R_2 + R_3 - D_3 - D_2 - T \times D_1 \tag{21}$$

Furthermore, each particle fitness value represents the calculation result of the objective function when using q_{ij}^t as the function input value; the best solution (fitness) that the particle i has achieved so far is called the individual optimal solution p_{ibest}; the overall best value (fitness), and its location, obtained thus far by any particle in the population is called the global optimal solution g_{best}.

According to the water dispatch problem and referring to the BMPSO algorithm, hence, in "Attraction phase" the velocity and position of each particle will be updated according to the following equations:

$$V_{ij}^{iter+1} = wV_{ij}^{iter} + c_1 r_{1j}^{iter} (p_{ibest}^{iter} - X_{ij}^{iter}) + c_2 r_{2j}^{iter} (g_{best}^{iter} - X_{ij}^{iter}) \qquad (22)$$

$$X_{ij}^{iter+1} = X_{ij}^{iter} + V_{ij}^{iter+1} \qquad (23)$$

In "Mutation phase", the velocity update equation is the same as the Eq. (20), but the position of each particle will be updated as following:

$$X_{ij}^{iter+1} = X_{ij}^{iter} + \sigma_{ij}' \cdot Betarnd_j() \qquad (24)$$

The calculation result shows the best particle the best monthly reservoir outflow.

The process for implementing the BMPSO algorithm on water dispatch problem can be described as follows:

Step1: Defining the water dispatch objective function and constraints.

Step2: Setting particle population size m, the solution hyperspace dimensionality D, the maximum iterations $iter_{max}$, other parameters of the PSO algorithm respectively c_1, c_2, r_1, r_2, etc.

Step3: Initialize particles with random positions and velocities in the search space with constrains of reservoir outflow, reservoir storage, reservoir water level and the water equilibrium equation (the specific steps is descried in section 4.2.3).

Step4: Evaluate the desired optimization fitness function value for each particle, and set p_{ibest} of each particle and its objective value equal to its current position and objective value, and set g_{best} and its objective value equal to the position and objective value of the best initial particle.

Step5: 1) Calculate the diversity of the swarm.

If $f(m)_{diversity} < d_{low}$ (often d_{low} can be set as $1/D$), update particles velocity and position using Eq. (10) and Eq. (13) respectively;

Else update particles velocity and position vectors using Eq. (10) and Eq. (11) respectively.

During iteration, if particles' velocities on each dimension are clamped to a minimum velocity v_{min} and a maximum velocity v_{max}. If the velocity calculation on one dimension exceeds v_{max} or is inferior to v_{min}. Then the velocity on that dimension is limited to the scope of velocity. The position calculation also needs to meet the scope of position.

2) Evaluate the objective function values.

3) Compare the particle's fitness value with p_{ibest}. If the current value is better, then let p_{ibest} and its objective function value equal to the current position and objective value.

4) Determine the best particle of the current swarm with the best objective function value. If the current best objective function value is better, then let g_{best} and its objective function value equal to the current best particle's position and objective value.

Step6: Evaluate the objective function values and update p_{ibest} and g_{best}.

Step7: Return to step (5) until a predetermined criterion is met.

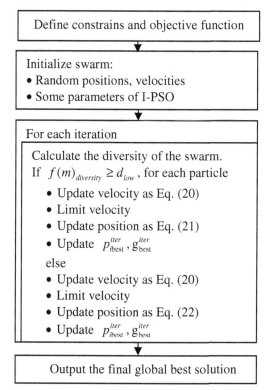

Fig. 7. The flowchart illustrates the steps and update equations of the I-PSO for the water dispatch problem

4.2.3   Constraint Handing

One of the most difficult parts encountered in practical engineering design optimizations is v constraint handing. In [42], Hu and Eberhart propose a kind of PSO to solve the nonlinear engineering optimization problems with constrains. In this paper they propose a generalized method to handing constrains.

For long term optimal hydraulic scheduling of MSMR interconnected systems problem, real-world limitations frequently introduce multiple, nonlinear and non-trivial constraints on a design. Constraints often limit the feasible solutions to a small subset of the design space. However, some constraint handing techniques can be employed to facilitate the optimization process and have the following features:

1. They need be simple. There is no need to consider all constrains or the complicated manipulation when calculating the fitness value. In addition, it is easy to understand the process of constraint handing.

2. They need be faster. The only part of methods dealing with constraints is to randomly initialize the particles according to the equality constrains. Thus it will make every particle satisfy all constraints and reduce the computation time when handing multiple or complicated constrains.

The following example of a reservoir illustrates this approach in details. Considering the water balance equation of the 1th reservoir at the first period:

$$v_{i1}^2 = v_{i1}^1 + T(I_{i1}^1 - q_{i1}^1)\Delta t - L_{i1}^1 \tag{25}$$

where v_{i1}^1 means the initial reservoir storage capacity, and it is a constant.

According to the boundary constraint conditions: $v_{i1}^2 \in \left[V_{1min}^2, V_{1max}^2\right]$, where V_{1max}^2, V_{1min}^2 are maximum and minimum reservoir storage capacity of the 1th reservoir at 2nd period, respectively

$$\sup : \min\{Q_{1max}^1, (v_{i1}^1 + T \times I_{i1}^1 - V_{1min}^2 - L_{i1}^1)/T\} \tag{26}$$

$$\inf : \max\{Q_{1min}^1, (v_{i1}^1 + T \times I_{i1}^1 - V_{1max}^2 - L_{i1}^1)/T\} \tag{27}$$

where Q_{1max}^1, Q_{1min}^1 are maximum and minimum reservoir outflow capability of the 1st reservoir at the 1st period respectively. Eq. (26), (27) can obtain a random initial reservoir outflow value q_{i1}^1, then using Eq. (25) calculates the random initial reservoir v_{i1}^2. Next period, repeat this process until find the fitness value.

5   Simulation and Results

5.1   Simulation for the Drought Evaluation

In order to verify the effectiveness and precision of the proposed model, a computer-based simulation is carried out by using the drought data of a middle south province of China, which includes 11 cities. These cities can be regarded as large regions as referred in the former sections and each of them can be divided to several smaller subregions. According to the administrative divisions of such cities, these 11 cities are separated to 98 subregions (namely, the Chinese counties or districts). The drought data of such areas include the total area of farmland and the drought effected area of farmland. All the data are collected at July 7, 2007 by the official drought monitoring department of this province. The simulation was conducted with MATLAB 2007a on a desktop computer with a dual-core 2.6GHz processor.

The whole simulation includes a training phase and a verification phase. In the training phase, the data of seven cities including 44 subregions are applied. In the verification phase, the remains (4 cities with 54 cities) are applied. To simplify the programming of the verification process, all the data of the 11 cities including 98 subregions are used in the verification process. The drought severity levels are the same as showed in Table. 1 with a tolerance $\Delta_{tol} = 0.1$. Table 2 shows the simulation results of the subregions (evaluated by the first level of the proposed model) without transferring to their literal meanings. Table 2 shows the overall results of both the subregions and their parent regions (evaluated by the whole proposed model) and the comparison between the simulation results and the actual values of the drought severity levels (the italic numbers in the column of simulation results are error evaluation results).

Table 2. Suffered areas and the corresponding severity classification with manual calculation and RBFNN model

Subregion number	Farmland area $(\times 10,000)$	Drought effected area $(\times 10,000)$	Drought effected percentage $(\%)$	Actual results	RBFNN results
1	0.9	0.4	39.2	0.30	0.40
2	88.1	23.4	26.6	0.30	0.30
3	38.2	9.8	25.7	0.30	0.29
4	77.9	12.4	16.0	0.10	0.13
5	122.3	16.3	13.3	0.10	0.08
6	3.7	0.3	6.7	0.10	0.08
7	10.2	0.05	0.5	0.10	0.10
8	5.8	1.3	22.5	0.30	0.26
9	56.7	7.5	13.3	0.10	0.08
10	27.0	0.8	2.8	0.10	0.15
11	24.9	7.8	31.3	0.30	0.31
12	22.2	4.6	20.7	0.30	0.23
13	13.8	1.7	12.6	0.10	0.07
14	17.0	2.1	12.5	0.10	0.07
15	2.3	0	0	0.10	0.08
16	43.5	24.8	57.1	0.50	0.50
17	35.7	20.0	56.1	0.50	0.51
18	41.5	18.9	45.6	0.50	0.41
19	22.3	8.4	37.8	0.30	0.32
20	57.3	19.9	34.8	0.30	0.31
21	61.6	16.6	26.9	0.30	0.30
22	71.1	17.3	24.3	0.30	0.28
23	22.8	5.4	23.6	0.30	0.27
24	57.5	12.1	21.1	0.30	0.23
25	24.7	4.6	18.7	0.10	0.19

Subregion number	Farmland area (×10,000)	Drought effected area (×10,000)	Drought effected percentage (%)	Actual results	RBFNN results
26	7.2	1.2	16.1	0.10	0.15
27	8.2	1.0	12.5	0.10	0.07
28	57.4	10.9	19.0	0.10	0.20
29	45.0	7.7	17.1	0.10	0.16
30	48.1	7.2	14.9	0.10	0.11
31	4.6	0.6	12.7	0.10	0.08
32	50.2	3.3	6.5	0.10	0.10
33	3.2	0.2	4.8	0.10	0.10
34	17.4	13.5	77.5	0.70	0.70
35	33.1	25.3	76.4	0.70	0.69
36	15.8	8.6	54.3	0.50	0.51
37	4.7	2.3	48.0	0.50	0.47
38	7.3	2.2	29.7	0.30	0.31
39	13.0	3.7	28.7	0.30	0.30
40	17.9	4.9	27.4	0.30	0.30
41	55.6	13.8	24.9	0.30	0.29
42	44.7	8.8	19.7	0.10	0.21
43	51.3	9.9	19.3	0.10	0.20
44	15.8	2.7	17.4	0.10	0.15
45	44.9	7.7	17.1	0.10	0.16
46	35.1	5.6	15.9	0.10	0.13
47	10.3	1.5	14.1	0.10	0.10
48	58.4	7.3	12.6	0.10	0.07
49	18.9	1.2	6.4	0.10	0.10
50	16.1	0.6	4.0	0.10	0.14
51	24.1	0.7	2.9	0.10	0.15
52	14.5	0.4	2.4	0.10	0.15
53	44.4	17.7	40.0	0.30	0.33
54	30.8	6.6	21.4	0.30	0.24
55	20.9	4.4	21.1	0.30	0.23
56	48.9	9.3	19.0	0.10	0.20
57	77.8	14.2	18.2	0.10	0.18
58	49.1	7.7	15.7	0.10	0.13
59	66.0	8.3	12.5	0.10	0.08
60	45.0	5.5	12.3	0.10	0.07
61	14.7	1.7	11.3	0.10	0.07
62	63.0	7.0	11.1	0.10	0.06
63	29.3	2.8	9.6	0.10	0.06
64	16.8	1.1	7	0.10	0.10
65	44.2	0.8	2	0.10	0.14
66	14.1	7.0	49.4	0.50	0.48

Subregion number	Farmland area ($\times 10,000$)	Drought effected area ($\times 10,000$)	Drought effected percentage (%)	Actual results	RBFNN results
67	51.3	19.7	38.4	0.30	0.32
68	140.4	49.9	35.6	0.30	0.31
69	85.1	22.7	26.7	0.30	0.30
70	83.7	21.2	25.4	0.30	0.29
71	148.9	36.7	24.6	0.30	0.28
72	18.1	4.1	22.4	0.30	0.26
73	63.4	10.2	16.1	0.10	0.13
74	47.9	7.1	14.9	0.10	0.11
75	55.3	8.0	14.4	0.10	0.10
76	83.1	54.5	65.6	0.70	0.42
77	49.6	16.0	32.2	0.30	0.31
78	25.1	4.3	17.2	0.10	0.16
79	20.5	3.1	15.0	0.10	0.11
80	10.3	1.5	14.1	0.10	0.10
81	27.5	3.1	11.1	0.10	0.07
82	78.0	8.6	11.1	0.10	0.06
83	38.6	3.9	10.2	0.10	0.06
84	34.9	1.5	4.3	0.10	0.14
85	33.0	1.4	4.3	0.10	0.14
86	31.6	0.5	1.6	0.10	0.14
87	145.4	73.1	50.2	0.50	0.48
88	102.5	44.7	43.6	0.50	0.38
89	39.6	9.1	23.1	0.30	0.26
90	30.5	4.5	14.8	0.10	0.11
91	18.6	2.4	12.8	0.10	0.08
92	34.6	4.4	12.6	0.10	0.08
93	31.7	3.9	12.2	0.10	0.07
94	55.7	6.2	11.1	0.10	0.06
95	29.4	3.0	10.2	0.10	0.06
96	10.6	0.8	7.3	0.10	0.08
97	44.5	3.2	7.3	0.10	0.09
98	30.6	1.9	6.2	0.10	0.10

Fig. 8 illustrates the evaluation results using the proposed model without the tolerance Δ_{tol}, and the evaluation level are unstable and fluctuate among the exact quantized levels. Fig. 9 illustrates the evaluation results after introduction the tolerance Δ_{tol} so that the evaluation performance can be observed more clearly by moving the unstable data to its nearest quantized level in the sense of the tolerance. In Fig. 8, four subregions are evaluated uncorrectedly and the error results made by the proposed model are one level up or below the expected level. The overall error rate can come to about 4.08%, which concludes that the proposed model can acquire evaluation levels in a high precision. Meanwhile, the total time for the training phase is between 0.5~0.6s based on the above data and hardware platform, and the time for the verification is about 0.15s, which does not increase significantly with the increasing number of data. It also shows that the proposed model is suitable for the real-time evaluation.

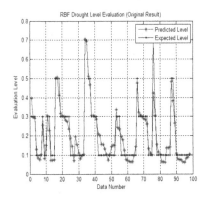

Fig. 8. The simulation results of the double-layered RBFNN model compared with the expected results without the tolerance Δ_{tol}

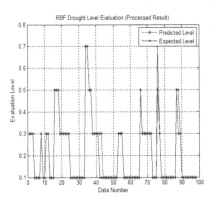

Fig. 9. The simulation results of the double-layered RBFNN model compared with the expected results with the tolerance Δ_{tol}

5.2   Simulation of the Water Resource Scheduling of Reservoirs

5.2.1   Curve Fitting

As we have defined the reservoir outflow as a particle (solution), but in fact we cannot directly control the reservoir outflow ability as it changes with the reservoir storage or elevation in real time. However water resources research indicates that we can acquire the reservoir storage capability and elevation to know the reservoir outflow ability. Therefore, it is necessary to make sure the relation between the reservoir storage capability and the reservoir elevation, also the relation between the reservoir elevation and the reservoir outflow.

Usually, we collect these data at regular intervals and then use mathematical method to find out relationship of them. As we know linear least-square method is the most commonly used methods to solve the curve fitting and obtain the mathematical function among them.

The Figs. 10-17 show the results, we can find the relations meeting secondary polynomial. Even thought the third polynomial looks like better than the secondary, but it may cause "curse of dimensionality" problem.

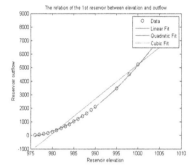

Fig. 10. The relation of the 1st reservoir between elevation and outflow

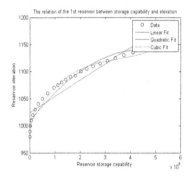

Fig. 11. The relation of the 1st reservoir between storage capability and elevation

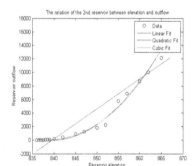

Fig. 12. The relation of the 2nd reservoir between elevation and outflow

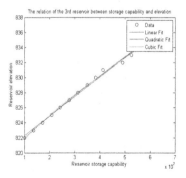

Fig. 13. The relation of the 3rd reservoir between elevation and outflow

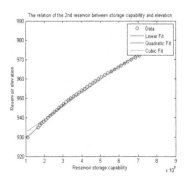

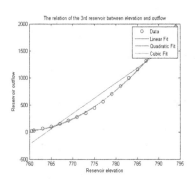

Fig. 14. The relation of the 2nd reservoir between storage capability and elevation

Fig. 15. The relation of the 3rd reservoir between storage capability and elevation

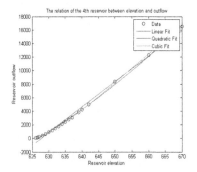

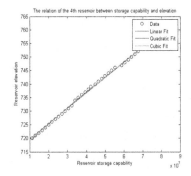

Fig. 16. The relation of the 4th reservoir between elevation and outflow

Fig. 17. The relation of the 4th reservoir between storage capability and elevation

5.2.2   Simulation Results

The implementation of the I-PSO was written and compiled in Matlab. All the simulations deploy the same parameter settings for the I-PSO. The population size (number of particles) is 50; the maximum velocity is set as the average-monthly maximum reservoir storage capability; the minimum velocity is set as the average-monthly minimum reservoir storage capability; the dynamic range for all elements of a particle is defined as the maximum ,minimum reservoir outflow capability, that is the particle cannot move out of this range in each dimension; the maximum number of iterations allowed is 1000. If the I-PSO cannot find an acceptable solution within 1000 iterations, it is claimed that the I-PSO fails to find the global optimum in this run.

Table 3. The result of simulating the scheduling model of water reservoir (Unit: m/s)

Reservoir	Jan.	Feb.	Mar.	Apr.	May	June	July	Aug.	Sept.	Oct.	Nov.	Dec.
1st	12.57	18.69	21.63	35.02	97.42	355.80	479.07	189.78	210.74	92.56	126.15	14.028
2nd	51.00	68.24	87.93	91.21	141.46	459.10	437.65	284.85	136.53	77.72	68.73	42.40
3rd	1.01	6.09	5.35	6.73	95.95	298.12	143.75	51.78	32.32	9.14	28.29	8.10
4th	0.62	15.47	15.31	25.15	50.13	234.76	285.358	61.68	73.20	6.27	28.83	43.97

Table 4. The result of simulating the scheduling model of water reservoir (Unit: m/s)

Reservoir	Jan.	Feb.	Mar.	Apr.	May	June	July	Aug.	Sept.	Oct.	Nov.	Dec.
1st	10.43	15.12	18.57	32.15	62.62	544.00	446.27	141.26	193.69	46.37	105.93	19.78
2nd	78.25	40.68	88.35	60.78	167.19	322.14	578.41	258.31	124.16	106.56	55.34	110.68
3rd	1.0	3.87	4.97	10.72	111.15	278.23	169.90	21.44	36.40	10.27	26.64	6.19
4th	2.01	18.55	14.40	26.05	47.07	213.82	371.57	2.20	51.46	12.92	34.70	43.86

Table 3 and Table 4 show twice computation results, overall result is successful, especially in April and September the water peak, by increasing the major issue of outbound traffic to meet demand. It means that using I-PSO to solve the water dispatch is feasible. Furthermore, compared with the same period of each reservoir outflow in these tables, we can find that the I-PSO has strong robustness. In addition, the I-PSO has the good convergence performance and easy realize advantage, thus it can be a solution method in water dispatch.

I-PSO is a random heuristic search algorithm with benefit of parallel calculation, strong robustness and positive feedback, etc. The parts of article propose an improved particle swarm algorithm to water dispatch problem. Constraint handing can improve the searching efficiency and easily find out the optimal solution, even could solve the "Dimension Disaster" problem.

Although, the scheduling model of water reservoir mainly considerate the collaborative scheduling problem, but no consideration of water dispatch rainy. In addition, due to the reservoir water dispatching problem largely depends on rivers runoff, but how to de predict rivers runoff in future also needs to do further research about it.

6 Conclusions

In this chapter, an integrated intelligent cooperative model for risk management of water-related natural disasters is firstly proposed, and then derived into two instances, which aim to solve problems in drought evaluation and water scheduling of reservoirs, respectively. The former is based on the radial base function neural network (RBFNN) and the later takes an improved particle swarm optimization (I-PSO) algorithm as its carrier for implementation. Simulation results demonstrate that the first model can make full use of the spatial and time data of the drought and high accuracy of evaluation and classification of the drought severity can therefore be acquired. The second model can distribute the water storage among the reservoirs timely and efficiently, which is of great significance in eliminating the damage of the seasonal droughts and floods occurred in the tributary.

Acknowledgements. This work was supported in part by the Key Project of the National Nature Science Foundation of China (No. 61134009), the National Nature Science Foundation of China (No. 60975059), Specialized Research Fund for the Doctoral Program of Higher Education from Ministry of Education of China (No. 20090075110002), Specialized Research Fund for Shanghai Leading Talents, Project of the Shanghai Committee of Science and Technology (Nos. 11XD1400100, 11JC1400200, 10JC1400200, 10DZ0506500).

References

1. Hubbard, D.: The Failure of Risk Management: Why It's Broken and How to Fix It, p. 46. John Wiley & Sons (2009)
2. Xuan, F.Z.: Artificial intelligence and integrated intelligent information systems. United States of America by Idea Group Publishing (February 2006)
3. Zadeh, L.A.: Fuzzy sets. Information and Control 8(3), 338–353 (1965)
4. Zhou, H.-C., Zhang, D.: Assessment Model of Drought and Flood Disasters with Variable Fuzzy Set Theory. Transaction of the Chinese Society of Agricultural Engineering (CSAE) 9, 56–61 (2009)
5. Moody, J., Darken, C.J.: Fast Learning in Networks of Locally Tuned Processing Units. Neural Computation 1, 281–294 (1989)
6. Wei, H.-K., Amari, S.-I.: Dynamics of Learning Near Singularities in Radial Basis Function Networks. Neural Networks 7, 989–1005 (2008)
7. Srinivasan, D., Ng, W.S., Liew, A.C.: Neural-Networked-Based Signature Recognition for Harmonic Source Identification. IEEE Transactions on Power Delivery 1, 398–405 (2006)
8. Lee, J., Sankar, R.: Theoretical Derivation of Minimum Mean Square Error of RBF Based Equalizer. Signal Processing 7, 1613–1625 (2007)
9. Rank, E.: Application of Bayesian Trained RBF Networks to Nonlinear Time-Series Modeling. Signal Processing 7, 1393–1410 (2003)
10. Broomhead, D.S., Lowe, D.: Radial Basis Functions, Multi-Variable Functional Interpolation and Adaptive Networks. RSRE-MEMO-4148. Defense Research Information Center, Orpington, England (1988)
11. Constantinopoulos, C., Likas, A.: An Incremental Training Method for the Probabilistic RBF Networks. IEEE Transactions on Neural Networks 4, 966–974 (2006)
12. Eberhart, R.C., Kennedy, J.: A new optimizer using particle swarm theory. In: Proceedings of the Sixth International Symposium on Micro Machine and Human Science, Nagoya, Japan, pp. 39–43 (1995)
13. Kennedy, J., Eberhart, R.C.: Particle swarm optimization. In: Proc. IEEE Int'l Conf. on Neural Networks, vol. IV, pp. 1942–1948. IEEE Press, Piscataway (1995)

14. Villanova, L., et al.: Functionalization of microarray devices: Process optimization using a multiobjective PSO and multiresponse MARS modeling. In: 2010 IEEE Congress on Evolutionary Computation, CEC (2010)

15. Alipoor, M., Parashkoh, M.K., Haddadnia, J.: A novel biomarker discovery method on protemic data for ovarian cancer classification. In: 2010 18th Iranian Conference on Electrical Engineering, ICEE (2010)

16. Goken, C., Gezici, S., Arikan, O.: Optimal signaling and detector design for power-constrained binary communications systems over non-gaussian channels. IEEE Communications Letters 14(2), 100–102 (2010)

17. Huang, H., Tsai, C.: Particle swarm optimization algorithm for optimal configurations of an omnidirectional mobile service robot. In: Proceedings of SICE Annual Conference 2010 (2010)

18. Zhao, B., Guo, C.X., Cao, Y.J.: A multiagent-based particle swarm optimization approach for optimal reactive power dispatch. IEEE Transactions on Power Systems 20(2), 1070–1078 (2005)

19. Kanakasabapathy, P., Swarup, K.S.: Evolutionary Tristate PSO for Strategic Bidding of Pumped-Storage Hydroelectric Plant. IEEE Transactions on Systems, Man, and Cybernetics. Part C: Applications and Reviews 40(4), 460–471 (2010)

20. Luo, Y., Wang, W., Zhou, M.: Study on optimal scheduling model and technology based on RPSO for small hydropower sustainability. In: International Conference on Sustainable Power Generation and Supply, SUPERGEN 2009 (2009)

21. Shi, Y., Eberhart, R.C.: A modified particle swarm optimizer. In: Proceedings of the IEEE International Conference on Evolutionary Computation, pp. 69–73. IEEE Press, Piscataway (1998)

22. Liang, J.J., Qin, A., Suganthan, K.P.N., Baskar, S.: Comprehensive learning particle swarm optimizer for global optimization of multimodal functions. IEEE Trans. Evol. Comput. 10(3), 281–295 (2006)

23. Li, H.R., Gao, Y.L.: Particle Swarm Optimization Algorithm with Adaptive Threshold Mutation. In: 2009 International Conference on Computational Intelligence and Security (2009)

24. Xie, X.F., Zhang, W.J., Yang, Z.L.: Adaptive Particle Swarm Optimization on Individual Level. In: International Conference on Signal Processing (ICSP), Beijing, China, pp. 1215–1218 (2002)

25. Li, X.D., Engelbrecht, A.P.: Particle swarm optimization: An introduction and its recent developments. In: Proc. Genetic Evol. Comput. Conf., pp. 3391–3414 (2007)

26. Shi, Y., Eberhart, R.C.: Empirical study of particle swarm optimization. In: Proc. of Congress on Computational Intelligence, Washington DC, USA, pp. 1945–1950 (1999)

27. Liu, H., Abraham, A., Zhang, W.: A fuzzy adaptive turbulent particle swarm optimation. International Journal of Innovative Computing and Applications 1(1), 39–47 (2007)

28. Pant, M., Radha, T., Singh, V.P.: A new diversity based particle swarm optimization with Gaussian mutation. J. of Mathematical Modeling, Simulation and Applications 1(1), 47–60 (2008)

29. Pant, M., Thangaraj, R., Abraham, A.: Particle swarm optimization using adaptive mutation. In: 19th International Conference on Database and Expert Systems Application, pp. 519–523 (2008)

30. Jiang, Y., Hu, T.S., Huang, C.C., Wu, X.N.: An improved particle swarm optimization algorithm. Applied Mathematics and Computation 193, 231–239 (2007)

31. Yang, X.M., Yuan, J.S., Yuan, J.Y., Mao, H.N.: A modified particle swarm optimizer with dynamic adaptation. Applied Mathematics and Computation 189, 1205–1213 (2007)

32. Ho, S.Y., Lin, H.S., Liauh, W.H., Ho, S.J.: Orthogonal particle swarm optimization and its application to task assignment problems. IEEE Trans. Syst., Man, Cybern. A, Syst., Humans 38(2), 288–298 (2008)

33. Liu, B., Wang, L., Jin, Y.H.: An effective PSO-based memetic algorithm for flow shop scheduling. IEEE Trans. Syst., Man, Cybern. B, Cybern. 37(1), 18–27 (2007)

34. Ciuprina, G., Ioan, D., Munteanu, I.: Use of intelligent-particle swarm optimization in electromagnetic. IEEE Trans. Magn. 38(2), 1037–1040 (2002)

35. Niu, B., Zhu, Y.L., He, X.X., et al.: An Improved Particle Swarm Optimization Based on Bacterial Chemotaxis. In: Proc. IEEE Congress on Intelligent Control, Dalian, China, pp. 3193–3197 (2006)

36. Thangaraj, R., Pant, M., Abraham, A.: A new diversity guided particle swarm optimization with mutation. In: World Congress on Nature & Biologically Inspired Computing, NaBIC 2009 (2009)

37. Engelbrecht, A.P.: Fundamentals of Computational Swarm Intelligence. John Wiley & Sons Ltd., England (2005)

38. Chang, G.W., et al.: Experiences with mixed integer linear programming based approaches on short-term hydro scheduling. IEEE Transactions on Power Systems 16(4), 743–749 (2001)

39. Xuan, Y., Mei, Y., Xu, J.: Research of Hydropower Stations Optimal Operation Based on the Discrete Differential Dynamic Programming - Progressive Optimization Algorithm Combination Method. In: 7th ACIS International Conference on Software Engineering Research, Management and Applications, SERA 2009 (2009)

40. Leite, P.T., Carneiro, A.A.F.M., Carvalho, A.C.P.L.: Hybrid Genetic Algorithm Applied to the Determination of the Optimal Operation of Hydrothermal Systems. In: Ninth Brazilian Symposium on Neural Networks, SBRN 2006 (2006)

41. Orero, S.O., Irving, M.R.: A genetic algorithm modelling framework and solution technique for short term optimal hydrothermal scheduling. IEEE Transactions on Power Systems 13(2), 501–518 (1998)

42. Hu, X.H., Eberhart, R.C.: Adaptive particle swarm optimization: detection and response to dynamic systems. In: Proceedings of the 2002 Congress on Evolutionary Computation, CEC 2002 (2002)

Chapter 19
Determining the Significance of Assessment Criteria for Risk Analysis in Business Associations

Omar Hussain, Khresna Bayu Sangka, and Farookh Khadeer Hussain

Digital Ecosystems and Business Intelligence Institute
Curtin University of Technology, GPO Box U1987, Perth, W.A., Australia
{O.Hussain,Farookh.Hussain}@cbs.curtin.edu.au,
khresnab@postgrad.curtin.edu.au

Abstract. Risk assessment in business associations is the process which determines the likelihood of negative outcomes according to a given set of desired criteria. When there is more than one desired criterion to be achieved in a business association, the process of risk assessment needs to be done by capturing the importance that each of the criteria will have on the successful completion of the business activity. In this paper, we present an approach that determines the significance of each criterion with respect to the goal of the business association and by considering the inter-dependencies that may exist between the different assessment criteria. This analysis will provide important insights during the process of risk management, where the occurrence of such negative outcomes can be managed, according to their significance, to ensure the successful completion of a business activity.

1   Introduction

Business collaborations are associations formed between two or more users for the achievement of specific desired outcome/s. These users can be software agents, individual users etc who utilize the constantly developing and advancing business facilitating architectures, such as cloud computing, for completing and achieving their business outcomes in a more efficient and less time consuming way. Apart from all the advantages that such interaction facilitating infrastructure provides, interacting users have to constantly watch for factors that will produce a 'negative' outcome. The occurrence of such negative outcome/s signifies the non-achievement of desired outcomes which are termed *'Risk'* in the business association.

The significance of analyzing and managing *'Risk'* in any form of activity has been discussed extensively in the literature. Specific to the domain of business activities, *'Risk'* is a bi-dimensional concept that expresses: a) the occurrence of those events that will lead to the non-occurrence of desired outcomes, and b) the level and magnitude of possible losses or consequences. The process by which *'Risk'* is identified, analyzed and managed in a business interaction is termed 'risk analysis' [1]. As shown in Figure 1, risk analysis is a multi-step, iterative process which deals with first identifying the risks in the business association, determining the level/s of their

J. Lu, L.C. Jain, and G. Zhang: Handbook on Decision Making, ISRL 33, pp. 403–416.
springerlink.com   © Springer-Verlag Berlin Heidelberg 2012

occurrence and the impact which they will have on the successful completion of the business interaction, evaluating the risks to ascertain which risks need to be managed, treating the risks and continually monitoring the risks to ensure a successful completion of the business activity. These steps of risk analysis are broadly categorized in the sub-steps of risk identification, risk assessment, risk evaluation and risk management.

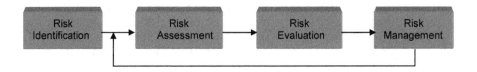

Fig. 1. Different stages in the process of risk analysis

One of the important factors that need to be captured during the steps of risk analysis is the importance or significance of each desired outcome in the business association. The significance of a desired outcome represents its importance in the successful completion of its business activity. The expected achievements from a business association (termed the 'expectations') may be a collection of more than one desired outcome (termed the 'assessment criterion'), the achievement of each of which is important for the successful completion of the business activity. However, this may not mean that all of the assessment criteria of the expectations are of equal importance to the interaction initiating agent (termed as the risk assessing agent). There may be some assessment criteria from the expectations which the risk assessing agent believes are vital or more important than others for the successful completion of the business activity, or there may be some assessment criteria which, even though the risk assessing agent may not consider them important, are significant for the successful achievement of the business activity due to the inherent dependencies between the assessment criteria. As explained in the next section, the significance value of the assessment criteria is an integral part of the various steps of risk analysis. So, such inter-dependencies between the assessment criteria need to be captured for determining the appropriate significance value of the assessment criteria, before reflecting it along the different steps of risk analysis.

Due to the importance of risk, various techniques for risk assessment have been proposed in the literature for business interactions [2-8], project management in different domains [9, 10], occupational safety and hazard management [11, 12] etc. We do not discuss those approaches as the focus of this paper is not on risk assessment and management but on determining the significance of the assessment criteria that are important in shaping the focus of the steps of risk assessment and management. Some of the current approaches do consider the significance of each assessment criteria, but they do not have a systemized process; nor do they capture the different types of inter-dependencies that may exist between them. In this paper, we aim to overcome that by proposing an approach that determines the significance of the assessment criteria of a business interaction by capturing the different levels of dependencies between them. The proposed method is explained in the next sections. In Section 2, we formulate and explain in detail the problem that is addressed in this paper. In Sections 3 and 4, we explain our approach for determining the significance of the assessment criteria in detail. In Section 5, we conclude the paper.

2   Significance of Assessment Criteria during Risk Analysis

The significance of assessment criterion/criteria in an activity has important inputs that determine how the process of risk assessment and management has to be carried out. To understand better, let us consider the business interaction scenario of a risk assessing agent (termed r1) who wants to achieve desired outcomes in a business association. Agent r1 has to make a decision about forming an interaction with either agent b1 or b2 (termed the 'risk assessed agents') in order to achieve desired outcomes, for which there are three assessment criteria (termed c1, c2 and c3).

Significance of Assessment Criteria during Risk Assessment Process: Agent r1 may consider assessment criteria c1 as vital to the successful achievement of its business interaction. By considering agent b1's past history, the probability of agent b1 committing to criteria c2 and c3 according to the expectations of agent r1 may be high but may also have a high probability of failure in committing to criterion c1. This may not be the case with agent b2, who can commit to the assessment criteria c1 with less probability of failure as compared to agent b1 and criteria c2 and c3 with a greater probability of failure as compared to agent b1. In such a case, even though the overall level of risk in forming an interaction with agent b1 may seem lower as compared to agent b2, in reality it may be higher due to the greater level of non-commitment of assessment criterion c1 by agent b1. This is determined by weighing the probability of failure of achieving each assessment criterion according to the significance or importance that it will have on the successful completion of the business activity.

Significance of Assessment Criteria during Risk Communication Process: After agent r1's interaction with agent b1, another risk assessing agent r2 may solicit recommendations from agent r1 about risk assessed agent b1's capability in committing to assessment criteria c1, c2 and c3. However, agent r2 may consider criteria c2 and c3 as significant to the completion of its business interaction. Subsequently, the level of risk communicated by agent r1 to r2 for agent b1 may be too focused on assessment criterion c1. This might give an incorrect impression for agent b1 to r2 as it considers assessment criteria c2 and c3 vital to the successful completion of its business interaction. This is addressed by communicating the significance of each assessment criterion clearly along with the recommendations for agent r2 to make its own assessment during risk analysis.

Significance of Assessment Criteria during Risk Management Process: The process of risk management is also dependent on the significance of each assessment criterion. In the above case of agent r1's interaction with agent b1, the process of risk management after the risk assessment and risk evaluation process will focus more on assessment criterion c1 as compared to assessment criterion c2 or c3 (depending on the output of the risk evaluation phase), as it is considered vital for the successful completion of the business activity. This may vary in agent r2's interaction with b1 as it considers criteria c2 and c3 to be more significant than criterion c1. So the significance of the assessment criteria play an important part in shaping the process of risk management.

2.1   Related Work

We divide the approaches from the literature that consider the significance value of the assessment criteria into two categories, namely quantitative and qualitative. This is according to how they represent the significance values of the assessment criteria. Quantitative approaches are those that utilize numerical values whereas qualitative approaches are those that utilize linguistic terms to represent the significance value of the assessment criteria. Among the quantitative approaches, a numerical range of either between 0 and 1[13, 14] or between 0 and 5 [3] is used. In that range, some approaches assign the significance values to the assessment criteria randomly, whereas some define different levels of significances, and the risk assessing agent assigns a criterion with a significance value according to the level of importance that it considers to have towards the successful completion of its business activity. This is similar to those approaches that use a qualitative scale for determining and assigning the significance value of the assessment criteria [15]. However, it can be argued that such process of assigning a significance value to an assessment criterion is in isolation with respect to the other criteria. What is needed is a systemized and informed process that assists the risk assessing agent in determining the level of significance of each assessment criterion by capturing the important factors such as: a) the relative importance and impact of each criterion to the successful achievement of the goal of the business interaction; b) the level of interdependence between the assessment criteria. In this paper, we will address this by proposing a method that accurately determines the significance of each assessment criterion by considering the different levels of inter-dependencies between them. For the sake of explanation we limit our discussion in this paper to the domain of business activities, but in reality it can be applied to the process of risk analysis in any domain. In the next sub-section, we define the problem that we address in this paper.

2.2   Problem Definition

Agent 'A' wants to form a business association with agent 'B' (which is a logistics company) in the context of transferring goods from Europe to Perth. The expectations of the business contract formed between the agents are as follows:

- Timely pickup of the goods from agent 'A' address in Europe on 26 March 2010 (Assessment Criterion C1)
- Meeting the required connections to leave Europe to reach and meet the delivery time in Perth (Assessment Criterion C2)
- Provision of a tracking facility during the goods transfer (Assessment Criterion C3)
- Delivery of the goods at agent 'A' address in Perth on or by 2 April 2010 (Assessment Criterion C4)
- Completion of the business activity in A$6000. Of this, agent 'A' has to pay A$4000 when the goods are picked up and the remaining A$2000 on delivery (Assessment Criterion C5).

Let us consider that the Easter holiday season in 2010 starts on 3 April 2010 and agent 'A' wants its goods to be delivered before that. So, according to the above expectations, it may consider assessment criterion C4 to be an important one compared to the other assessment criteria. If the risk assessed agent does not fulfill assessment criterion C4 according to the expectations, but fulfils others, then it may lead to agent 'A' assigning a high level of risk to agent 'B'. Other scenarios are possible where the risk assessing agent 'A' may want agent 'B' to pick up the goods at the specified time and hence may give more importance to that criterion (C1) when ascertaining its performance risk. In this scenario, if this assessment criterion is not completed, then agent 'B' might be assigned with a high performance risk value. So, depending upon the goal of the interaction and the inter-dependencies between the different assessment criteria, how is the significance of each assessment criterion, in terms of the successful completion of the business activity, determined? Mathematically, the problem is defined as:

$$\text{Performance Risk} = \sum_{i=1}^{n} Sc_n * \text{Uncommit } C_n$$

where: n represents the number of assessment criteria,
C_n represents the assessment criterion,
Sc_n represents the significance of each assessment criterion, and
Uncommit C_n represents the level of un-commitment of assessment criterion Cn.

Our aim is to determine the significance Sc_n of each assessment criterion according to (a) its importance in the successful completion of the business activity, and (b) considering the different level/s of inter-dependence between the assessment criteria. We will explain our approach to determine that in the next sections.

3   Analytic Hierarchy Process to determine the Significance of the Assessment Criteria

Analytic Hierarchy Process (AHP) is a quantitative technique that assists in making structured decisions about the weights to be given to the multiple criteria with respect to the goal to be achieved. It achieves this by performing a pair-wise comparison among the different assessment criteria and transforming the analysis to numerical representation by which the relative significance of assessment criteria are determined [16]. AHP has been applied in several applications like Balanced Scorecard [17, 18], Decision-Making [19] Geography Information System (GIS) [20] etc in different domains including education, business, social sciences and natural sciences to determine the importance of different factors with respect to the goal of analysis. For performing pair-wise comparison, AHP requires a hierarchy to be built among the criteria as goals, criteria, sub-criteria and alternatives whose significance values have to be determined with respect to the goal [18]. To build the hierarchy among criteria in our problem, the goal to be achieved has to be decomposed into different subfactors. This is achieved by the risk assessing agent forming the expectations of its interaction with the risk assessed agent and defining the assessment criteria that come under the goal as shown in Figure 2. It should be noted here that there might be other assessment criteria from the expectations which come under the goal of the interaction, but only those assessment criteria that are dependent on the risk assessed agent and which will be used by the risk assessing agent during the risk analysis process are considered while forming the hierarchy among the criteria.

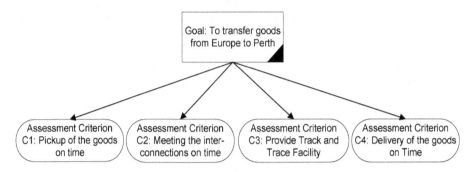

Fig. 2. Assessment Criteria with respect to the goal of the interaction

Once the hierarchy of the assessment criteria has been formed, a pair-wise comparison between them should be carried out to determine the significance of each with respect to the goal. But before that, the factors on the scale against which the criteria will be compared in a pair-wise fashion need to be defined. As our objective in this paper is to determine the relevant significance of each assessment criteria with respect to the goal, we define the levels which can assist in comparing any two assessment criteria and then determine which of them is more significant than the other and to what extent. To achieve that, we use a 4 level scale which is *Equal, Weakly More Important, Slightly More Important* and *Absolutely More Important* that compares the importance of the assessment criteria. It should be noted that these levels are defined according to the problem objective and they will change when the problem objective varies.

Pair-wise comparison between the assessment criteria is done by the risk assessing agent by using these levels of importance and its own critical judgment. For pair-wise comparison, the defined levels of importance are represented as a numeric number on a numeric scale. Each number represents a different level of importance. Pair-wise comparison among the criteria can be carried out by two approaches, namely conventional and fuzzy analysis. Conventional analysis requires the selection of an appropriate level of importance of an assessment criterion over the other, from a given scale during pair-wise comparison. The corresponding reciprocal value is assigned to the reverse comparison between the assessment criteria. For example, if assessment criterion C1 is considered important as compared to assessment criterion C2 for the successful completion of the goal, and an importance value of 5 (on a scale of 1-5, with 1 being the least important and 5 being the most important) is assigned to C1→C2, then for the reverse comparison C2→C1, a value of 1/5 is assigned as the relative importance. One drawback of this approach is that it has a degree of uncertainty associated with it as it is difficult to map qualitative preferences to point estimates [21].This is overcome by using a fuzzy analysis of pair-wise comparisons, which also assigns values of importance during pair-wise comparison, but uses fuzzy sets that have membership functions and a degree of membership to capture the uncertainty. One way to express the fuzzy sets to represent the level the importance on a given scale is by using a triangular fuzzy number that has three parameters *(l,m,u)*, with *l* representing the smallest possible level of importance, *m* representing the most possible level of importance and *u* representing is largest possible value of importance. If a value of *(l,m,u)* is assigned as the significance value of C1 over C2, then for reverse comparison the reciprocal of that value i.e. $(\frac{1}{u},\frac{1}{m},\frac{1}{l})$ is assigned. This process is more appropriate and effective than conventional AHP to capture the uncertainties during the pair-wise comparisons between the assessment criteria. In our approach, we will utilize the fuzzy-based approach to determine the weights of the assessment criteria. The series of steps to be followed are:

1. Define a linguistic range (Equal, Weakly More Important, Slightly More Important, and Absolutely More Important) to classify the different levels of importance and significance shown in Figure 3.
2. Define a triangular fuzzy scale to determine the different fuzzy sets of importance within the fuzzy scale shown in Table 1.
3. Determine the triangular fuzzy reciprocal value for the different linguistic fuzzy sets defined in step 2.
4. Construct an nXn matrix from pair-wise comparisons of the different assessment criteria with respect to the final goal to be achieved in the business activity. 'N' represents the number of assessment criteria in the business activity.
5. Utilize the fuzzy synthetic extent approach to determine the degree of possibility of a given assessment criterion being more important than the others.
6. The analysis of step 5 will represent the degree of importance of the different assessment criteria with respect to the goal of the interaction.

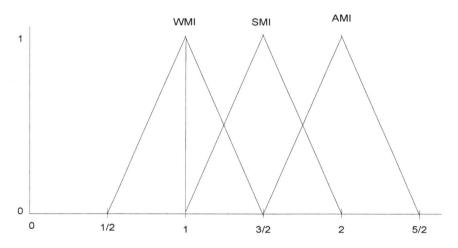

Fig. 3. The Linguistic scale for measuring the relative importance of assessment criteria

Table 1. Linguistic Scale for Importance of the Assessment Criteria

Linguistic Scale for Importance	Triangular Fuzzy Scale	Reciprocal Fuzzy Value
Equal	(1,1,1)	(1, 1, 1)
Weakly more Important - WMI	(1/2, 1, 3/2)	(2/3, 1, 2)
Slightly more Important -SMI	(1, 3/2, 2)	(1/2, 2/3, 1)
Absolutely more Important - AMI	(3/2, 2, 5/2)	(2/5, 1/2, 2/3)

Various techniques have been proposed in the literature that utilize the pair-wise comparison results and ascertain the relative importance of each criterion (Step 5). In this paper, we utilize the approach proposed by Chang [22] that calculates the priority vectors from the pair-wise comparisons and then determines the possibility of a fuzzy number being greater than the other by using the following formulas:

$$P\,(V_1 \geq V_2) = \qquad 1 \qquad \text{if } m_1 > m_2 \text{ or}$$

$$P\,(V_1 \geq V_2) = \frac{l_2 - u_1}{(m_1 - u_1) - (m_2 - l_2)} \qquad \text{if } m_2 > m_1$$

Interested readers are encouraged to look at Chang [22] for a detailed explanation of the process. Extending our example from Section 2, let us consider agent 'A' wants the goods to be delivered at the destination address by 2nd April 2010 and considers that assessment criterion as the most significant and important one for the successful achievement of its business activity. Determining the pair-wise comparison of each assessment criterion of Figure 2 with respect to the other assessment criteria, we obtain the 4 X 4 comparison matrix as shown in Table 2. Using the fuzzy synthetic approach, the weights of each assessment criteria determined are shown in the last column of Table 2.

Table 2. Pair-wise comparison matrix and the weights of the Assessment Criteria

Assess-ment Criteria (C)	C1	C2	C3	C4	Significance Value
C1	(1,1,1)	(3/2,2,5/2)	(3/2,2,5/2)	(2/3,1,2)	0.362
C2	(2/5,1/2, 2/3)	(1,1,1)	(1/2,2/3,1)	(2/5,1/2,2/3)	0.060
C3	(2/5,1/2,2/3)	(1,3/2,2)	(1,1,1)	(2/5,1/2,2/3)	0.197
C4	(1/2,1,3/2)	(3/2,2,5/2)	(3/2,2,5/2)	(1,1,1)	0.381

From the analysis, it can be seen that assessment criterion C4 is the most important of the expectations followed by C1, C3 and C2. The weights of these assessment criteria are determined by the risk assessing agent in comparison with the other criteria and hence, the analysis gives an informed representation of their importance for the successful completion of its business interaction. Consistent with the obtained analysis, existing approaches may assign the highest level of significance to assessment criteria C4, but they do not make any informed representation of the significance value that has to be assigned to the other assessment criteria C1- C3. This is addressed by using the AHP to ascertain the significance of each assessment criterion.

Approaches in the literature utilize AHP to determine the weight of each index or criterion in e-business [23]. However, a limitation of AHP when used to find the weights of the assessment criteria is that it works well only when there is no inter-dependence between the assessment criteria. In other words, the AHP process works well when all of the assessment criteria are independent of others and do not have an effect on the successful achievement of the other assessment criteria. For example, consider the following objective of forming a business association with the expectations as shown in Figure 4. In such activity, each assessment criterion is important to the successful achievement of the business goal but none of them has an effect on the other assessment criteria to be achieved successfully. In other words, there is no implied relationship between them.

In contrast, there may be certain business activities where the assessment criteria are inter-dependent. For example, considering the logistics interaction scenario mentioned in Figure 2, in order for the goods to be delivered on time, a sequence and series of steps must be completed beforehand, such as the goods being picked up on time, the different connections for transport to be met before the deadline which ensure that the goods reach the destination before the holiday closedown period. Also, let us consider that for the logistic company to activate the track and trace facility, the goods should reach the logistic company's warehouse from the remote location when they are picked up as shown in Figure 5.

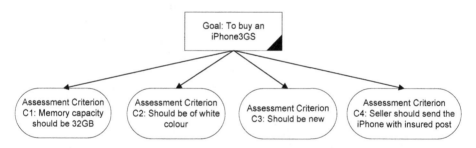

Fig. 4. Expectations and assessment criteria without dependency between them

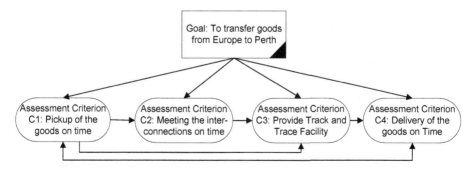

Fig. 5. Assessment criteria with respect to the goal and the inter-dependencies between them

On a scale of either [0-1] or [0-5], the risk assessing agent may assign the highest importance value to the assessment criterion 'delivering the goods on time' (C4) as it considers this to be the most important assessment criterion. This, however, does not take into consideration the level of dependency that the assessment criteria C1and/or C2 has on the assessment criterion C4, in order to meet that criteria. For example, if the first assessment criterion of picking up the goods on time (C1) is not met, then it is quite difficult for agent 'B' to meet the connections in time (C2) which will, in turn, have an effect on delivering the goods on time (C4). The risk assessing agent assigning a very high significance value to assessment criterion C4 and a low significance value to criteria C1 and C2 does not capture the level of dependency that C4 has on those criteria. Thus, a technique is needed which, aside from considering the importance of each assessment criterion with respect to the goal, also considers the level of dependencies between the different assessment criterion and ascertains their significance values accordingly. To address this in our approach, we utilize the Analytic Network Process (ANP) to ascertain the different weights of the expectations according to the inner dependencies between them. The approach is presented in the next section.

4   Analytic Network Process to Ascertain the Significance of the Assessment Criteria

Analytic Network Process (ANP) determines the impact of a given set of factors on the goal to be achieved by considering the levels of inter-dependencies between them. This is in contrast to Analytic Hierarchy Process (AHP) that considers only the weight of the factors on the goal. Many decision-making problems cannot be structured hierarchically because they involve the interaction and dependence of higher-level elements on lower-level elements [24]. When network system relationships exist between the criteria, the ANP can be used to make a better decision. ANP models the relationships between the criteria in two parts: the first models the control hierarchy among the criteria according to the goal, whereas the second models the different sub-networks of influences among the criteria, for each criterion [25]. The inter-relationships between the decision levels and attributes are modeled in ANP by determining the composite weights through the development of a supermatrix [16]. The supermatrix is actually a partitioned matrix, where each matrix segment represents a relationship between two components or clusters in a system.

As with AHP, the level of inter-dependence between the criteria can be modeled by conventional or fuzzy techniques. In our approach, we will use the fuzzy ANP process, which can tolerate vagueness and ambiguity in the information, thereby assisting in the decision-making process. The steps for the ANP are an extension of the AHP steps mentioned in the previous section. At first, the pair-wise comparison of each assessment criterion to the other criteria with respect to the goal is determined and their importance is ascertained as shown in Table 2. Then the following extension steps are followed:

1. Determine the level of inter-dependencies among the different assessment criteria and compute a pair-wise comparison matrix for each dependant criterion according to the linguistic scale defined in step 2 of the AHP.
2. Repeat steps 2- 6 mentioned in the previous section to obtain the relative inter-dependent importance weights for each assessment criteria over the other.

3. Multiply the weights obtained in step 6 of the AHP with the ones obtained in step 2 of the ANP to ascertain the final weights of each assessment criterion with respect to the goal and the inter-dependencies between them.

To explain with an example, let us consider the interaction scenario mentioned in Figure 5 and the inter-dependencies between the assessment criteria as shown in Figure 6. It can be seen from the figure that assessment criteria C4 is dependent on C1 and C2, C3 too is dependent on C1 and C2, C2 is dependent on C1 and there is inter-dependency between C2 and C4.

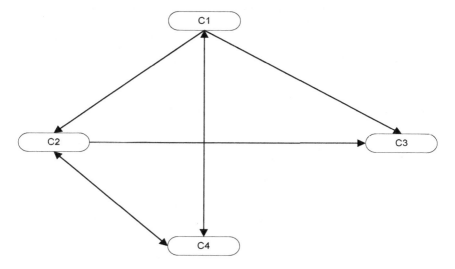

Fig. 6. Level of dependency between the Assessment Criteria of the Logistics Interaction

Once the importance of the assessment criteria with respect to the goal (shown in Table 2) is determined, then the inter-dependencies between the assessment criteria should be captured for ascertaining their appropriate significance value. This is done by using the scale of Figure 3 and determining the inner dependence matrix of each assessment criterion by using pair-wise comparisons. The inner dependence matrix represents the level of importance between any two or more assessment criteria with respect to another criterion by considering the level of dependence between them. Table 3 represents the level of inter-dependence between assessment criteria C1 and C2 with respect to assessment criterion C4. The inner dependence matrices for the other assessment criteria are shown in Tables 4 –5.

Table 3. The Inner dependence matrix for Assessment Criterion C4 on other criteria

Inner dependence matrix with respect to Assessment Criterion C4			
C4	C1	C2	Weight
C1	(1,1,1)	(1/2,2/3,1)	0.314
C2	(1,3/2,2)	(1,1,1)	0.685

Table 4. The Inner dependence matrix for Assessment Criterion C2 on other criteria

Inner dependence matrix with respect to Assessment Criterion C2			
C2	C1	C4	Weight
C1	(1,1,1)	(3/2,2,5/2)	0.833
C4	(2/5,1/2,2/3)	(1,1,1)	0.167

Table 5. The Inner dependence matrix for Assessment Criterion C3 on other criteria

Inner dependence matrix with respect to Assessment Criterion C3			
C3	C1	C2	Weights
C1	(1,1,1)	(1,3/2,2)	0.693
C2	(1/2,2/3,1)	(1,1,1)	0.307

Once the level of inter-dependencies between the assessment criteria has been determined, the final weights of the assessment criteria are determined by:

Dependencies within   X   Weight of the assessment   =   Significance of the
the assessment   criteria with respect to the goal   assessment criteria
criteria

Multiplying the interdependence weights of the assessment criteria with the individual weights of the criteria with respect to the goal determined in Table 2:

$$
\begin{array}{cccc}
C1 & C2 & C3 & C4
\end{array}
$$

$$
\begin{array}{c}
C1 \\ C2 \\ C3 \\ C4
\end{array}
\begin{bmatrix}
1 & 0.833 & 0.693 & 0.314 \\
1 & 1 & 0.307 & 0.685 \\
0 & 0 & 1 & 0 \\
1 & 0.617 & 0 & 1
\end{bmatrix}
X
\begin{bmatrix}
0.362 \\ 0.060 \\ 0.197 \\ 0.381
\end{bmatrix}
=
\begin{bmatrix}
0.668 \\ 0.381 \\ 0.098 \\ 0.753
\end{bmatrix}
$$

Normalizing the determined weights, we obtain the final weights of the assessment criteria as shown in Table 6.

Table 6. Significance values of the assessment criteria after considering the inter-dependencies

Assessment Criteria	Previous Significance Value	Significance value after considering inter-dependencies
C1	0.362	0.334
C2	0.060	0.190
C3	0.197	0.098
C4	0.381	0.377

As can be seen from Table 6, the significance values of the assessment criteria change considerably after considering the inter-dependencies between them. The importance of assessment criteria in descending order is now C4, C1, C2 and C3 as compared to C4, C1, C3 and C2 previously. Thus, it can be concluded that assessment criterion C2 is an important step for agent 'B' to commit to as compared to criterion C3 if agent 'A' has to achieve its 'goods delivered on time at the destination address' criterion. ANP helps to consider such level of inter-dependencies among the assessment criteria and correspondingly determines their significance values with respect to the goal of the interaction.

5 Conclusion

The significance of an assessment criterion is an important factor to be considered during the process of risk analysis in any activity. Although various techniques for assessing and managing risks have been proposed, a systematic approach is needed by which the importance of each assessment criteria on the successful completion of the business activity is determined. In this paper, we proposed an approach which utilizes the Analytic Hierarchy Process (AHP) to address this. AHP is a comprehensive framework that models the uncertainty associated in multi-objective or multi-criterion goals by structuring it and doing pair-wise comparisons to determine their significance. To determine the level of inter-dependence between the different assessment criteria and to model the network structure we used the Analytic Network Process (ANP). Such a structured and systematic process to ascertain the significances of assessment criteria helps to model the steps of risk assessment and management accurately for the successful achievement of the goal. The proposed approach can be applied to any activity in any domain to determine the significance of the assessment criteria while performing risk assessment and management. Once the significance of each assessment criterion has been determined, the risk assessing agent can then utilize our previously proposed approaches [3-5] to assess and manage the different levels of risk in the activity.

References

1. Joint Technical Committee OB-007 on Risk Management: AS/NZS ISO 31000:2009 Risk Management— Principles and Guidelines. In: Australia, S. (ed.): Sydney, p. 37 (2009)
2. Bohnet, I., Zeckhauser, R.: Trust, risk and betrayal. Journal of Economic Behavior & Organization 55, 467–484 (2004)
3. Hussain, O.K., Chang, E., Hussain, F.K., Dillon, T.S.: A methodology to quantify failure for risk-based decision support systems in digital business ecosystems. Data & Knowledge Engineering 63, 597–621 (2007)
4. Hussain, O.K., Dillon, T., Hussain, F., Chang, E.: Probabilistic Assessment of Financial Risk in E-Business Association Simulation Modelling Practice and Theory 19, 704–717 (2010)
5. Hussain, O.K., Dillon, T.S., Hussain, F.K., Chang, E.: Transactional Risk-based Decision Making System in E-Business Computing. Computer Systems Science and Engineering 25, 15–25 (2010)

6. Xiong, L., Liu, L.: Peertrust: Supporting Reputation-Based Trust for Peer-to-Peer Electronic Communities. IEEE Transactions on Knowledge and Data Engineering 16, 843–857 (2004)

7. Gourlay, I., Djemame, K., Padgett, J.: Reliability and Risk in Grid Re-source Brokering. In: Second IEEE International Conference on Digital Ecosystems and Technologies (DEST 2008), pp. 437–443. IEEE, Phitsanulok (2008)

8. Griffiths, N.: Enhancing peer-to-peer collaboration using trust. Expert Systems with Applications 31, 849–858 (2006)

9. Al-Bahar, J.F., Crandall, K.C.: Systematic Risk Management Approach for Construction Projects. Journal of Construction Engineering and Management 116, 533–546 (1990)

10. Fairley, R.: Risk management for software projects. IEEE Software 11, 57–67 (1994)

11. University of Queensland: Occupational Health & Safety Risk Assessment and Management Guideline (2010), http://www.uq.edu.au/ohs/pdfs/ohsriskmgt.pdf

12. Safety & Risk Management Charter:
http://www.rex.com.au/AboutRex/pdf/
Safety%20&%20Risk%20Management%20Charter.pdf

13. Wang, Y., Wong, D.S., Lin, K.-J., Varadharajan, V.: Evaluating trans-action trust and risk levels in peer-to-peer e-commerce environments. Information Systems and E-Business Management 6, 25–48 (2008)

14. Li, Y., Li, N.: Software Project Risk Assessment Based on Fuzzy Linguistic Multiple Attribute Decision Making. In: IEEE International Conference on Grey Systems and Intelligent Services, pp. 1163–1166. IEEE, China (2009)

15. Altenbach, T.J.: A Comparison of Risk Assessment Techniques from Qualitative to Quantitative. In: Proceedings of the Joint ASMEIJSME Pressure Vessels and Piping Conference, Honolulu, pp. 1–25 (1995)

16. Saaty, T.L.: The Analytic Network Process (2008)

17. Searcy, D.L.: Aligning the Balanced Scorecard and a Firm's Strategy Using the Analytic Hierarchy Process. Management Accounting Quarterly 5 (2004)

18. Jovanovic, J., Krivokapic, Z.: AHP In Implementation Of Balanced Scorecard. International Journal for Quality Research 2 (2008)

19. Wu, H.-H., Shieh, J.-I., Li, Y., Chen, H.-K.: A Combination of AHP and DEMATEL in Evaluating the Criteria of Employment Service Outreach Pro-gram Personnel. Information Technology Journal 9, 569–575 (2010)

20. Duc, T.T.: Using GIS and AHP Technique for Land-use Suitability Analysis. In: International Symposium on Geoinformatics for Spatial Infrastructure Development in Earth and Allied Sciences (2006)

21. Yu, C.-S.: AGP-A HP method for solving group decision-making fuzzy AHP problems. Computers & Operations Research 29, 1969–2001 (2002)

22. Chang, D.-Y.: Applications of the extent analysis method on fuzzy AHP. European Journal of Operational Research 95, 649–655 (1996)

23. Wang, L., Zeng, Y.: The Risk Identification and Assessment in E-Business Development. In: Wang, L., Jin, Y. (eds.) FSKD 2005. LNCS (LNAI), vol. 3614, pp. 1142–1149. Springer, Heidelberg (2005)

24. Saaty, T.L.: Time dependent decision-making; dynamic priorities in the AHP/ANP: Generalizing from points to functions and from real to complex variables. Mathematical and Computer Modelling 46, 860–891 (2007)

25. Saaty, T.L.: Multi-decisions decision-making: In addition to wheeling and deal-ing, our national political bodies need a formal approach for prioritization. Mathematical and Computer Modelling 46 (2007)

Chapter 20
Artificial Immune Systems Metaphor for Agent Based Modeling of Crisis Response Operations

Khaled M. Khalil, M. Abdel-Aziz, Taymour T. Nazmy, and Abdel-Badeeh M. Salem

Faculty of Computer and Information Science Ain Shams University Cairo, Egypt
{kmkmohamed,mhaziz67}@gmail.com, ntaymoor@yahoo.com,
absalem@asunet.shams.edu.eg

Abstract. Crisis response requires information intensive efforts utilized for reducing uncertainty, calculating and comparing costs and benefits, and managing resources in a fashion beyond those regularly available to handle routine problems. This paper presents an Artificial Immune Systems (AIS) metaphor for agent based modeling of crisis response operations. The presented model proposes integration of hybrid set of aspects (multi-agent systems, built-in defensive model of AIS, situation management, and intensity-based learning) for crisis response operations. In addition, the proposed response model is applied on the spread of pandemic influenza in Egypt as a case study.

Keywords: Crisis Response, Multi-agent Systems, Agent-Based Modeling, Artificial Immune Systems, Process Model.

1   Introduction

The challenge of crisis response is reducing the influence crises cause to society, the economy, and the lives of individuals and communities. This challenge is extreme in several dimensions. The demand is highly diverse and largely unpredictable in terms of location, time, and specific resources needed. Moreover, the urgency associated with crisis has many implications, such as the need to rapidly identify information about the developing situation, and to have the capability to make good decisions in the face of an inevitable degree of uncertainty and incompleteness of information. An efficient crisis response is of paramount importance, because if not responded to promptly and managed properly, even a small mishap could lead to a very big catastrophe with significantly severe consequences. Being equipped with a profusion of resources does not ensure a successful response to the crisis situation. Thus, the key to the successful response necessitates an effective and expedited allocation of the requested resources to the emergency locations. Such complexity suggests the use of intelligent agents for adaptive real-time modeling of the crisis response operations [16]. Multi-agent Systems are computational systems where software agents cooperate or compete with each other to achieve an individual or collective task [30].

In order to build multi-agent architecture, and increase the effectiveness of the crisis response operations, a similar metaphor is required that mimics the crisis

J. Lu, L.C. Jain, and G. Zhang: Handbook on Decision Making, ISRL 33, pp. 417–428.

response operations. AIS metaphor is selected in this study. AIS are a computational systems inspired by the principles and processes of the biological immune system [5] [9]. AIS represent an area of vast research over the last few years. For example, developed AIS in a variety of domains, such as machine learning [14], anomaly detection [4] [10], data mining [21], computer security [20] [6], adaptive control [24] and fault detection [7]. The biological immune system is a robust, complex, adaptive system that defends the body from foreign pathogens. It is able to categorize all cells (or molecules) within the body as self-cells or non-self cells. It does this with the help of a distributed task force that has the intelligence to take action from a local and also a global perspective using its network of chemical messengers for communication [6]. A more detailed overview of the immune system can be found in many textbooks [23] [26]. The immune system combines a priori knowledge with the adapting capabilities of a biological immune system to provide a powerful alternative to currently available techniques for pattern recognition, learning and optimization [22]. It uses several computational models and algorithms such as Bone Marrow Model, Negative Selection Algorithm, and Clonal Selection Algorithm [13].

In this paper we propose a multi-agent based model for crisis response. The proposed model architecture and operations process are adopted from AIS. Then the proposed response model is applied on controlling pandemic influenza in Egypt. Section 2 provides the proposed response model, while section 3 presents design of the proposed model for pandemic influenza in Egypt. Section 4 includes experiments. Finally, section 5 includes conclusions.

2   The Proposed Response Model

The view of the biological immune system provides the basis for a representation of AIS as systems of autonomous agents which exist within a distributed and compartmentalized environment [29]. In what follows, we present the multi-agent model based on the AIS metaphor for crisis response operations.

2.1   Proposed Hierarchical Architecture for Multi-agent Response Model

The architecture of the AIS can be abstracted into hierarchy of three levels (cells, tissue, and host) (see Fig. 1). Cells are able to interact with their environment and communicate and coordinate their behavior with other cells by synthesizing and responding to a range of molecules. Cells within the body aggregate to form tissue, such as muscle or connective tissue. Tissues themselves combine to form hosts, such as the heart, brain, or thymus. Hosts work together to form the immune system.

The proposed multi-agent architecture follows the same hierarchical architecture of the biological immune systems with mapping of the functionalities of cells to agents and adopting crisis response domain attributes and operations levels (operational, tactical and strategic levels [2] [3]) (see Table 1). Pathogens represent source of danger to the body entity, in which immune systems antibody cells tries to detect and kill. Pathogens are mapped to danger sources or undesired situations in the crisis domain, in which agents have to detect and overcome. Cells are represented by agents working as first responders and voluntaries. Cells contain different type of receptors which affect their capability to match pathogens and to kill them. Thus, receptors are mapped to agents skills or resources required to overcome danger. Agents are working in groups belonging to certain organization (tissue) which provide help by other agents teams dedicated in the tactical level. Host represents the grouping of different tissues working together. This can be mapped to emergency operations center (EOC) of different working divisions and each division contains specialized teams.

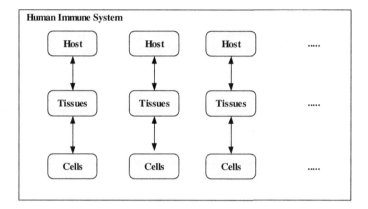

Fig. 1. Hierarchical architecture of human immune system

Table 1. Mapping biological immune system levels to crisis response operations and levels

Biological Immune System Level	Crisis Response Level	Crisis Domain
Pathogens	-	Danger sources – harmful situations
Cellular	Operational	First Responders and Voluntaries
Tissue	Tactical	Helper Agents staff
Host	Strategic	Emergency Operations Centers
System	-	Human Society and Important Properties

2.2   AIS Operational Architecture for Multi-agent Model

However, the AIS layered structure which is adopted in previous section is not complete from the conceptual framework perspective, which is required to allow effective algorithms to be developed [28]. Brownlee [1] said that "The acquired immune system provides a general pattern recognition and defense system that modifies itself, adapting to improve its capability with experience".

Decision Making Process of AIS (Conceptual Model Formalization)
The effectiveness of the system is due to a set of internal strategies to cope with pathogenic challenges. Such strategies remodel over time as the organism develops, matures, and then ages. Towards determining the decision making process of AIS, rational reconstructions approach is used. Rational reconstructions operate so as to

transform a given problematic philosophical scientific account-particularly of a terminological, methodological or theoretical entity-into a similar, but more precise, consistent interpretation [8]. Proposed rational reconstruction of AIS follows the same steps of Grant et al. work [12]. Grant et al. steps include: definition of requirements of the model, definition of the top-level use-cases, selection of notation, formalization of the process model by walking through use-cases, implementation, and evaluation.

Definition of requirements of the proposed response model follows crisis response systems design requirements [17] [18]. The following requirements were defined as:

- The model processes are concurrent.
- Support multiple instances of agents.
- Agents located in different layers should share required information only.
- Allow continuous monitoring of the situation and available resources.
- Permit relationships between agents to be collaborative.
- Allow separate behavior of each agent based on its role and objectives.
- Allow removal and adding of new agents and components at run time.
- Allow continuous planning/re-planning.
- Integrate planning and learning processes.
- Each process defined can be done by one or more agents. Agents with different roles differ by their allocated processes.

Definition of the top-level use cases is as follows (see Fig. 2):

- Use-case (0): no change in environment. This use-case applies when no pathogens found in the environment.
- Use-case (1): antibody found a pathogen. This use-case applies when the antibody finds a pathogen in the environment.
- Use-case (2): antibody receptors detect the required response for pathogens. This use-case applies when the antibody receptors match the pathogen, and can provide required response.
- Use-case (3): antibody receptors cannot detect the type of the pathogen thus failed to provide required response. This use-case applies when the antibody failed to match the pathogen and failed to provide response.
- Use-case (4): antibody cell asks for help for handling the unknown pathogen. This use-case applies when antibody failed to response to the pathogen. Antibody sends signals to activate the adaptive response.
- Use-case (5): adaptive cells mutate to match the pathogen and generate required receptors. This use-case applies when the adaptive cells mutate to match the pathogen and generate the required receptors.
- Use-case (6): required receptors are cloned to be applied to pathogen. This use-case applies when the required receptors for response are cloned to provide response to pathogen.
- Use-case (7): successful response is sustained as memory cells. This use-case applies when a successful response is executed; the receptors are stored in memory cells for later usage. Go to use-case (0).
- Use-case (8): failed response is ignored. This use-case applies when a failed response is gained; the receptors are ignored and not stored in memory. Go to use-case (5).

Selection of Notation includes selection of process notation for representing the process model. Integrated DEFinition Methods Technique (IDEFS0) [15] represents a common notation used in process modeling which is highly suited to specifying systems in terms of functional processes [12].

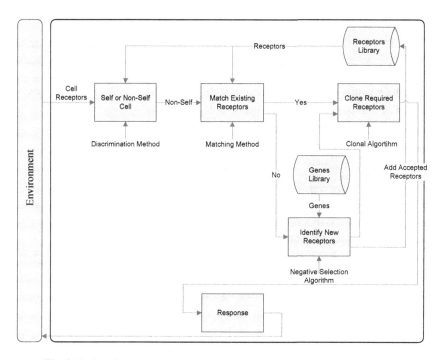

Fig. 2. Rationally reconstructed AIS model, formalized using SADT notation

The rationally reconstructed model (see Fig. 2) can be explained by starting at the environment and considering the activity of a typical Cell. Antibody cell checks other cells in the environment. Antibody cell determines if the examined cell is self or non-self based on the available receptors in the system using specific discrimination method. If cell is identified as non-self, cell immediately tries to match the receptors of the cell with existing receptors for response using matching method. If response receptors are found, clonal of the antibodies using clonal algorithm is applied to attack the pathogen cell. In case of new pathogen receptors not currently recognized, cell activates the adaptive response mechanism. Cells in the adaptive response mechanism mutate to match the non-self receptors using available genes library and negative selection algorithm. When reach an acceptable receptors form, the generated antibodies are cloned to provide response to the pathogen cell. Then, the newly generated receptors are added to the receptors library.

Table 2 shows comparison between the AIS rational reconstructed model and other process models such as OODA, RPDM, and Rasmussen models. Comparison criteria and OODA, RPDM, and Rasmussen models values are presented by Grant et al. [11].

Table 2. Comparing proposed AIS, OODA, RPDM, and Rasmussen models

Criteria/Process Model	OODA	Rasmussen	RPDM	AIS Model
Control Loop	√	√	√	√
Detailed	×	×	√	√
Tempo (fast decision making)	√	×	×	×
Planning	×	√	×	√
Learning	×	×	×	√

Formalizing the process model by walking through use-case. Use case (8) represents failing to response to pathogen cells. To formalize the process model, steps of the use case are presented as follows (see Fig. 3):

1. Antibody cell examines other cells looking for pathogens.
2. Antibody cell detects that cell is non-self.
3. Antibody cell tries to find proper receptors to kill the pathogen.
4. Antibody cell failed to response.
5. Antibody cell activates the adaptive system to generate proper receptors.
6. Adaptive cells mutate to match the pathogen and generate required receptors.
7. Required receptors are cloned and response is provided.
8. Failed response is reported as the pathogen is detected again which backs to step 1.

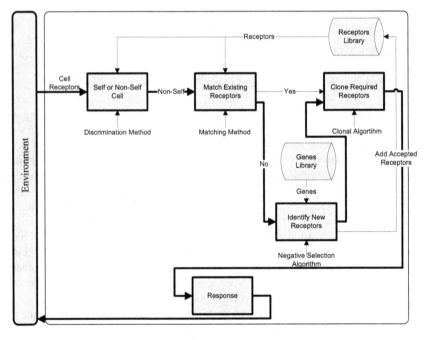

Fig. 3. Walk through use case (8), failed response to pathogen

Mapping AIS Operational Model to Crisis Response Multi-agent Model

After definition of the conceptual model of the AIS, mapping the proposed model to multi-agent model for crisis response is a straight forward process (see Tables 3 and 4). An agent can examine environment searching for danger sources or undesired feature. This can be mapped to operational agents. When agent finds an un-desired situation in environment, agent tries to handle the problem using available procedures. If the situation exceeds the available routine procedures, agents ask for help from tactical agents. Tactical agents check available memory for similar situations and check if the old experienced situation can be adopted for the current situation or not. If an old situation matches the current state of the environment, apply the course of actions coupled with the experienced situation. Otherwise, tactical agent asks for decision making (strategic) agent help. Using nearest matched situation, decision making agent mutates different course of actions to handle the current situation till reach an acceptable course of actions (intensity-based learning [25]). Decision making agent allocates course of actions to tactical agents to be deployed (cloned). During the execution of actions, operational agents report status of tasks execution, and in case of failed task re-planning is presented. Finally, generation and death of agents are related to the application domain of the proposed model. In crisis response death of cells and generation of new cells can be mapped to deployment of effective actions and neglecting others, or removal and adding new responders to the response field.

Table 3. Mapping AIS operational model to crisis response multi-agent model

Biological Immune System Level	Crisis Response Domain
Cells	Operational agent
Helper Cells	Tactical agents
Cells mutation	Decision making agents
Memory Cells	Case Memory

Table 4. Mapping AIS Operational Model Processes to agents' roles

Biological Immune System Process	Crisis Domain	Agent Role
Self or non-self discrimination	Reporting un-desired situation	Operational
Matching existing receptors	Situation Recognition	Tactical
Identify new receptors	Planning (Situation Assessment)	Decision making
Clone required receptors	Allocation of plan tasks	Tactical
Response	Executing plan tasks	Operational
Cells generation and death	Deployment of effective actions and neglecting non-effective actions	

3   Crisis Response to Pandemic Influenza in Egypt

This section presents the design, and implementation of the proposed response model for pandemic influenza in Egypt. Agent environment includes two parts: the pandemic model which is implemented in the previous work [19], and the available resources (control strategies or actions). Agent retrieves the current pandemic situation based on the agents' health states. Each control strategy is represented by (resource type, amount, cost, from/to date, and efficiency [19]). Agents' roles argue that each agent has its own profile and roles which specify its responsibilities and skills. For example decision making agent has the role of making decisions and adding new memory cases, while tactical agent has the role of processing information and allocating tasks to operational agents. Table 5 shows different agent roles, role responsibilities, and number of agents allowed per emergency operations center. Added here a new role titled tactical communication agent. Actually tactical communication agent is a tactical agent which is specialized for managing communication among the EOC parties. Tactical communication agent has to receive reports from operation agents, send reports to other tactical agents, deliver reports to crisis decision maker, and deliver plans back to the tactical agents.

Crisis decision making agent follows the BDI (Belief-Desire-Intention) cognitive model [27]. While, tactical and operational agents are recognized as reactive helper agents for the decision making agents, and do not have believes nor desires. Decision making agents need to work with other agents in EOC through collaboration to deliver resources information and to deploy actions. Decision making agents need to make decisions to control the spread of pandemic influenza using available effective resources. Decision making agents need to determine the course of actions to be deployed to control the pandemic.

Table 5. Agent roles and number of agents per EOC

Agent Role	Responsibilities	# of Agents
Decision Making Agent	Providing course of actions	1
	Adding new cases to memory	
Tactical Communication Agent	Receiving reports from operational agents	1
	Receiving resources reports from tactical agents	
	Sending reports to decision making agent	
	Receiving plan from decision making agent	
	Assigning plan tasks to tactical agents	
	Receiving status of plan execution	
	Sending plans status to decision making agent	
Tactical Agent	Handling reports from communication agent	1 or more
	Processing available resources	
Operational Agent	Reporting current environment state	1 or more
	Deployment of course of actions	

Design of the AIS Planning Methodology in the Proposed Response Model

Design of AIS planning methodology follows Stepney et al. [28] structure of AIS engineering. Pandemic situation is represented as a record of the total number of agents based on agents' health states. For example, situation can be represented by: (Susceptible: 50 agents, In-Contact: 10 agents, Infectious: 20 agents, Isolated Infected: 10 agents, Recovered: 1 agent, and Dead: 3 agents). The required course of actions to control given pandemic situation is represented as follows: (identifier, successfulness of the course of actions, and current pandemic situation). In addition the course of actions is coupled with deployed actions. The entry of course of actions and its coupled actions constructs a memory case.

The city block distance is used here to find the similarity between situations. For example: distance between situation 1 (Infectious: 2 agents, Isolated Infectious: 1 agent) and situation 2 (Immunized: 31 agents) is: $|0-2|+|0-1|+|31-0|=33$. Immune algorithms present the mutation (dynamic) behavior of AIS. Bone marrow algorithm, positive selection algorithm, clonal selection is used in the proposed model implementation.

4   Experiments

The main goal of experiments scenarios is to validate the proposed response model. Scenarios basically include simulation of pandemic influenza in a closed population of 1000 agents and initially three infected agents located in Cairo. Cairo EOC contains 3 operational agents, 2 tactical agents, and one decision making agent. The duration of the simulation round is 50 days. All simulation rounds involve randomly generated resources pool. Each action in the resource pool has efficacy of (0.75). The basic simulation round with no control strategies gives pandemic peak on day 10 with %60.8 infected agents [19].

Fig. 4 shows the flow of control during a simulation round, and the total cost of deployed actions (total cost = 2821.4 and plan certainty = 0 due to there are no previous plans in the system). While, Fig. 5 shows the stored case memory for the finished simulation round (case id = 174, successfulness = 0.001198). It is found that the pandemic peak is shifted to day 16 with %55 infected agents of the population.

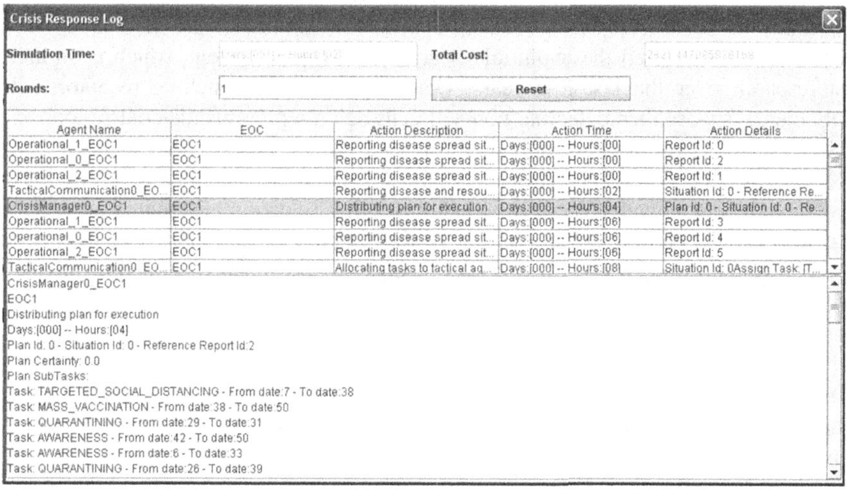

Fig. 4. Round 1 - Crisis Response Log

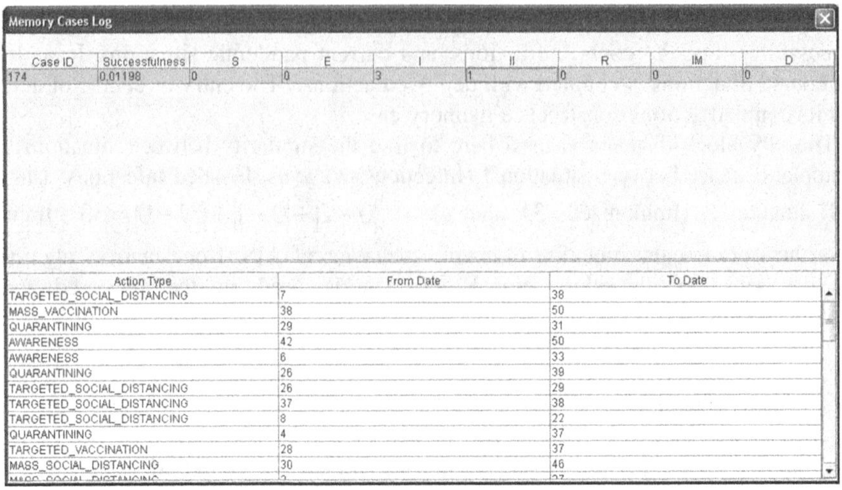

Memory Cases Log

Case ID	Successfulness	S	E	I	II	R	IM	D
174	0.01198	0	0	3	1	0	0	0

Action Type	From Date	To Date
TARGETED_SOCIAL_DISTANCING	7	38
MASS_VACCINATION	38	50
QUARANTINING	29	31
AWARENESS	42	50
AWARENESS	6	33
QUARANTINING	26	39
TARGETED_SOCIAL_DISTANCING	26	29
TARGETED_SOCIAL_DISTANCING	37	38
TARGETED_SOCIAL_DISTANCING	8	22
QUARANTINING	4	37
TARGETED_VACCINATION	28	37
MASS_SOCIAL_DISTANCING	30	46
MASS_SOCIAL_DISTANCING	2	37

Fig. 5. Round 1 - Case Memory

5   Conclusions

Currently, multi-agent architecture is the essence of response systems. The original idea comes out from agent characteristics in MAS, such as autonomy, local view of environment, capability of learning, planning, coordination and decentralized decision making. The incorporation of multi-agent systems can be clarified by discussing major disciplines involved during crisis response operations; situation management, and decision making and planning.

The study of biological systems is of interest to scientists and engineers as they turn out to be a source of rich theories. They are useful in constructing novel computer algorithms to solve complex engineering problems. Immunology as a study of the immune system inspired the evolution of artificial immune system, which is an area of vast research over the last few years. Detecting non-self, matching receptors, clone antibodies, response then detecting non-self cells again represents the control loop in the AIS metaphor. This process loop (control loop) is the core of the decision making process in AIS metaphor. The AIS model allows learning by mutating genes to generate acceptable receptors and store them in memory cells. Selection of required antibodies to be cloned and the cloning process represents the planning methodology of AIS. Artificial immune systems represent an interesting metaphor for building effective based and defensive multi-agent crisis response operations model. The proposed architecture proposed by AIS, promises effective operations and system architecture for crisis response.

According to experiments scenarios, AIS model shows slow learning process but it is very fast in handling desired or un-desired situations. Number of cases represents the growth of the system. While, scenarios results show that the effectiveness of response operations and utilization of resources are improved within the growth of the response model by ignoring low successfulness memory cases while deliberating new course of actions. The simulation process is very slow and consumes a lot of computation power.

Each round takes at least 10 minutes to complete. It is recommended to implement the model using high performance computing to enable fast growth of case memory.

References

1. Brownlee, J.: A Hierarchical Framework of the Acquired Immune System. Technical Report, Complex Intelligent Systems Laboratory (CIS), Swinburne University of Technology, Victoria, Australia (2007)
2. Builder, C.H., Bankes, S.C., Nordin, R.: Command Concepts A Theory Derived from the Practice of Command and Control. RAND Corporation's National Defense Research Institute (1999) ISBN/EAN: 0-8330-2450-7
3. Chen, R., Sharman, R., Rao, H.R., Upadhyaya, S.: Design Principles for Critical Incident Response Systems. Journal of Information Systems and E-Business Management 5(3), 201–227 (2007)
4. Dasgupta, D., Forrest, S.: Novelty detection in time series data using ideas from immunology. In: ISCA 5th International Conference on Intelligent Systems (1996)
5. Dasgupta, D., Attoh-Okine, N.: Immunity-Based Systems: A Survey. In: The Proceedings of the IEEE International Conference on Systems, Man and Cybernetics, Orlando, Florida, vol. 1, pp. 369–374 (1997)
6. Dasgupta, D.: Immune-based intrusion detection system: A general framework. In: Proceedings of the National Information Systems Security Conference, pp. 147–160 (1999)
7. Dasgupta, D., Krishnakumar, K.T., Wong, D., Berry, M.: Negative Selection Algorithm for Aircraft Fault Detection. In: Nicosia, G., Cutello, V., Bentley, P.J., Timmis, J. (eds.) ICARIS 2004. LNCS, vol. 3239, pp. 1–13. Springer, Heidelberg (2004)
8. Davia, G.A.: Thoughts on a Possible Rational Reconstruction of the Method of "Rational Reconstruction". In: Twentieth World Congress of Philosophy, Boston, Massachusetts, U.S.A., pp. 10–15 (1998)
9. Golzari, S., Doraisamy, S., Sulaiman, M.N.B., Udzir, N.L.: A Review on Concepts, Algorithms and Recognition-Based Applications of Artificial Immune System. CCIS, vol. 6, pp. 569–576. Springer, Heidelberg (2008)
10. Gonzalz, F., Dasgupta, D.: Anomaly detection using real-valued negative selection. Genetic Programming and Evolvable Machines, 383–403 (2004)
11. Grant, T.J., Kooter, B.M.: Comparing OODA and Other Models as Operational View C2 Architecture. In: Proceedings, 10th International Command and Control Research and Technology Symposium (ICCRTS 2005), Washington DC, USA (2005)
12. Grant, T.: Unifying Planning and Control using an OODA-based Architecture. In: Proceedings of Annual Conference of South African Institute of Computer Scientists and Information Technologists (SAICSIT), White River, Mpumalanga, South Africa, pp. 111–113 (2005)
13. Hofmeyr, S.A.: An interpretative introduction to the immune system. In: Cohen, I., Segel, L.A. (eds.) Design Principles for the Immune System and other Distributed Autonomous Systems. Oxford University Press, New York (2000)
14. Hunt, J.E., Cooke, D.E.: Learning using an artificial immune system. Journal of Network Computing Applications 19, 189–212 (1996)
15. Integrated DEFinition (IDEF) methods IDEF0 – Function Modeling Method, http://www.idef.com/IDEF0.htm

16. Khalil, K.M., Abdel-Aziz, M.H., Nazmy, M.T., Salem, A.M.: The Role of Artificial Intelligence Technologies in Crisis Response. In: Mendel Conference, 14th International Conference on Soft Computing (2008)

17. Khalil, K.M., Abdel-Aziz, M.H., Nazmy, M.T., Salem, A.M.: Multi-Agent Crisis Response systems – Design Requirements and Analysis of Current Systems. In: Fourth International Conference on Intelligence Computing and Information Systems, Cairo, Egypt (2009)

18. Khalil, K.M., Abdel-Aziz, M.H., Nazmy, M.T., Salem, A.M.: Bridging the Gap between Crisis Response Operations and Systems. Annals of University of Craiova, Mathematics and Computer Science Series 36(2), 141–145 (2009) ISSN: 1223-6934

19. Khalil, K.M., Abdel-Aziz, M.H., Nazmy, M.T., Salem, A.M.: An agent-based modeling for pandemic influenza in Egypt. In: INFOS 2010: 7th International Conference on Informatics and Systems, Cairo, Egypt (2010)

20. Kim, J., Bentley, P.: Toward an artificial immune system for network intrusion detection: An investigation of dynamic clonal selection. In: Proceedings of the 2002 Congress on Evolutionary Computation, Honolulu, Hawaii, pp. 1244–1252 (2002)

21. Knight, T., Timmis, J.: AINE: An immunological approach to data mining. In: IEEE International Conference on Data Mining, pp. 297–304 (2001)

22. Krishnakumar, K.: Artificial Immune System Approaches for Aerospace Applications. Technical Report, AIAA (2003)

23. Kubi, J.: Immunology, 5th edn. W H Freeman (2002)

24. Krishna, K.K., Neidhoefer, J.: Immunized adaptive critic for an autonomous aircraft control application. In: Artificial Immune Systems and Their Applications, ch. 20, pp. 221–240. Springer-Verlag, Inc. (1999)

25. Lewis, L.: Managing Computer Networks: A Case-Based Reasoning Approach. Artech House (1995)

26. Perelson, A.S., Weisbuch, G.: Immunology for physicists. Reviews of Modern Physics 69, 1219–1267 (1997)

27. Shendarkar, A., Vasudevan, K., Lee, S., Son, Y.: Crowd simulation for emergency response using BDI agent based on virtual reality. In: WSC 2006: Proceedings of the 38th Conference on Winter Simulation, pp. 545–553 (2006)

28. Stepney, S., Smith, R.E., Timmis, J., Tyrrell, A.M.: Towards a Conceptual Framework for Artificial Immune Systems. In: Nicosia, G., Cutello, V., Bentley, P.J., Timmis, J. (eds.) ICARIS 2004. LNCS, vol. 3239, pp. 53–64. Springer, Heidelberg (2004)

29. Twycross, J., Aickelin, U.: Information Fusion in the Immune System. Information Fusion 11(1), 35–44 (2010)

30. Weiss, G.: Multiagent Systems: A Modern Approach to Distributed Artificial Intelligence, pp. 1–23. MIT Press (2000)

Chapter 21
Mobile-Based Emergency Response System
Using Ontology-Supported Information Extraction

Khaled Amailef and Jie Lu

Decision Systems & e-Service Intelligence (DeSI) Lab
Centre for Quantum Computation & Intelligent Systems (QCIS)
School of Software
Faculty of Engineering and Information Technology
University of Technology Sydney
P.O. Box 123, Broadway, NSW 2007, Sydney, Australia
{kamailef,jielu}@it.uts.edu.au

Abstract. This chapter describes an algorithm within a Mobile-based Emergency Response System (MERS) to automatically extract information from Short Message Service (SMS). The algorithm is based on an ontology concept, and a maximum entropy statistical model. Ontology has been used to improve the performance of an information extraction system. A maximum entropy statistical model with various predefined features offers a clean way to estimate the probability of certain token occurring with a certain SMS text. The algorithm has four main functions: to collect unstructured information from an SMS emergency text message; to conduct information extraction and aggregation; to calculate the similarity of SMS text messages; and to generate query and results presentation.

1   Introduction

Within the context of electronic government (e-Government), mobile government (m-Government) services offer greater access to information and services for citizens, businesses, and non-profit organizations through wireless communication networks and mobile devices such as Personal Digital Assistants (PDAs), cellular phones, and their supporting systems [1]. m-Service is one of m-Government dimension in the public sector, and an example of one the most promising services is crisis communication. In Great Britain, for example, people with a hearing disability can send a text message to an emergency number to assist an emergency centre to locate the person in need and respond to them quickly [2]. In emergency situations such as the 9/11 terrorist attack, information communication technology (ICT) in the form of mobile telecommunications and the Internet have the potential to provide more capable and effective services in warning and response activities [1].

As stated in [3], the objective of emergency response systems is to minimize the impact of a disaster on several organizations including government agencies, businesses, and volunteer organizations, as well as experts, and residents. Disasters can be

J. Lu, L.C. Jain, and G. Zhang: Handbook on Decision Making, ISRL 33, pp. 429–449.
springerlink.com   © Springer-Verlag Berlin Heidelberg 2012

classified into two main categories: man-made and natural [4]. In a large-scale terrorist, information sharing and incompatible technology are significant problems for all such organizations [5]. Emergency response systems, as one of the new m-Government services, have the facilities to access the electronic public services from anywhere and at any time in dealing with an emergency. The goal of this study is to make mobile devices valuable tools for emergency response systems under the platform of m-Government.

Literature shows that mobile-based information systems can be a solution that benefits responders in a variety of ways. First, the mobility devices allow first responders to perform better rapid and actionable decision-making. Second, the development of wireless/mobile technology and communication trends provides a ubiquitous environment for the deployment of mobile devices in a variety of fields [6]. In addition, using mobile devices will create interactive communication mechanisms that can facilitate just-in-time communication and collaboration among a large number of residents and responders [5]. The MERS makes use of mobile technologies to assist government to obtain information and respond to disasters. In this way, every mobile user has the ability to report emergency information and to rapidly information that may safe his or her life. This research mainly considers situations involving man-made disasters, such as bomb attacks, non-natural events, fire, chemical attacks, hijacking and so on, all of which have common features in response system development.

The chapter is organized as follows: Section 2 briefly discusses the mobile-based emergency response system (MERS) within the m-Government dimension. Section 3 presents information extraction and text aggregation concepts which are used for the MERS. Section 4 proposes system architecture for SMS text extraction and aggregation. Section 5 provides an ontology-based representation for unstructured Short Message Service (SMS) text. Section 6 gives an illustrated example. Section 7 describes system implementation. Finally, concluding remarks and further work are discussed in Section 8.

2   Emergency Response System within m-Government Dimensions

Recent advances in Internet technologies and services have allowed governments to provide a new way to deal with citizens and businesses through mobile platforms. The demand for better, more efficient and more effective government services will put serious pressure on the government with regard to m-Government (i.e. transparency, access, affordability, and participation) [7]. m-Government can be seen as 'a subset of e-Government. It stands for the use of mobile and wireless communication technology within the government administration and in its delivery of services and information to citizens and firms' [8]. According to [9, 10], four types of m-Government services are identified and categorized as Government to Government (G2G), Government to Employee (G2E), Government to Citizens (G2C), and Government to Businesses (G2B). This study mainly focuses on the G2C category but also involves some part of G2G.

As stated in [8, 11], m-Government applies to four main dimensions in the public sector. (1) m-Communication: Providing information to the public is not a trivial activity. It is the foundation of citizen empowerment. Without relevant information citizens are unable to form intelligent opinions and are consequently unable to act meaningfully on the issues before them. (2) m-Services: Providing a channel of communication between citizens and government via SMS, and also enabling G2C transactions. Some examples of existing m-Services include m-Parking, m-Teacher, m-library and crisis communication (3) m-Democracy: m-Voting and the use of SMS and mobile devices for citizen input to political decision-making is an m-Government application with tremendous potential to enhance democratic participation. (4) m-Administration: m-Government also provides opportunities to improve the internal operation of public agencies.

2.1   The MERS Conceptual Framework

The MERS proposed in this study is an important new m-Service under m-Government platform, which aims to provide a new function and service of m-Government [12]. A MERS under an m-Government platform is a mobile-based information system designed to enable people to get help from and provide information to the government in an emergency situation. The MERS conceptual framework proposed in this study consists of four main components: inputs, processes, outputs, and outcomes as illustrated in Figure 1.

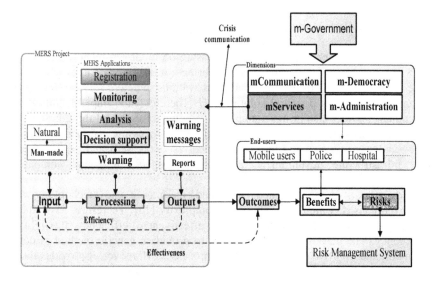

Fig. 1. Emergency response system within m-Government dimension

Input stands for the elements to be entered into the system. Examples of inputs are a mobile user's details and emergency request data. Processing represents all the elements necessary to convert or transform inputs to outputs, for example, a process may include registration application; monitoring application, analysis application; decision support application, and warning application. Output is defined as the consequences of being in the system, for example, warning messages. Outcomes can take either or both of two forms. Benefits and risks, each should be well planned.

The MERS supports five major applications:

- **Registration Application:** This application keeps track of all entities of interest in the disaster area such as relief organizations, experts, mobile users, etc. The information includes mobile users' personal details, and the data also includes a comprehensive list of expert's relief organizations, and volunteer organizations. In addition, it captures the essential services provided by each organization.
- **Monitoring Application:** This application constantly monitors crisis and reports quickly report in charge of analysis application. The module collects complete and accurate disaster data via mobile communication devices.
- **Analysis Application:** The functions of this application are to support crisis classification, to identify experts to contact, and to search for background information.
- **Decision Support Application:** This application is used to facilitate information gathering, generate quick data queries (assessment report), facilitate information sharing (convert data into usable information), and provide consultation information.
- **Warning Application:** This application is used to identify the parties to be notified, and to facilitate the generation and distribution of warning information.

2.2   Risk and Risk Management System

A MERS risks represent another type of pressure on m-Government organization. A MERS system (as s one of m-Services) will bring a series of challenges which shared by the m-Government efforts. Like many innovations, wireless services may present as many challenges as potential solutions [1]. Therefore, the implementation of m-Government services also brings a number of challenges. Some of these issues shared with e-Government implementation, such as limited processing and memory capacity, privacy and security, cost and lack of financial resources, staff and expertise, acceptance, infrastructure development, accessibility, legal issues, and compatibility [13, 14].

The complexity of government agencies and emerging of a new technology will introduce more risks, which may add to its complexity [15]. Therefore, Risk Management (RM) system is required to guarantee effectiveness, efficiency, flexibility and transparency. The purpose of RM system is to provide a generic method to defined, implement, and improve the MERS project processes. Moreover, it provides the public organizations with appropriate support in the planning and management phases and it guarantees integration between different projects carried out in the area of m-Government project. These goals can only be achieved through an effective management system or framework. RM frameworks set out the relationship between the processes of risk identification, evaluation and management

In the following sections, we will focus on the analysis application which is one of the most important applications within the MERS system. In particular, we will present an algorithm within the analysis application to automatically extract information from SMS emergency text messages.

3   An Information Extraction and Text Aggregation

In this section, we lay out the background of the related research area of this study. We will provide related concepts of information extraction (IE), named entity recognition (NER), and the maximum entropy model.

3.1   Information Extraction

In order to define and understand the IE concept, we will consider the following examples. In the first example, a government is looking for a summary of the latest data available about a natural disaster that occurred in the last five years. The second example concerns the investigation of general trends in terrorist attacks all over the world. A government has a huge database of news and emails and wants to use these to obtain a detailed overview of all terrorist events and natural disasters in a particular region. These two different examples share in the following points:

- The request for information;
- The answer to this request is usually presented in unstructured data sources such as text and images;
- It is impractical for humans to process huge data sources and;
- Computers are not able to directly query an unstructured data format.

IE is a significant research field for solving this kind of problem within the Artificial Intelligence area because it tries to extract related information out of huge amounts of data [16-18]. IE stands for '*A technology for finding facts in plain text, and coding them in a logical representation such as a relational database*' [18, 19]. Moens [20, p. 4] sees IE as:

> ' *the identification, and consequent or concurrent classification and structuring into semantic classes, of specific information found in unstructured data sources, such as natural language text, making the information more suitable for information processing tasks*'.

As illustrated in Figure 2, IE consists of five major subtasks including [21]:

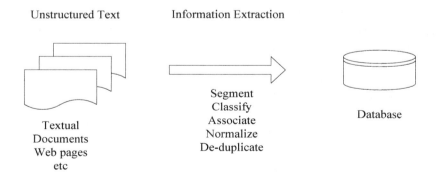

Fig. 2. Information extraction subtasks

- Segmentation finds the boundaries of the text snippets that will fill a database field.
- Classification determines which database field is the correct destination for each text segment.
- Association determines which fields belong together in the same record.
- Normalization puts information in a standard format by which it can be reliably compared.
- De-duplication collapses redundant information and remove duplicate record in your database.

3.2   Information Aggregation Definitions

Aggregation aims to combine a finite number of observed values into a single output [22]. Many aggregation operators have been developed to aggregate information with a wide range of properties. There are various linguistic operators and it is not a simple task to select a suitable operator for a particular system [23, 24]. To aggregate numeric values, we use the simplest and most common way to aggregate which is an average aggregation function. Mathematically we have (Equation (1)):

$$M(x_1, x_2, ..., x_n) = \frac{1}{n} \sum_{i=1}^{n} x_i \text{ , where } x_i \text{ is a numeric value} \tag{1}$$

To aggregate linguistic values, we conduct a linguistic aggregation operation as follows (Equation (2)):

$S\alpha, S\beta \in \overline{S}, \lambda \in [0,1]$, and their operational rules can be defined as follows:

$$S_\alpha \oplus S_\beta = S_{\alpha+\beta}, \lambda S_\alpha = S_{\lambda\alpha} \text{ , where} \tag{2}$$

$$\overline{S} = \{ S_{-2} = very\ low, S_{-1} = low,$$
$$S_0 = medium, S_1 = high,$$
$$S_2 = very\ high \}$$

Then, we compute the union operation between them, so that we have, $\eta = \bigcup R_i$, where R_i is a given SMS test message such that $R_i = \{r_1, r_2, ... r_m\}$, m is the number of words in message R_i, and r_j is the jth word in message R, $i=\{1,2,...,n\}$, n is the number of SMS text messages.

3.3   Named Entity Recognition

Named entity recognition (NER) classifies every word in a document into predefined categories [23]. In [25], the authors defined NER as the identification of text terms referring to items of interest. NER is important for semantically oriented retrieval tasks, such as question answering [26], biomedical retrieval [24], trend detection, and event tracking [27]. Much research has been carried out on NER, using both knowledge engineering and machine learning approaches [28, 29].

Several approaches have been proved to successfully employ NER development. Hidden Markov Model (HMM) [30], Maximum Entropy (ME) [31, 32], Conditional Random Field (CRF) [32], and Support Vector Machine (SVM) [33] are the most commonly used techniques. In this chapter, a ME statistical model is used for name-entity recognition. The ME is a flexible statistical model which assigns an outcome for each token (word) based on its history and features. For example, in a SMS text including a title element such as "St" after proper name, a proper name is the name of a street. In the sentence: "Car bomb attack in Oxford St". The term "Oxford" is recognized as the name of a street because it is a proper name preceded by term which belongs to contextual information ("St"). In the above example, we can directly conclude that the term 'Oxford' is a proper street name because of its features (spelling with an upper case in the beginning and followed by the element "St".

3.4   Maximum Entropy Model

The ME model has been applied in many areas of science and technology such as natural language processing, text classification, and machine learning [34]. It produces a probability for each category of a nominal candidate conditioned on the context in which the candidate occurs. It is a flexible statistical model which assigns an outcome for each token based on its history and features. We can compute the conditional probability (Equations (3-6)) as:

$$p(o/h) = \frac{1}{Z(h)} \prod_{j=1}^{k} \alpha_j^{f_j(0,h)} \tag{3}$$

$$t = \arg\max \; p_j(o \setminus h) \tag{4}$$

$$Z(h) = \sum_h \prod_{j=1}^{k} \alpha_j^{f_j(o,h)} \tag{5}$$

where o refers to the outcomes and h is the history or the context, history can be viewed as all the information derivable from the training corpus relative to the current token[35]. $Z(h)$ is a normalization function and is equal to 1 for all h. $f_j(j=1,2,... k)$ are known as features that are helpful in making predictions about the outcome. Parameter α_j is the weight of feature f_j where $f_j(o,h) \rightarrow \{0,1\}$. For instance, for the problem of emergency SMS text categorization, we can define a feature function as follows:

$$f(o,h) = \begin{cases} 1 & \textit{iff } o = disaster\ location \\ & and\ CurrentWord(h) \in DIC; \\ 0 & otherwise. \end{cases} \tag{6}$$

In this section, we describe the features used in MERS. Since the features are critical to the success of machine learning approaches [21], we will discuss them in more detail. The features used in our model come from reference [22]:

- **Slipping Window:** Typically, word windows are of size 5 consisting of the current token, two tokens to the left and two tokens to the right (e.g. w_{-2} w_{-1} w_0 w_1 w_2). This feature is useful for determining the class entity to which the current token belongs.
- **Previous Token Tags:** These features are helpful for capturing information on the interdependency of the label sequence.
- **Prefix and Suffix:** These features provide the composition of the word, which usually yields useful information on its semantic.
- **Capitulation and Digit Information:** Semantic or syntactic type information can provide useful information.
- **Dictionary Lookup Features:** This feature is used to decide whether or not a token phrase belongs to a certain prebuilt dictionary.
- **Pattern Feature:** This feature is used to determine a set of decision rules that will produce a positive example

4   System Architecture for SMS Text Extraction and Aggregation

To extract and aggregate SMS text messages received from mobile phone users in an emergency situation, we propose the following processing steps. First, unstructured information is pre-processed by storing it in a relational database. The information extraction algorithm is then applied. The result of this information extraction algorithm will be stored in another relational database. For decision makers, the structured information relating to tables and messages can be used to assist in responding to an emergency situation. The main components used in the SMS text extraction and aggregation process and their interactions are depicted in Figure 1.

Below are the main components of the system:

1) **Pre-processing (lookup resources):** Obviously, the first task in the IE system is to collect the data. In our IE scenarios, the relevant data is given as SMS text messaging. The collection of SMS emergency text messages is acquired and is deposited in a relational database. We use the following set of attributes to represents SMS messages:

    - Words – sequences of alpha or numeric characters in the message text.
    - Word bi-grams – sequences of n words in a window of 5 words preceding the current word.

    An example of an SMS text message:
    "There is a car bomb attack, near UTS Uni Building 10, king George St, about 10 students <u>were</u> killed and 20 students <u>were</u> injured".

2) **Information Processing:** Information processing is the task of finding specific pieces of information from unstructured or semi-structured documents. It can be seen as a process that aims to fill the database from unstructured SMS text messages. Information extraction involves five major subtasks:

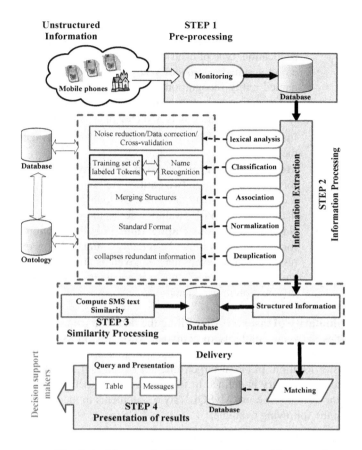

Fig. 3. Architecture for SMS text extraction and aggregation

- **Lexical Analysis:** The first step in handling the SMS text is to divide the stream of characters into words or more precisely, tokens. Identifying the tokens is an essential task for the extraction of information from SMS text. Each token represents an instance of a type. As an example, in the above example of an SMS text message, the two underline tokens ("were") represent instances of a type "were". Once a text has been segmented into a sequence of tokens, the next step is to convert each token into a standard form. For example *"UTS Uni"* is converted to UTS University. The final step in lexical analysis is to use the stopping technique to remove information that is not useful for categorization (so-called stop-words). For example, remove articles, prepositions, pronouns and other functional words that are not related to the content. A list of predefined stop words must be developed first. The application will then identify and remove all the stop words in the SMS text based on the predefined list.

- **Name Entity Recognition:** Once a name entity (NE) is identified, it is classified into predefined categories of SMS text messages such as location, type of disaster, physical target, etc. We use a maximum entropy statistical model with various predefined features for the named entity recognition. From a mathematical point of view, the classification is a function that maps words or tokens to labels, $f : w \rightarrow L$, where w is a vector of words and L is a set of labels. In our case, the label represents a goal that is potentially related to the words.

- **Merging Structure:** This step is referred to as relation extraction for the case in which two entities are being associated. For example, a disaster event may described by multiple names. Extraction must determine which filed values refer to the same disaster event.

- **Normalization and Duplication:** Puts information into a standard format by which it can be reliably compared. For example, the status of a disaster event may be given as "in-progress", and in another case as "in progress", thus it is necessary to avoid getting redundant information in the database.

3) **Similarity Measures:** After filling the database from unstructured SMS text messages, the similarity of composite SMS text messages is calculated. The system measures the similarity based on four different types of features: disaster location, physical target, disaster event, and disaster weapon used. To do this, a query is defined to filter the resulting set of resources (SMS text). For example, we use MySQL[1] style statements to measure similarity as seen in the following example:

```
Example :Java Code
rs  =  stmt.executeQuery("SELECT     *,     COUNT(*)     from
smsmgs.smstextextraction   GROUP  BY  Disaster_Event,Street_Name,
Post_Code, State, Physical_Target, Weapon_Used");
```

In this example, the search is carried out in the MySQL table of each retrieved resource, matching the classification with preferences (disaster event, location physical target and weapon used). This example is used to list all the data in the specified MySQL table column based on certain specified conditions.

4) **Presentation of Results:** Provides decision makers and mobile users with a query related to the emergency situation. A simply query, for example, would allow mobile phone users to receive a warning emergency text message. A query also allows decision support makers to analyze the current emergency situation.

In the following, we propose an algorithm for SMS text extraction and aggregation:

---

[1] http://www.mysql.com/

Algorithm 1. SMS Text Classification

Input: Emergency SMS text messages
Output: SMS text classification - tag $_{a_i}$

Purpose: To identify all pre-defined of the structured information in the
unstructured text, and, to populate a database of these entities.

1: **Input**
    C= {c_1, c_2......, c_n} // the category set
    S= {s_1, s_2......, s_m} // SMS messages set
2: **Collect unstructured information stage:**
4: **For** every received SMS text message **DO**
5:    database ($_{field_i}$)$_{\leftarrow s_i}$ // store each SMS text into the relational database
6: **Endfor**
7: **Tokenization stage:**
8: **Initialize:**
    Set currentPosition to 0
    Set delimiterSet1 to {, . ; ! ? ( ) + " space}
    Set delimiterSet2 to { articles, propositions,
            pronouns and other functional words }
9: **Procedure getToken:**
10:   C=cuurentPostion; ch=charAt(C)
11:   **If** ch = endofSMS then return -; **endif**
12:   **While** ch $\neq$endofSMS nor ch $\notin$ delimiterSet1
13:       token=token+ch C++; ch=chart(C)
14:   **Endwhile**
15:   **If** token $\notin$ delimiterSet2 then
16:     ts_i = token // ts, all the tokens in the SMS text, where j is the number
      of SMS text
17:   **Endif**
18: **Name Entity Recognition and merging**
    Structure stage:
19:   **Initialize:**
    F_k= {f_1, f_2......, f_n} // is the feature set, where k type number of feature
               // $f_i(h,t) \in \{0,1\}$, t is defined as training data and h is
               // history available when predicting t
20: **Procedure NameRecognition:**
21:   **Do** for every ts_i
22:     Compute a maximum entropy model

$$p(o/h) = \frac{1}{Z(h)} \prod_{j=1}^{k} \alpha_j^{f_j(0,h)}$$, where α_j is weight of feature f_j,

and Z(h) is a normalization constant

23:     $t = \arg\max p_j(o/h)$
24: **Enddo**

Algorithm 2. SMS Text Aggregation

1: **Input:** tag a_i // i=1, 2,..., n
2: **Output** database record
3: **For** i=1 **to** n
4:     **If** a_i exits in database
5:         **Do case:**
6:             **Case** *numeric values*
                Conduct average aggregation functions
7:             **Case** *linguistic values*
                conduct a linguistic aggregation operation
8:         **Endcase**
9:     **Endif**
10: database ($field_i$) $\leftarrow a_i$ // store each tag in the correct database field
       Destination
11: **Endfor**

5   Ontology-Based Representation for Unstructured SMS Text

Ontology has been used in many information extraction systems [36]. In this chapter, the ontology for the SMS text messages domain has been developed for the purpose of experimental text classification. In the domain ontology, we construct an SMS text message domain ontology which consists of seven main entities including: physical target; disaster location; human target; weapon used; stage of execution; and disaster event using

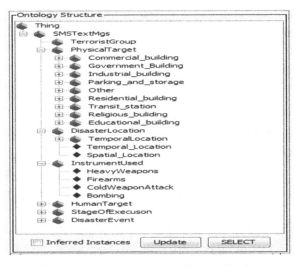

Fig. 4. Ontology structure of a disaster situation

Protégé tool1[2]. All entities are dividing into categories and sub-categories. We define the classes of words in an SMS emergency text message as shown in Figure 4.

To support the understandability of our extraction method and its implementation, we explain by using an example. For this purpose we have chosen the domain of disaster location, because the domain is widely known and its characteristics are suitable for extraction. Figure 5 shows a possible conceptualization of the domain containing only relevant aspects of the domain in which we are interested. In this sense, we can think of the ontology as the task specification where the properties of concepts represent the properties for which our extraction method has to determine appropriate values from the input data.

Fig. 5. Graphical representation of disaster spatial location

6   An Illustrated Example

Suppose our system receives the following SMS emergency text message: "A chemical attack is being carried out in UTS, Building 10 Level 4, 3 people were killed"

Step 1: Collect unstructured information of each individual SMS emergency text message and store it a relational database as shown below in Table 1. We store sender phone number if applicable, SMS text message, date and time.

---

[2] http://protege.stanford.edu/

Table 1. SMS emergency text message database

SMSID#	Time	Date	SMS Text	Sender Phone
1	10:25am	01.01.11	A chemical attack is being carried out in UTS, Building 10 Level 4, 3 people were killed	0422080991

Step 2: Apply information extraction algorithm. The first step is to break the text into tokens as shown in Table 2. The tokenization process consists of splitting the words and removing all unnecessary characters such as "," and "." from the list

Table 2. Words after tokenization

Tokenized Word
A/ chemical /attack /is /being /carried /out /in /UTS /Building /10 /level /4 /3 /people /were /killed

The next step after tokening all SMS text is to remove all stop words (see Table 3). A stop word is a type of word that is useless for search and retrieval purpose that it means; it has no great significance in the SMS text collection and must be eliminated in order to extract adequate key terms. Some of the most common stop words include: a, an, the, in, of, on, are, be, if, into, which.

Table 3. List of stopping word

Stop Word
A / is / being / out / in / were

Table 4 shows the outcome of the entity name recognition process based on an iterated feature argument of the current word. Eventually, we get the maximum value as the result and tag the current word. For example, disaster location can be calculated as follows:

Given a set of answer candidate h= {chemical, attack, UTS, Building, 10, level, 4, 3, people, killed}. We have: o="Disaster Location", and a current word "UTS" is an element of dictionary with an upper case in the beginning. $f_1(h,o)$ and $f_2(h,o)$ are a set of binary valued features, which are helpful in making a prediction about the outcome o. Officially, we can represent these features as follows:

$$f_1(h,o) = \begin{cases} 1 & \textit{iff Current} - \textit{Token}(h) \in \textit{DICTIONARY} \\ 0 & \textit{Otherwise} \end{cases}$$

$$f_2(h,o) = \begin{cases} 1 & \textit{iff Current} - \textit{Token} - \textit{First} - \textit{Letter} - \textit{Caps}(h) \\ 0 & \textit{Otherwise} \end{cases}$$

Consequently, $f_1(h,o)=1$ and $f_2(h,o)=1$. Suppose the model parameter $\alpha_1=0.5$ and $\alpha_2=0.2$ are weight of feature $f_1(h,o)$ and $f_2(h,o)$ respectively, and $Z(h)=1$ is a normalization constant. From Equation (2), we can compute the conditional probability of word or token "UTS" as $P(o\backslash h)= \alpha_1 \times f_1(h,o)+ \alpha_2 \times f_2(h,o)=0.5 \times 1+0.2=0.7$. The probability is given by multiplying the weights of active features (i.e., those $f(h,o) = 1$). We then compare the result with the threshold value.

Table 4. Tokens to feature vectors

Entity Name Type	Feature used	Example
DisasterLocation	Dictionary lookup features, Prefix/Suffix and digit information	UTS Building 10 Level 4
StageOfExecution	none	none
HumanTarget	Prefix/Suffix and digit information	3 killed
WeaponUsed	none	none
PhysicalTarget	Dictionary lookup features	University
DisasterEvent	Dictionary lookup features	chemical attack

Step 3: Query and information presentation. The last step is to generate a warning message based on the above extracted information. The following is an example of the warning message that will automatically be sent to mobile users:

"UTS building emergency at Building 10, level 4, if you are off campus, do not enter the campus. Wait for further instructions"

7   System Implementation

Automatic extraction of SMS text is evaluated using the standard information retrieval measures precision, recall, and F measure. The recall score measures the ratio of the correct information extracted from the texts against all the available information present in the SMS text messages. The precision score measures the ratio of correct information that was extracted against all the information that was extracted. F measure is a combined measure of precision and recall [37].

Let us assume that N is a collection of SMS text messages. In this collection, n represents all correct specific relevant information we want to extract (Disaster location, Weapon used, Physical target Stage of execution, Stopped words), and r all correct specific relevant information extracted manually. IE system recognizes a collection of SMS text messages, in this collection, k represents all results extracted. Then the recall (R), precision (P), and F-measure formulae used are given by Equations 7, 8 and 9:

$$\mathrm{Re}\,call\,(R) = \frac{a}{a+c} \tag{7}$$

$$\mathrm{Pr}\,ecision(P) = \frac{a}{a+b} \tag{8}$$

$$F - measure = \frac{2RP}{R + P} \tag{9}$$

Table 5 summarizes the relationships between the system classifications and the expert judgments in terms of a binary classification.

Table 5. Contingency table of classification decisions

Expert / IE system	Relevant	Not-relevant	Total
Matched	a	b	a+b=k
Not-matched	c	d	c+d=n-k
Total	a+c=r	b+d=n-r	a+b+c+d=n
Overall accuracy (OA)= (a/b)/n			

Where n = number of classified objects, k = number of objects classified into the class Ci by the system, r = number of objects classified into the class Ci by the expert.

7.1   Text Collection

A total text collection of 100 SMS is manually generated. In this collection, we avoid the use of unnecessary information, keep messages brief and comply with the 160 character limitation, use direct, straightforward language and communicate all necessary actions. The SMS text collection is composed of 2851 words.

7.2   Results

We ran the system with information consisting of five classes: disaster location, human target, physical target, weapon used, and disaster event. The results of two types of information are summarized in Tables 6 and 7 respectively. To compute precision and recall, we first compared the extracted instances automatically.

Table 6. Evaluation metrics for disaster location

Expert / IE system	Relevant	Not-relevant	Total
Matched	96	2	98
Not-matched	6	0	6
Total	102	2	104
Recall	0.94		
Precision	0.97		
F-measure	0.95		

Table 7. Evaluation metrics for weapon used

Expert / IE system	Relevant	Not-relevant	Total
Matched	42	0	42
Not-matched	14	0	14
Total	56	0	56
Recall	0.75		
Precision	1.00		
F-measure	0.85		

Our system suggests that overall the entity extraction system correctly locates and classifies the disaster location entity. Precision exceeds recall 97 percent as shown in Table 6. The standard measures In Table 7, our system suggests that precision exceeds recall 100 percent for the weapon used entity, and this variance is statistically significant. The standard measures for evaluating the performance of our system are shown in Figure 6.

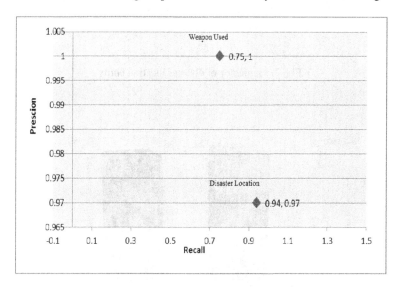

Fig. 6. Frequency of disaster location entity

Figure 7 shows our system classification results side by side. The right bar represents all instances of the disaster location entity. The left bar represents the ratio of our accurate results and the different types of classification. Figure 7 illustrates the number of correctly found and classified entities (black section of left bar), the number of correctly classified and misclassified entities (gray section of left bar), and the number of false positives with respect to identification and classification (white of left bar). On average, our system correctly classified 92.30 percent of the instances of the disaster location entity in the data, while 1.92 percent of the entities suggested by our system are false positive. At the same time, our system fails to correctly classify 5.76 percent of the disaster location entity in the data.

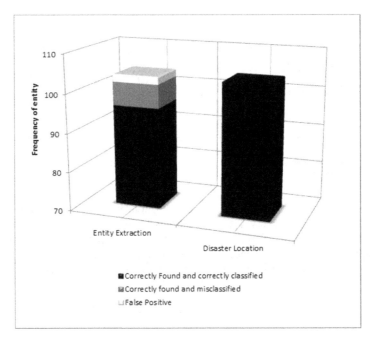

Fig. 7. Frequency of disaster location entity

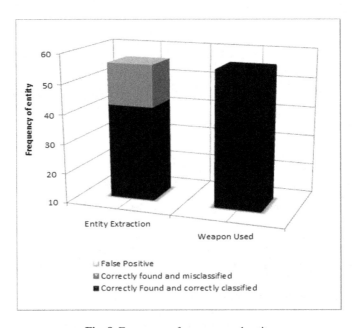

Fig. 8. Frequency of weapon used entity

Another example of entity extraction is illustrated in Figure 8. On average, our system correctly classified 75 percent of the instances of the weapon used in the data, and zero percent of the entities suggested by our system are false positive. At the same time, our system fails to correctly classify 25 percent of the weapon used entity in the data.

8   Conclusion

The key aspect of the MERS framework and system is to enable people get help from the government in an emergency situation. It makes use of mobile technologies to assist the government to get information and make decisions in response to disasters at anytime and anywhere. We have described the mobile-based MERS framework and its main components. As one of the most important MERS applications, this chapter introduced the components of the information extraction and aggregation process. In addition, the proposed algorithm can extract many kinds of semantic elements of emergency situation information such as disaster location, disaster event, and status of disaster.

In a future study, we will develop an intelligent knowledge management approach for the MERS system. This approach is concerned with getting the right information knowledge to the right people at the right time in the right form. The approach also aims to support a need for rapid decision making under pressured conditions in which information is partial, conflicting and often overloaded in an emergency disaster situation.

Acknowledgment. The work presented in this chapter was supported by Australian Research Council (ARC) grant under Discovery Project DP0880739.

References

[1]   Moon, M.J.: From e-government to m-government? Report, IBM Center for The Business of Government, Washington (2004)

[2]   Huijnen, C.: Mobile tourism and mobile government - an inventory of european projects. Working paper, European Center for Digital Communications (2006)

[3]   Careem, M., et al.: Demonstration of sahana: Free and open source disaster management. In: Proceedings of the 8th Annual International Conference on Digital Government Research: Bridging Disciplines & Domains, Philadelphia, PA (2007)

[4]   NCRIS, 2008 strategic roadmap for australian research infrastructure, Department of Innovation, Industry, Science and Research (DIISR), Canberra (2008)

[5]   Jaeger, P.T., et al.: Community response grids: E-government, social networks, and effective emergency management. Telecommun. Policy 31, 592–604 (2007)

[6]   Kim, S., et al.: Mobile analytics for emergency response and training. Information Visualization, Houndmills 7, 77–88 (2008)

[7]   Rossel, P., et al.: "Mobile" e-government options: Between technology-driven and user-centric. The Electronic Journal of e-Government 4, 79–86 (2006)

[8]   El-Kiki, T., Lawrence, E.: Government as a mobile enterprise: Real-time, ubiquitous government. In: Proceedings of 3rd International Conference on Information Technology: New Generations, Las Vegas, Nevada, pp. 320–327 (2006)

[9]   Kim, Y., et al.: Architecture for implementing the mobile government services in korea. In: First International Workshop on Digital Government: Systems and Technologies (DGOV 2004). LNCS, pp. 601–614 (2004)

[10] Saldhana, A.: Secure e-government portals - building a web of trust and convenience for global citizens. Presented at the Paper Presented to the W3C Workshop on e-Government and the Web, Washington DC, USA (2007)

[11] Wang, H., Rong, Y.: Case based reasoning method for computer aided welding fixture design. Computer-Aided Design 40, 1121–1132 (2008)

[12] Amailef, K., Lu, J.: M-government: A framework of mobile-based emergency response system. Presented at the The International Conference on Intelligent System and Knowledge Engineering, Xiamen, China (2008)

[13] Antovski, L., Gusev, M.: M-government framework. In: EURO mGov 2005, pp. 36–44. University of Sussex, Brighton (2005)

[14] Kushchu, I., Kuscu, H.: From e-government to m-government: Facing the inevitable. In: Proceedings of European Conference on E-Governemnt (ECEG 2003), Trinity College, Dublin (2003)

[15] El-Kiki, T., et al.: A management framework for mobile government services. In: Proceedings of CollECTeR, Sydney, Australia (2005)

[16] Yildiz, B., Miksch, S.: OntoX - A Method for Ontology-Driven Information Extraction. In: Gervasi, O., Gavrilova, M.L. (eds.) ICCSA 2007, Part III. LNCS, vol. 4707, pp. 660–673. Springer, Heidelberg (2007)

[17] Soysal, E., et al.: Design and evaluation of an ontology based information extraction system for radiological reports. Computers in Biology and Medicine 40, 900–911 (2010)

[18] Zhou, G., et al.: Tree kernel-based semantic relation extraction with rich syntactic and semantic information. Information Sciences 180(8), 1313–1325 (2010)

[19] Roman, Y., et al.: Extracting information about outbreaks of infectious epidemics. In: Proceedings of HLT/EMNLP on Interactive Demonstrations, Vancouver, BC, Canada (2005)

[20] Moens, M.-F.: Information extraction: algorithms and prospects in a retrieval context. Springer (2006)

[21] Bahora, A.S., et al.: Integrated peer-to-peer applications for advanced emergency response systems. Part ii. Technical feasibility. In: Systems and Information Engineering Design Symposium, pp. 261–268. IEEE (2003)

[22] Zhu, H., et al.: Information aggregation – a value-added e-service. In: Proceedings of the International Conference on Technology, Policy, and Innovation: Critical Infrastructures, The Netherlands (2001)

[23] Saha, S.K., et al.: A composite kernel for named entity recognition. Pattern Recognition Letters 31(12), 1591–1597 (2010)

[24] Li, L., et al.: Two-phase biomedical named entity recognition using crfs. Computational Biology and Chemistry 33, 334–338 (2009)

[25] Yang, Z., et al.: Exploiting the contextual cues for bio-entity name recognition in biomedical literature. Journal of Biomedical Informatics 41(4), 580–587 (2008)

[26] Min-Kyoung, K., Han-Joon, K.: Design of question answering system with automated question generation. In: Proceedings of the 4th International Conference on Networked Computing and Advanced Information Management, vol. 02, pp. 365–368 (2008)

[27] Valentin, J., et al.: Named entity normalization in user generated content. In: Proceedings of the 2nd Workshop on Analytics for Noisy Unstructured Text Data, Singapore (2008)

[28] Hai Leong, C., Hwee Tou, N.: Named entity recognition with a maximum entropy approach. In: Proceedings of the 7th Conference on Natural Language Learning at HLT-NAACL, Edmonton, Canada, vol. 4, pp. 160–163 (2003)

[29] Kozareva, Z., Ferrández, Ó., Montoyo, A., Muñoz, R., Suárez, A.: Combining Data-Driven Systems for Improving Named Entity Recognition. In: Montoyo, A., Muñoz, R., Métais, E. (eds.) NLDB 2005. LNCS, vol. 3513, pp. 80–90. Springer, Heidelberg (2005)

[30] Shaojun, Z.: Named entity recognition in biomedical texts using an hmm model. In: Proceedings of the International Joint Workshop on Natural Language Processing in Biomedicine and its Applications, Geneva, Switzerland, pp. 84–87 (2004)

[31] Saha, S.K., et al.: Feature selection techniques for maximum entropy based biomedical named entity recognition. Journal of Biomedical Informatics 42(5), 905–911 (2009)

[32] Sun, C., et al.: Rich features based conditional random fields for biological named entities recognition. Computers in Biology and Medicine 37(9), 1327–1333 (2007)

[33] Takeuchi, K., Collier, N.: Bio-medical entity extraction using support vector machines. Artificial Intelligence in Medicine 33(2), 125–137 (2005)

[34] Javed, A.A., et al.: The maximum entropy method for analyzing retrieval measures. In: Proceedings of the 28th Annual International ACM SIGIR Conference on Research and Development in Information Retrieval, Salvador, Brazil (2005)

[35] Kumar, S.S., et al.: Feature selection techniques for maximum entropy based biomedical named entity recognition. Journal of Biomedical Informatics 42, 905–911 (2009)

[36] Tran Quoc, D., Kameyama, W.: A proposal of ontology-based health care information extraction system: Vnhies. In: 2007 IEEE International Conference on Research, Innovation and Vision for the Future, pp. 1-7 (2007)

[37] Katharina, K., et al.: How can information extraction ease formalizing treatment processes in clinical practice guidelines? Artif. Intell. Med. 39, 151–163 (2007)

Author Index

Editors

Professor Jie Lu is the Head of School of Software in the Faculty of Engineering and Information Technology, and the Director of the Decision Systems and e-Service Intelligence Research Laboratory in the Centre for Quantum Computation & Intelligent Systems at the University of Technology, Sydney (UTS). She received her PhD from Curtin University of Technology in 2000. Her main research interests lie in the area of decision making modeling, decision support system tools, uncertain information processing, recommender systems and e-Government and e-Service intelligence. She has published five research books and 270 papers in refereed journals and conference proceedings. She has won five Australian Research Council (ARC) discovery grants, an Australian Learning & Teaching Council grant, and 10 other research and industry linkage grants. She received the first UTS Research Excellent Medal for Teaching and Research Integration in 2010. She serves as Editor-In-Chief for *Knowledge-Based Systems* (Elsevier), Editor-In-Chief for *International Journal on Computational Intelligence Systems* (Atlantis), editor for book series on *Intelligent Information Systems* (World Scientific) and proceedings series on *Computer Engineering and Information Science* (World Scientific), and has served as a guest editor of six special issues for international journals, as well as having delivered five keynote speeches at international conferences.

Professor Lakhmi C. Jain is a Director/Founder of the Knowledge-Based Intelligent Engineering Systems (KES) Centre, located in the University of South Australia. He is a fellow of the Institution of Engineers Australia.

His interests focus on the artificial intelligence paradigms and their applications in complex systems, art-science fusion, e-education, e-healthcare, unmanned air vehicles and intelligent agents.

 Guangquan Zhang is an Associate Professor and co-Director of the Decision Systems and e-Service Intelligence Research Laboratory in the Centre for Quantum Computation & Intelligent Systems in Faculty of Engineering and Information Technology at the University of Technology Sydney (UTS), Australia. He has a PhD in Applied Mathematics from Curtin University of Technology, Australia, and been awarded an Australian Research Council (ARC) QEII Fellow (2005-2010). He was with the Department of Mathematics, Hebei University, China, from 1979 to 1997, as a Lecturer, Associate Professor and Professor. His main research interests lie in the area of multi-objective, bilevel and group decision making, decision support system tools, fuzzy measure, fuzzy optimization and uncertain information processing. He has published four monographs, four reference books and over 200 papers in refereed journals and conference proceedings and book chapters. He has won four ARC Discovery grants and over 10 other research grants. He has served as a guest editor of three special issues for international journals.

Printed in the United States
By Bookmasters